COLOUR ENGINEERING

Wiley-SID Series in Display Technology

Editor:
Anthony C. Lowe
The Lambent Consultancy, Braishfield, UK

Display Systems:
Design and Applications
Lindsay W. MacDonald and Anthony C. Lowe (Eds)

Electronic Display Measurement:
Concepts, Techniques and Instrumentation
Peter A. Keller

Projection Displays
Edward H. Stupp and Matthew S. Brennesholz

Liquid Crystal Displays:
Addressing Schemes and Electro-Optical Effects
Ernst Lueder

Reflective Liquid Crystal Displays
Shin-Tson Wu and Deng-Ke Yang

Colour Engineering:
Achieving Device Independent Colour
Phil Green and Lindsay MacDonald (Eds)

Published in Association with the
Society for Information Display

COLOUR ENGINEERING

Achieving Device Independent Colour

Edited by

Phil Green

Colour Imaging Group, London College of Printing, UK

and

Lindsay MacDonald

Colour & Imaging Institute, University of Derby, UK

JOHN WILEY & SONS, LTD

Other Wiley Editorial Offices

John Wiley & Sons Inc., 111 River Street, Hoboken, NJ 07030, USA

Jossey-Bass, 989 Market Street, San Francisco, CA 94103-1741, USA

Wiley-VCH Verlag GmbH, Boschstr. 12, D-69469 Weinheim, Germany

John Wiley & Sons Australia Ltd, 33 Park Road, Milton, Queensland 4064, Australia

John Wiley & Sons (Asia) Pte Ltd, 2 Clementi Loop #02-01, Jin Xing Distripark, Singapore 129809

John Wiley & Sons Canada Ltd, 22 Worcester Road, Etobicoke, Ontario, Canada M9W 1L1

British Library Cataloguing in Publication Data

A catalogue record for this book is available from the British Library

ISBN 0-471-48688-4

Typeset in 10/12 Times by Laserwords Private Limited, Chennai, India
Printed and bound in Great Britain by Biddles Ltd, Guildford and King's Lynn
This book is printed on acid-free paper responsibly manufactured from sustainable forestry
in which at least two trees are planted for each one used for paper production.

Contents

Colour plate captions

Plate 1 (Chapter 5) (a) CIELUV colour picker; (b) Munsell colour picker; (c) NCS colour picker; (d) TekHVC colour picker

Plate 2 (a) ISO 12640 target (IT8.7/3) CMYK data file for characterizing printers (described in Chapters 6, 10 and 18); (b) ISO 12641 target (IT8.7/1) on photographic media for characterizing input devices (Chapters 6, 8 and 18); (c) gamut target CMYK data file for determining the gamut boundary of printers and photographic media (Chapter 10)

Plate 3 (Chapter 9) An image in different stages of colour processing. (a) Raw CFA data stored by the DSC. (b) After linearization, dark current subtraction, flare subtraction, and channel balancing. The image data is still called raw CFA data after these operations have been performed. (c) Color filter pattern interpolated to producec three fully populated color channels. This stage is denoted as raw device RGB. Software supporting raw DSC image data exchange must be aware which operations need to be performed after exchange. (d) After a matrix transformation has been applied to go from device RGB to ITU-R BT.709 based RGB (scene-referred image data). (e) After color rendering has been applied to map scene colorimetry to the sRGB virtual display in a preferred manner (standard output-referred).

Plate 4 (Chapter 9)
(a) A photographic reproduction model is used to capture a high-key scene. The reproduction appears too dark because the model assumes that the mean scene luminance should be placed at an 18% gray reflectance.
(b) The result obtained using a more sophisticated proprietary reproduction model. The model used still applies a global transform, but the nature of the transform is determined using several image-based statistical parameters.
(c) A photographic reproduction model is used to capture a high dynamic range scene. Shadows are blocked up in the reproduction because the model assumes a scene dynamic range of 160:1, and this scene is 1400:1.
(d) A global transform determined using the proprietary reproduction model is applied. This transform compresses the dynamic range of the scene to that of the output medium.
(e) A simple video reproduction model is used to capture a 180:1 scene. Although this scene does not have an extremely high dynamic range, it is severely clipped because the model system gamma of about 1.2 limits the range captured.
(f) A global transform determined using the proprietary reproduction model is applied. Detail is maintained throughout the entire scene dynamic range and the color reproduction is still pleasing.

(g) A colorimetric reproduction of a 24:1 scene. Even though the colorimetry of the print is representative of the original scene (system gamma of unity), the reproduction is not optimally pleasing. (h) A global transform determined using the proprietary reproduction model is applied. The more pleasing color reproduction may indicate that the human visual system includes scene dynamic range adaptation.

Plate 5 (Chapter 13) (a) Effect of reproduction (printer) gamut size on performance of GMA types. (b) Two images and their image gamuts (solid) in CIELAB when seen on a given CRT (mesh)

Plate 6 (Chapter 13) (a) An original, (b) a simple gamut-clipped, (c) a simple gamut-compressed, and (d) an advanced gamut-compressed reproduction to the same reproduction gamut. (e) CRT gamut (solid) and gamut of block dyes representing the gamut of theoretically possible surface colours (mesh), which was obtained by independently varying spectral reflectances at 16 wavelengths whereby generating spectral reflectance curves, which were then combined with the spectral power distribution of CIE Standard Illuminant D50

Plate 7 (Chapter 14) Rendering of large dynamic range scene optimized for (a) foreground, and (b) background

Plate 8 (Chapter 14) (a), (b) Two different renderings of a scene. In (a) the colorimetry of the rendered image closely matches that of the original scene. In (b) the rendered image is not colorimetrically accurate, but the resulting image generally would be judged to have improved color reproduction. (c) In this rendering of the image shown on Plate 7, a digital 'dodge-and-burn' operation has been used to produce a print where both the foreground and the background are properly rendered

Contributors

Professor Roy Berns
Munsell Color Science Laboratory
Chester F. Carlson Center for Imaging
    Science
Rochester Institute of Technology
54 Lomb Memorial Drive
Rochester
NY 14623-5604
USA
Tel: +1-716-475-2230
E-mail: berns@cis.rit.edu
http://www.cis.rit.edu:80/people/faculty/berns/
Munsell lab: www.cis.rit.edu/research/mcsl/

Edward J. Giorgianni
Imaging Research and Advanced
    Development
Eastman Kodak Company
1700 Dewey Avenue
Rochester
NY 14650-1816
USA
E-mail: 57526n@isbgate.kodak.com

Phil Green
Colour Imaging Group
London College of Printing
Elephant & Castle
London
SE1 6SB
UK
Tel: +49-20-7514-6759
E-mail: pj.green@lcp.linst.ac.uk
http://www.digitalcolour.org

Lawrence R. Hanlon
Hewlett-Packard Company
1501 Page Mill Road
MS 2U-18, Palo Alto
CA 94304
USA
E-mail: larry_hanlon@hp.com

Jack Holm
Hewlett-Packard Company
1501 Page Mill Road
MS 2U-19, Palo Alto
CA 94304
USA
Tel: +1-650-236-2436
E-mail: jack_holm@hp.com

Paul M. Hubel
Hewlett-Packard Company
1501 Page Mill Road
MS 2U-18, Palo Alto
CA 94304
USA
Tel: +1-650-857-7338
E-mail: paul_hubel@hp.com

Professor Tony Johnson
Colour Imaging Group
London College of Printing
Elephant & Castle
London
SE1 6SB
UK
Tel: +49-20-7514-6759
E-mail: tony@colouruk.demon.co.uk
http://www.digitalcolour.org

James C. King
A Principal Scientist
Adobe Systems Incoporated
Mail Stop: W14
345 Park Avenue
San Jose
CA 95110-2704
USA
Tel: +1-408 536-4944
E-mail: jking@adobe.com

Naoya Katoh
Sony Corporation
MNC/PIC/Dept 1
2-15-3 Konan
Minatu-Ku
Tokyo 108-6201
Japan
E-mail: naoya@color.sony.co.jp

Wolfgang Lempp
Computer Film Company
19–23 Wells Street
London
W1T 3PG
UK
E-mail: wolf@cfc.co.uk

Professor M. Ronnier Luo
Colour & Imaging Institute
University of Derby
Kingsway House
Kingsway
Derby
DE22 3HL
UK
E-mail: m.r.luo@derby.ac.uk

Professor Lindsay W. MacDonald
Colour & Imaging Institute
University of Derby
Kingsway House
Kingsway
Derby

DE22 3HL
UK
E-mail: L.W.MacDonald@derby.ac.uk

Marc Mahy
Manager Color Imaging Technologies
Graphic Systems Business Group
GS/R&D/Color Technology Center
Agfa-Gevaert N.V.
Septestraat 27
B-2640 Mortsel
Belgium
Tel: +32-3-444 3990
E-mail: marc.mahy.mm@belgium.agfa.com

David Q. McDowell
Standards Consultant
51 Parkwood Lane
Penfield
NY 14526
USA
Tel: +1-716-383-1706
E-mail: mcdowell@kodak.com

Ján Morovic
Lecturer in Digital Colour Reproduction
Colour & Imaging Institute
University of Derby
Kingsway House
Kingsway
Derby
DE22 3HL
UK
Tel: +44-1332-593113
E-mail: j.morovic@derby.ac.uk
http://colour.derby.ac.uk/~jan/

Leonardo Noriega
Colour and Imaging Institute
University of Derby
Kingsway House
Kingsway
Derby
DE22 3HL
UK
E-mail: L.Noriega@derby.ac.uk

Peter A. Rhodes
Colour & Imaging Institute
University of Derby
Kingsway House
Kingsway
Derby
DE22 3HL
UK
E-mail: P.A.Rhodes@derby.ac.uk

Danny C. Rich
86 Joni Avenue
Hamilton Square
NJ 08690
USA
E-mail: drich7@softhome.net

Kevin E. Spaulding
Imaging Research and Advanced
      Development
Eastman Kodak Company
1700 Dewey Avenue
Rochester
NY 14650-1816
USA
E-mail: kevin.spaulding@kodak.com

Sabine Süsstrunk
Assistant Professor
Swiss Federal Institute of Technology
      (EPFL)
Laboratory for Audiovisual
      Communications (LCAV)

Communication Systems Department
      (DSC)
Ecublens INR-140
CH-1015 Lausanne
Switzerland
Tel: +41-21-693 66 64
E-mail: Sabine.Susstrunk@epfl.ch
web: lcavwww.epfl.ch

Arthur Tarrant
11 Preston Close
Twickenham
Middx
TW2 5RU
UK

Ingeborg Tastl
Hewlett-Packard Company
1501 Page Mill Road
MS 2U-7, Palo Alto
CA 94304
USA
Tel: +1-650-857-4559
E-mail: ingeborg_tastl@hp.com

Dawn Wallner
6785 Rainbow Drive
San Jose
CA 95129
USA
E-mail: dawn@dawnlink.com

Series Preface

The tenth Color Imaging Conference, organised jointly by the Society for Imaging Science and Technology and the Society for Information Display, will be held this year. How appropriate it is that "Colour Engineering", this new volume in the SID-Wiley series in Display Science and Technology, should be published in the same anniversary year.

The object of each book in the series is to provide information, which is of real practical value to practitioners in its field. This volume amply achieves that purpose. As an increasing variety of electronic image generation, capture, storage and rendering devices becomes available and as film and print media continue to be developed, the need for an understanding of the complex field of colour management has never been greater.

By their use of different expert authors for each chapter, the Editors have created a remarkably authoritative volume, which addresses all key areas of the subject. Numerical treatment achieves a balance of providing the necessary quantitative tools whilst offering sufficient derivation and explanation to satisfy those who require an understanding of the basis of the suject. Those who require more will find their needs satisfied by extensive bibliographies. The book begins with chapters, which introduce to the general concepts of colour and its measurement and explain colorimetry and colour space. These are followed by section, which describe the methods for characterising input and output devices, the determination of colour gamut and how these techniques can be used in the attempt to create device independent colour management systems. It concludes with an extensive description of standards.

I found "Colour Engineering" fascinating in a way that technical books rarely are. I finished it with the feeling that display developers ignore the subject at their peril and that this book should be required reading for anyone involved in the development or the advanced use of an image input, output or storage device. This book provides an extremely readable and practical means to raise the general level of understanding of colour engineering and colour management throughout the diverse field of information display.

Anthony C. Lowe
Braishfield, UK 2002

Preface

At the first Color Imaging Conference in Scottsdale, Arizona, organised in 1993 by the Society for Imaging Science and Technology (IS&T) and the Society for Information Display (SID), James King of Adobe Systems spoke eloquently on the need for 'color engineering' as a synthesis of the disciplines of colour science and computer engineering. He called for more education in the subject, and for greater awareness of its importance in the design and development of colour imaging products. Since then several postgraduate courses in colour imaging (such as those at the Rochester Institute of Technology, London College of Printing and the University of Derby) have been launched and a number of excellent textbooks have appeared. This book has arisen from the need for a specialist text that brings together key developments in colour management technology and findings from the colour engineering research community.

Many contributions to scientific endeavour are first published in the form of conference papers, and later refined as journal articles. The peer-reviewed book chapter can be seen as continuation of that process, one which allows a more detailed and thorough treatment of the subject. The opportunity to write at greater length enables the author to provide additional background and a more detailed discussion of the issues, making the resulting piece useful to those who wish to review the material in depth and also more accessible to readers new to the subject. While the single-authored textbook may have a more consistent authorial 'voice' than the multi-contributor text, the latter may be more comprehensive and be a vehicle for conveying the most important trends of industrial research.

We have been extremely fortunate to obtain agreement from the distinguished authors who have contributed to the book, generously giving up their time to write chapters and then revise them following peer review. We were also privileged to obtain the services of a panel of experts, all well known in the field, who anonymously reviewed the draft chapters. Where necessary we have added material to ensure a reasonably complete treatment of the main themes. Each chapter remains a distinctive and specialist contribution, however, and we have not attempted to integrate them into a consistent whole through excessive editing. Instead we have limited editing to minor alterations, with the aim of minimising unnecessary duplication. Two chapters on device characterisation have previously been published in a special issue of the journal *Displays*. Although first published some time ago they remain amongst the best descriptions of display and scanner characterisation.

The book is organised into four sections. The first deals with fundamental colour science and with methods of defining colours and their appearance, and quantifying differences between them. The second section focuses on the characterisation of colour imaging devices. The third covers some of the current themes in colour management, while in the fourth section contributors describe implementation issues through their experience in some of the leading imaging companies and the standards community.

Our main goals have been to provide an overview of the theoretical basis and the practical implementation of colour science to imaging technology, and to ensure that a wide range of theoretical perspectives are included. We asked contributors to address underlying issues rather than the current form of their solution, and we believe that the final text gives a good account of the current 'state of the art' in colour engineering.

The images in the colour plate section were supplied by the respective authors in RGB colour spaces with associated input profiles (with the exception of the characterization targets in Plate 2). Conversion to CMYK for printing was performed using the perceptual rendering intent of a reference CMYK profile for sheet-fed offset printing. The profile was part of a set developed for commercial offset printing by the Colour Imaging Group at LCP, using reference printing data available from FOGRA and IFRA. The perceptual intents of these profiles were optimised through a collaboration between the Colour Imaging Group and the Digital Imaging Group of the Association of Photographers, and the profiles are publicly available at http://www.digitalcolour.org.

We would like to offer our sincere thanks to all the authors and reviewers whose efforts have made this book a reality, and whose public contributions to research continue to sustain the colour engineering community. We would also like to acknowledge the continuing support and patience of Ruth and Sandra, to whose better natures we have frequently to appeal.

All royalties from this publication are being donated to a trust fund for students of colour engineering, to be administered by the Colour Group of Great Britain.

Phil Green and Lindsay MacDonald
April, 2002

1

Light and colour

Arthur Tarrant

1.1 LUMINOUS FLUX

In any engineering study of light and colour, we need to have clearly definable measures of light. Suppose a beam of light is emitted from a projector with an incandescent lamp (Figure 1.1). The beam contains a certain amount of light – close to the projector it occupies only a small area and further away it is spread over a larger area, but the total number of photons in any cross-section of the beam is the same. We can define the *flux* of the light as the rate of flow (i.e. number of photons per second) per unit cross-sectional area normal to the direction of the flow.

Let us now imagine that we can disperse the light of the projector beam into a spectrum (Figure 1.2). The spectrum will not appear equally bright at all points – the yellow and green near the centre will appear brighter than the red and the blue at the extremities, for example. That is for two reasons: first the lamp in the projector does not emit an equal amount of power at all wavelengths; and second our eyes are not uniformly sensitive to light of all wavelengths. The relative amount of power emitted at each wavelength by the lamp is shown in Figure 1.3a, and the relative sensitivity of our eyes is shown in Figure 1.4.

The curves in Figure 1.4 show the relative sensitivity of the eye for both the light-adapted (photopic) and dark-adapted (scotopic) states. Note that these are relative curves, which show the sensitivity at any wavelength relative to the maximum sensitivity at

Colour Engineering, edited by P.J. Green and L.W. MacDonald.
© 2002 John Wiley & Sons Ltd.

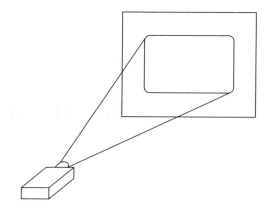

Figure 1.1   Luminous flux emitted by a projector.

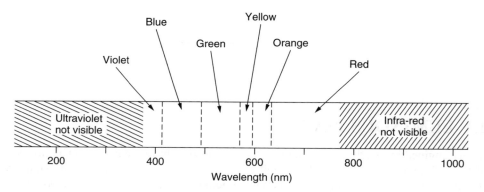

Figure 1.2   Apparent brightness as a function of wavelength.

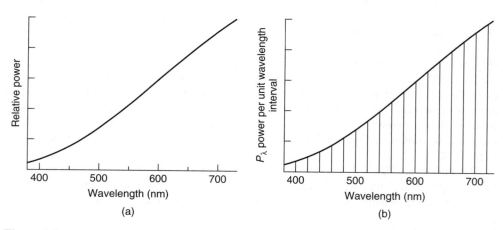

Figure 1.3   Relative spectral power distribution of a tungsten lamp; (b) Estimation of power per unit wavelength.

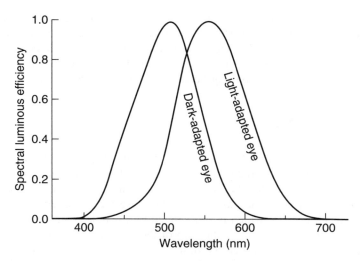

Figure 1.4 Relative spectral luminous efficiency curves for the light-adapted (photopic) and dark-adapted (scotopic) eye.

555 nm and 508 nm respectively. In practice we are nearly always concerned with photopic vision for typical image viewing conditions in daylight, domestic and office situations. The spectral luminous efficiency curves have been derived from studies on a large number of observers, and have been defined by the CIE [1]. Precise tabulated values can be found also in many textbooks [2]. The curve for photopic vision is often referred to as 'the visibility function', or simply as 'the $V(\lambda)$ function'.

It should be noted that the relative sensitivity curves are defined with reference to the CIE 'standard observer', which is a statistical average of the population with normal colour vision. In practice there is a certain amount of deviation from the standard observer, and it unlikely that any one individual will have exactly the same response as the standard observer or that any two individuals will have responses identical with each other [3].

Fortunately the eye works in an integrative manner, effectively summing all the stimuli that it receives from light of different wavelengths. So to quantify the amount of light visible in our projector beam, we have to take the amount of power at each wavelength, determine its light-producing effect in the eye by multiplying by the value of the visibility function for that wavelength, and add up the contributions across all wavelengths. Imagine that we divide the spectral power distribution of the light source into a series of narrow wavelength intervals, as in Figure 3b. Suppose that the power per unit wavelength interval is P_λ, and that the width of each wavelength band is $\delta\lambda$. Then the light-producing effect of the power in that band can be quantified as $P_\lambda V_\lambda \delta\lambda$. If we sum the effect across all wavelength bands we obtain $\Sigma P_\lambda V_\lambda \delta\lambda$ which represents the total light-producing effect, i.e. the total amount of light visible in the beam, called the *luminous flux*. The unit of luminous flux is the *lumen* and is specified by:

$$F = k_{\mathrm{m}} \Sigma P_\lambda V_\lambda \delta\lambda \qquad (1.1)$$

where F stands for the luminous flux, and k_m is a constant relating units of flux to units of power, equal to 683 lumens per watt. The constant k_m arises because the unit of light, the lumen, was established long before the SI system of units was conceived. The familiar 60 watt incandescent bulb that we use in our homes produces about 750 lumens; a five-foot fluorescent tube may produce anything between 3000 and 5000 lumens.

1.2 ILLUMINANCE

Luminous flux can be thought of as a measure of light in passage from place to place, but we are usually more concerned with the amount of light arriving on a surface - that is the amount of luminous flux falling on a unit area of the surface. That quantity is called *illuminance* and is defined as follows. Suppose we consider a small element of the surface, as in Figure 1.5, which is illuminated by a single point source, and subtends only a small cone from that point. If the area of the element is s, and the amount of flux contained within the cone is F, then the illuminance of the element will be the ratio F/s as the element is made vanishingly small. In mathematical terms, if we call the value of illuminance E, then:

$$E = \lim_{s \to 0} F/s \tag{1.2}$$

The unit of illuminance is the *lux*, defined as that produced when luminous flux of one lumen falls on an area of one square metre. The US unit of illuminance is the *foot-candle*, equal to 10.76 lux.

In practice we are usually concerned with surfaces that are lit from all directions rather than from a single point, but the same principle applies. Notice that illuminance is specified *at a point* on a surface. In most cases the value of illuminance varies considerably from point to point; and in fact it is very difficult to achieve even an approximation to uniform illuminance over an area of more than a few square centimetres in extent. The lighting in a typical office may produce an illuminance of anywhere between 150 lux and 500 lux on the working surfaces of the desks. On an overcast day in temperate latitudes the illuminance on the earth's surface may be approximately 5000 lux; in bright sunlight it can rise to 50,000 lux.

For vision, the important factor is *retinal illuminance*, measured in *trolands*. When the eye views a surface of uniform luminance, the number of trolands is equal to the product of the area in square millimetres of the pupil times the luminance of the surface

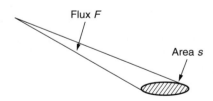

Figure 1.5   Small element of a surface illuminated by a point source of light.

in candelas per square metre. It should be noted that, under very high illumination levels, some of the neurones in the retina begin to saturate, and colour 'wash-out' begins to occur. At low illumination levels, photopic vision (cone photoreceptors) begins to be replaced by scotopic vision (rod photoreceptors). The transition stage is called mesopic vision, which occurs at adapted illumination levels from about 10 lux (dusk) down to about 0.1 lux (moonlight), in which colour vision ceases to be normal. Note that some common viewing conditions are mesopic, such as cinema or the projection of slides in a darkened room. For any location where accurate assessments of colour are needed, the illuminance level should not be below 50 lux or above 10,000 lux.

Recommended values of illuminance for all manner of visual tasks are published by the Chartered Institution of Building Services Engineers [4] (CIBSE) in the UK, and by the Illuminating Engineering Society in the USA [5]. These are average values – in any real room the illuminance in the corners is invariably lower than in the centre. Readers may sometimes find references to the *ambient* illuminance in a room, which is an almost meaningless phrase. This is usually taken to mean the average value of illuminance on a horizontal plane in the room at the height of bench or desk tops, however it is not an internationally agreed term. Lighting engineers have properly formulated methods of describing the average illuminance in a given space [5].

We now consider what happens when we tilt the surface upon which the light falls. Suppose as before that the light is emitted by a point source in a narrow cone and falls normally onto the surface, as in Figure 1.6a. If we call the illuminance E_0, then:

$$E_0 = F/s \tag{1.3}$$

If we now tilt the surface so that its normal is at an angle θ relative to the beam direction, as in Figure 1.6b, the same cone is spread over a larger area $s\cos(\theta)$, and the illuminance becomes:

$$E_\theta = F\cos\theta/s \tag{1.4}$$

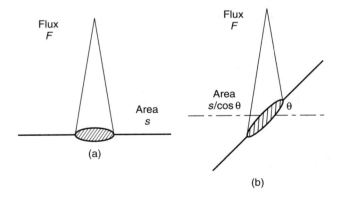

Figure 1.6 (a) Light falling normally onto a surface; (b) Light falling onto a tilted surface.

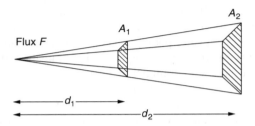

Figure 1.7 A_1 Surface illuminated by a near source; A_2 Surface illuminated by a more distant source.

and hence:

$$E_\theta = E_0 \cos \theta \qquad (1.5)$$

This means that when light from a point falls on a surface obliquely, the illuminance it produces depends on the cosine of the angle of incidence. This relationship is sometimes called the cosine law of illumination. Instruments made to measure illuminance, i.e. photometers, have to be designed to allow for this effect (see Chapter 2).

It is also necessary to consider the distance between the light source and the surface. Suppose again that we have a point light source illuminating a surface, as in Figure 1.7a, but this time we consider the light emanating from the source within a narrow square cone. Suppose the luminous flux is F, the area of the surface on which it falls is A_1, and the distance from the source to that surface is d_1. Then the illuminance E_1 on the surface will be:

$$E_1 = F/A_1 \qquad (1.6)$$

Suppose that we now move the receiving surface a position further away from the source, as in Figure 1.7b, in which a larger area A_2 is illuminated by the cone. The illuminance E_2 will now be:

$$E_2 = F/A_2 \qquad (1.7)$$

so the ratio of E_2/E_1 will be $A_1/A_2, = d_1^2/d_2^2$, and hence:

$$E_2/E_1 = (d_1/d_2)^2 \qquad (1.8)$$

This relationship is usually spoken of as the 'inverse square law'. It implies that, for a surface illuminated by a point source, the illuminance falls off as the inverse of the square of the distance. Thus if the distance is doubled the illuminance will fall to one quarter of its value.

1.3 LUMINOUS INTENSITY

The concept of luminous flux provides us with a measure of the light in passage from one place to another. Illuminance provides a measure of the amount of light falling onto a surface. But we also need to have a measure of the amount of light leaving a light source, called the *luminous intensity*.

In most practical light sources, the amount of light emerging varies greatly for different directions. For example, a plain incandescent bulb with a diffusing envelope – i.e. the ordinary domestic light bulb – hanging from a flex may emit significantly different amounts in the horizontal and downward directions, but it will emit none in the vertically upward direction, because the lamp-holder gets in the way. Likewise, a slide projector by design may emit all of its output in a narrow cone only a few degrees wide, and none in any other direction. Thus when we specify the luminous intensity of a source, we must also specify the precise direction in which it is measured.

Luminous intensity is specified in this way: suppose we have a point source of light, which emits luminous flux F in a narrow cone in the direction of measurement. Suppose also that this cone subtends a solid angle of ω. Then the luminous intensity I is defined as the limiting value of the ratio F/ω as the cone is made vanishingly small. In mathematical terms we may say:

$$I = \lim_{s \to 0} F/s \tag{1.9}$$

The unit of luminous intensity is the *candela*, which occurs when a source radiates one lumen into a solid angle of one steradian. The domestic lamp quoted above may typically have a luminous intensity of 80 to 100 candelas; a car headlamp may well have a luminous intensity of 20,000 candelas in the centre of its beam.

1.4 LUMINANCE

A quantity of great importance in illumination engineering is the light emission from a surface or visual field, for which we use the concept of *luminance*. Consider a small element of the surface, of area s, as if it were a light source in its own right, and determine the luminous intensity per unit area. If in a given direction, that element has a luminous intensity of I, then the luminance L is given by I/s. At a single point of the surface, we can say:

$$L = \lim_{s \to 0} I/s \tag{1.10}$$

The unit of luminance is the *candela per square metre*, usually abbreviated to cd/m^2. (In some European countries this unit is called the *nit*, but this is not an internationally agreed term. The US unit of luminance is the *foot-Lambert*, equal to $3.426\,cd/m^2$.) The luminance value of the walls in most rooms may be between 30 and $100\,cd/m^2$. Computer displays may have a white luminance of around $100\,cd/m^2$. The surface of a fluorescent lamp may have a luminance of $1800\,cd/m^2$. The tungsten filament in an incandescent lamp may have a luminance as high as $50,000\,cd/m^2$.

Note that when we specify the luminance of a surface, we have to specify both the point and the direction in which the measurement is made. Obviously the luminance of any reflecting surface depends on the illuminance of the light falling upon it. As illuminance is rarely uniform over a surface, its luminance usually varies as a function of position.

Most real surfaces do not emit light uniformly in all directions, moreover. They usually have some kind of texture – woven fabrics for example – and the luminance may vary widely with the angle of viewing. Glossy or shiny surfaces will reflect more strongly in the specular direction. This must be remembered when specifying the luminance of a display screen – most visual displays show marked directional characteristics, especially those using liquid crystal devices.

A Lambertian surface provides uniform diffusion of the incident radiation such that its luminance is the same in all directions from which it can be measured. Reflection from a diffuse Lambertian surface obeys the cosine law by distributing reflected energy in proportion to the cosine of the reflected angle. A Lambertian surface of area A that has a luminance of $1.0 \, \text{cd/m}^2$ will radiate a total of πA candelas into a hemisphere of π steradians. Conversely, if you were to illuminate a Lambertian surface uniformly with an illuminance of $3.1416 \, \text{lux}$, then you would measure a luminance on that surface of $1.0 \, \text{cd/m}^2$ if it were 100% reflective.

1.5 UNITS

1.5.1 Obsolete units

In some countries, international SI units are not yet in universal use. The foot-candle is defined as the illuminance produced when luminous flux of one lumen falls on an area of one square foot. Since there are 10.76 square feet in one square metre, an illuminance of one foot-candle is equivalent to that of 10.76 lux.

The foot-Lambert is a unit of luminance widely used in the USA which is defined in slightly different mathematical terms from the SI unit, the candela per square metre. Although mathematically related, it is easiest for the user to remember that one foot-Lambert is equivalent to 3.426 candela per square metre.

1.5.2 Radiometric units

In all of the preceding sections we have been concerned with units involving visible light. In some cases one must consider the total amount of radiation – i.e. including all of the infra-red and ultraviolet components, without any consideration of the visibility function. This would be necessary, for example, when considering the response of a photo-detector, such as a charge-coupled device (CCD), which may have a spectral sensitivity different to the human observer and may respond quite strongly to wavelengths outside the visible spectrum. There is an analogous series of units for such radiation measurements, defined in exactly the same way, except that the visibility factor $V(\lambda)$ is not taken into account. Thus *radiant flux* corresponds to luminous flux, *irradiance* corresponds to illuminance, *radiant intensity* to luminous intensity and *radiance* to luminance. The units of these radiometric measurements correspond to photometric units with the *watt* replacing the lumen.

1.6 TRANSMISSION MEASUREMENTS

1.6.1 Direct transmission

We now consider what happens when light falls passes through a translucent material. The simplest case to consider is that of a polished glass surface, as in Figure 1.8. Some small amount of light is reflected at the surface, the specular component, but most of the incident light enters the glass at a steeper angle, following the physical laws of *refraction*. If the angle of incidence is ι, the angle of refraction is r, and the refractive index of the glass is n, then Snell's Law applies:

$$\sin \iota = n \sin r \tag{1.11}$$

To be strictly correct, we should call the medium outside the glass medium 1, and refer to the glass as medium 2, with respective indices of refraction n_1 and n_2, so that:

$$n_1 \sin \iota = n_2 \sin r \tag{1.12}$$

In practice the refractive index of air is so near to unity ($n = 1.000\,017$) that we usually assume n_1 to be 1.0.

Now suppose that we have light passing through a glass plate with polished and parallel surfaces, as in Figure 1.9. At the first surface, some light will be reflected but most will be refracted and enter the glass. At the second surface, there is again a small loss by reflection, but most of the light will emerge, refracted back to a direction parallel to its original path. In passing through the glass some light will be absorbed, and the extent of this absorption is dependent on the wavelength of the light. Most glass has a slightly greenish tinge, because some of the longer (red) and shorter (blue) wavelengths are absorbed in their passage through the glass.

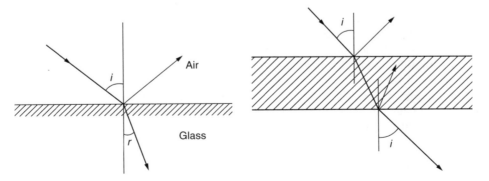

Figure 1.8 Reflection and refraction at a glass surface in air.

Figure 1.9 Passage of light through a polished glass plate.

Now consider the reflected light. It can be shown that for light entering the glass normally the amount reflected is given by the equation:

$$I_r = I_0(n - 1)^2/(n + 1)^2 \tag{1.13}$$

Where I_r represents the reflected intensity and I_0 the incident intensity. Since n for glass and most transparent solid materials is approximately 1.5, it can be calculated that in most cases I_r is about 4% of I_0. This means that on entering any transparent solid like glass, 4% of the light will be reflected at the surface. The same thing happens when light emerges from glass. Hence the amount of light transmitted by a sheet of glass cannot be greater than about 92% of the incident amount. It is true that by applying carefully designed coatings to the glass this reflection loss can be reduced, but it is important to remember that for all plain surfaces, these surface reflection losses are inescapable. It should also be noted that the proportion of light reflected from a surface in this way increases considerably as the angle of incidence is moved away from the normal and becomes more oblique.

Thus in passage through a transparent substance the light which emerges in the original direction will have suffered losses due to both absorption and reflection. The intensity ratio of the emergent to the incident light is called the *transmittance factor*. For any material the transmittance factor will vary with wavelength – when we speak of the transmittance factor at a single wavelength the phrase *spectral transmittance factor* is used.

1.6.2 Diffuse transmission

The case of direct transmission through perfectly transparent glass, described in the section above, is somewhat idealised. In practice, all transmitting materials scatter light to some extent – meaning that some of the transmitted light emerges in directions quite different from that of the incident light. Scattering of the light can occur: (a) through scattering at supposedly smooth polished surfaces; (b) through irregularities of shape or texture at the surfaces; or (c) through encountering inclusions of material with a different refractive index.

Scattering of type (a) occurs because the normal polishing process produces a myriad of micro-scratches on the surface of the glass. Scattering of type (b) can be seen in ground glass, or rough-rolled glass, both of which have irregular surfaces. Scattering of type (c) can be seen in materials such as opal glass, where the glass sheet contains inclusions of a glass with a different refractive index. In the case of highly polished optical glass, the components of (a) and (b) will be very small, and of (c) will be virtually zero, so that the transmission will nearly all be truly direct. In the case of a material like tissue paper, there will be large amounts of scattering of types (b) and (c), so that the transmission will nearly all be diffuse, with very little direct transmission.

1.6.3 Transmittance

The intensity ratio of the total light transmitted (in all directions) to that of the incident light is called the *transmittance*. The incident light need not all be incident in one direction, as for example in the case of skylight falling onto a translucent roofing material. Where the incident light and the transmitted light lie within defined cones the term *transmittance factor* is used; if these cones are complete hemispheres, the term *transmittance* is used. It should be noted that the term 'transmittance' is also used by chemists in a different sense, as described in the following section.

1.6.4 Optical density and absorbance

When measuring the transmittance factor for direct transmission, it is often more convenient to work on a logarithmic scale than a linear one, using the concept of *optical density*. If the transmittance factor is T, then the optical density D is defined as:

$$D = \log_{10}(1/T) = -\log_{10} T \qquad (1.14)$$

For example, if we have a material with a transmittance factor of 10%, i.e. $T = 0.1$, then the optical density will be 1.0. For a material transmitting 50%, the optical density would be 0.301. For a material transmitting only 1%, it would be 2.0.

Optical density measurements are widely used in the photographic industry and in chemical work of all kinds. Chemists primarily work with liquids in solution, and absorption measurements provide a very convenient method of analysis. Problems arise because liquids have to be contained in cells transparent to the wavelengths involved, and the solvents used also have intrinsic absorption characteristics.

Measurements on solutions are therefore made by comparing light which has passed through a cell containing the solution with light passing through an identical cell containing the solvent alone. This relativistic technique ensures that all absorption effects due to the cells or solvent are cancelled out, and the term 'transmittance' is frequently used for measurements made in this way. However since the results are invariably required for chemical purposes on a logarithmic scale, the transmittance values are always converted to *absorbance* values, and the term 'transmittance' is rarely used. Absorbance values A are related to transmittance values T' measured in this way in similar fashion to optical density:

$$A = \log_{10}(1/T') = -\log_{10} T' \qquad (1.15)$$

1.6.5 The Beer–Lambert law

It was discovered independently by the English chemist Beer and the French physicist Lambert that when light was passed through a liquid solution containing molecules of an absorbing species that the absorbance was proportional to the concentration of molecules of that species. The absorbance at each wavelength is in fact proportional to the number of absorbing molecules that the beam encounters, and thus depends on the intrinsic absorptivity $a(\lambda)$ of the solution, the path length b through the solution and its concentration c:

$$A(\lambda) = \log_{10}(1/T(\lambda)) = a(\lambda)bc \qquad (1.16)$$

If several species of absorbing molecules are present in a solution and the absorption characteristics of each are known, then by making absorbance measurements at several wavelengths, the concentration of each species can be determined – without the need to separate them chemically. Consequently the Beer-Lambert law is widely used in chemical analysis, but also finds many applications in the dyestuff, dyeing and printing industries.

1.7 REFLECTION MEASUREMENTS

1.7.1 Specular and diffuse reflection

When light falls on a surface, it may be reflected in one of two ways. A glossy painted surface provides a good example, as in Figure 1.10. At the smooth outer surface of the paint, which resembles a liquid, there is a mirror type reflection, as described in Section 1.6 above. The reflected light leaves the surface at an angle equal to the angle of incidence, and the spectral composition of the light is little altered. This type of reflection is called *specular* and shows the gloss on shiny surfaces.

Most of the incident light, however, penetrates the paint surface, where it encounters the particles of pigment in the paint, as shown at (b). These are of irregular shape, and will reflect the light in a myriad of different directions – any one ray may suffer multiple reflections before it escapes or is absorbed. This light emerges from the surface in all directions, and because the pigment particles absorb some wavelengths more strongly than others, the spectral composition of the light will be altered, i.e. there will be a colour change. This type of reflection is called *diffuse* and shows the 'body colour' of the

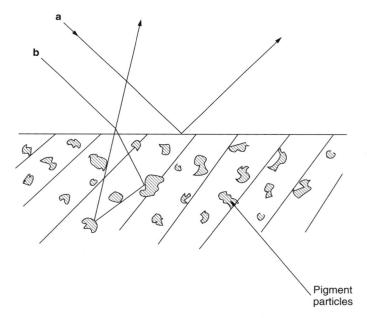

Figure 1.10   Specular and diffuse reflection from a painted surface.

material. An ideal matt paint would have only diffuse reflection. Most painted surfaces exhibit both specular and diffuse reflection, depending on the degree of gloss, whereas a mirror exhibits all specular and no diffuse component.

1.7.2 Surface texture

Reflecting surfaces which have a 'texture' show different reflection characteristics in different directions. Examples are all woven and knitted fabrics. The appearance of a dress fabric can change both when viewed from different directions and when lit from different directions. The 'stripes' on a newly mown lawn occur because the blades of grass have been rolled in different directions. To obtain reliable measurements of the amount of light reflected from a surface, therefore, we have to be careful in specifying both the illumination and viewing conditions. The CIE has established a series of standard 'geometries' for reflection measurements. Full details of these are given in CIE Publication 15.2 and are reviewed in Chapter 2.

It should also be noted that reflection from many structured surfaces causes some alteration to the polarisation characteristics of the light involved. Whilst this effect is rarely significant visually, it should not be forgotten because it may cause trouble with instrumental measurements.

1.7.3 Reflectance

Most reflection measurements involve comparing a sample with a *perfect reflecting diffuser*, an ideal surface that reflects all of the light of all wavelengths that fall upon it, equally strongly in all directions. A number of paints and materials have been developed which approximately meet these requirements. A typical white tile used as a reference in spectrophotometric measurement incorporates a mass of tiny crystals of barium sulphate ($BaSO_4$) in a transparent medium, and reflectance factors of over 99% can be achieved.

The *reflectance factor* of a surface is the ratio of light reflected from it to that reflected from a perfect diffuser under the defined geometrical conditions. We are usually concerned with its value at a single wavelength, when the term *spectral reflection factor* is used. The reflectance factors of everyday white materials such as paper are commonly in the 80–90% range. Strongly coloured surfaces show large variations of reflectance factor with wavelength (see Figure 1.11). Achromatic surfaces show little variation with wavelength, and the reflectance factors for grey and black are much lower than for whites – typically 20–60% for greys and 3–5% for blacks.

It is very difficult to produce a totally absorbing surface, i.e. an ideal black. Most of the so-called 'matt black' paints have reflectance factors of a few percent (optical density of less than 2), and it should be noted that at grazing incidence the reflectance factor of 'matt black' surfaces often becomes much higher. The interiors of optical instruments are usually painted matt black to minimise stray reflections, but this treatment is by no means always successful. Better black surfaces can be obtained by the use of black velvet or black flock wallpaper. If a total absorber is needed to provide as near as possible to zero reflectance, a black velvet lined cavity with a small aperture is recommended.

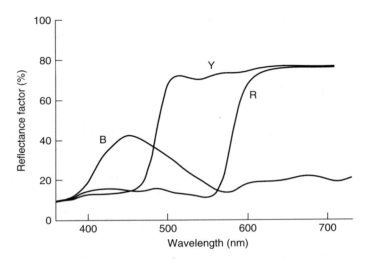

Figure 1.11 Spectral reflectance factors of three materials. The red material (R) absorbs strongly in the medium and short wavelengths and reflects only the long wavelengths; the yellow material (Y) absorbs strongly in the short wavelengths but reflects most of the medium and long wavelengths; the dark blue material (B) reflects only moderately the short wavelengths and absorbs all others.

1.8 COLOUR RENDERING AND METAMERISM

1.8.1 Colour rendering

The colour of an object as we see it is determined primarily by the *spectral power distribution* of the light that passes from the object to the eye. By 'spectral power distribution' we mean the relative proportion of radiant power at each different wavelength in that light. Many other factors, such as the surrounding colours, the viewing geometry and the observer's state of chromatic adaptation, affect the colour perceived, but the spectral power distribution (usually abbreviated to SPD) is by far the most important.

The SPD of the light reflected by a particular surface is the product of the spectral power distribution of the light incident upon the surface and the spectral reflectance factor at each point of the surface. By way of example, Figure 1.12 shows the spectral reflectance factor of a pink surface plotted against wavelength. Figure 1.13 shows the SPD of light from an incandescent lamp, and Figure 1.14 shows the SPD for a typical fluorescent lamp. Figure 1.15 then shows the SPD of light reflected from the pink surface when lit by the incandescent lamp, and Figure 1.16 the SPD when it is lit by the fluorescent lamp. Since the SPD of Figure 1.15 is quite different from that of Figure 1.16, we would expect the appearance of the surface to be quite different under the two lamps; and it is common experience that colours viewed under incandescent light often do look different when seen in fluorescent light. Perhaps the more surprising thing is that the coloured surface can be recognised as the same under both types of illumination, because the

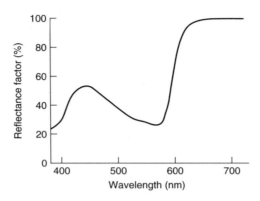

Figure 1.12 Spectral reflectance distribution of a pink coloured surface.

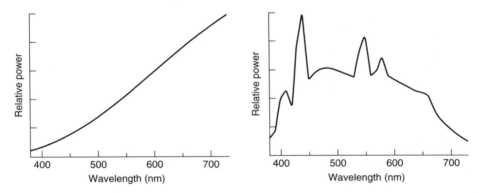

Figure 1.13 Spectral power distribution of   **Figure 1.14** Spectral power distribution of
an incandescent lamp.   a fluorescent lamp.

human vision system is able to adapt substantially from one condition to another (see also Chapter 4).

Many types of light source are in use today, and whilst some produce a satisfactory rendering of a wide range of colours, others do not. For example the simplest type of fluorescent tube – commonly referred to as 'white' – emits very little power in the red part of the spectrum, with the result that most red surfaces look rather dull when illuminated by it. Some light sources, such as the low pressure sodium lamps used in street lighting, emit only a single narrow band of wavelengths, and their rendering of colours is so poor that some colours are unrecognisable. We therefore need to have some measure of the colour rendering properties of light sources.

The CIE has produced a 'general colour rendering index', which can be thought of as a figure of merit for the colour rendering properties of a lamp. Its definition is based on the way the lamp renders the colour of a whole range of samples, and it is given the symbol

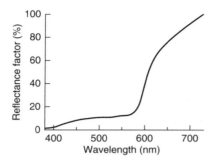 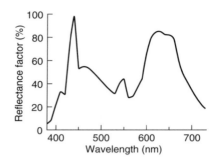

Figure 1.15 Spectral power distribution of the light reflected from the surface of Figure 1.12 when illuminated by the lamp of Figure 1.13.

Figure 1.16 Spectral power distribution of the light reflected from the surface of Figure 1.12 when illuminated by the lamp of Figure 1.14.

R_a. An ideal lamp would have an R_a of 100; a lamp with an R_a of zero would greatly distort colours. Details of minimum acceptable values of R_a for various applications are published by the CIBSE, but for colour engineering purposes it may be said that for any work where good colour rendering is required the light source should have an R_a of at least 80, and wherever accurate colour matching is required it should be at least 90. Many modern types of lamp meet these requirements.

1.8.2 Metamerism

Colours may be produced whose spectral reflectance curves are not the same, but will match under a specified illuminant. However if they are then viewed under an illuminant with a different SPD, they may no longer match. This phenomenon is known as *metamerism*, or more properly as *illuminant metamerism*, and most commonly occurs with near-neutral colours such as olive greens, browns and greys. For a pair of samples to exhibit metamerism, their two spectral reflectance curves must have crossover points within each of the three wavelength bands to which the ρ, γ and β cone cells in the observer's retina are sensitive [2].

A different form of metamerism is *observer metamerism*, in which two surfaces having different spectral reflectance curves viewed under a given illuminant may appear to one observer to match but to another observer to be different. This arises because of differing spectral sensitivities of the three channels between the observers, due to natural variance across the human population. As noted at the beginning of the chapter, it is rare for any individual to have exactly the same visual sensitivity as the CIE standard observer, and so observer metamerism may cause differences between the colour calculated using the CIE system of colorimetry and the colour actually observed. Observer metamerism can also occur between different trichromatic systems, for example a human observer and a digital camera [3].

REFERENCES

1. *Colorimetry*, CIE Publication No. 15.2, 2nd Ed., (1986).
2. R. W. G. Hunt, *Measuring Colour*, 3rd Ed., Fountain Press, Kingston-upon-Thames, (1998).
3. Berns, R. S., *Principles of Color Technology*, 3rd Ed., John Wiley & Sons, (2000)
4. *CIBSE Code for Interior Lighting*, The Chartered Institution of Building Services Engineers (CIBSE), London, (1994).
5. *IES Lighting Code*, Illuminating Engineering Society, USA

Publications of the CIE (Commission Internationale de L'Éclairage) may be obtained in the UK from the Chartered Institution of Building Services Engineers, 222 Balham High Road, London SW12 9BS.

Instruments and methods for colour measurement

Danny Rich

2.1 INTRODUCTION

2.1.1 The need for colorimetry

In the reproduction of colour and coloured images, the assessment of the colour appearance or colour match has historically been the job of experienced, trained artisans. They were schooled and apprenticed by a master colourist or artist who taught them how to hold the specimen or image, how to describe the appearance of the colour or colour difference and how to select the correct set of dyes, pigments or inks to create a workable recipe for the desired colour match. They were also taught a bias for certain colours and methods for making any secret recipes.

The master colourist could match any colour within the gamut of his or her known primaries and could find the most visually acceptable or pleasing near-match to any colour that was outside this gamut. This approach worked well when the coloured product, be it an illuminated manuscript, a fine textile or the paint to decorate and protect the woodwork of a stately house, was conceived, reproduced and sold locally. In today's fast-moving, global marketplace, more objective and rigorous methods are required for creating, matching and reproducing coloured materials and images. This is especially true

Colour Engineering, edited by P.J. Green and L.W. MacDonald.
© 2002 John Wiley & Sons Ltd.

in the application of electronic imaging where the coloration process occurs in milliseconds rather than hours. Colorimetry is the technology that attempts to capture the essence of the visual perception of the light reflected from or transmitted through coloured images or emitted by sources. Colorimetry then converts that essence in an objective nomenclature to communicate the colour or colour-differences to someone in a different place and time and still obtain the same level of fidelity and aesthetics. The following chapters describe the engineering approaches to the creation and manipulation of coloured images and to many of the issues raised here, especially electronically communicating and reproducing coloured images, but *all* applications of colour engineering must begin with a basis in colorimetry.

2.1.2 The principles of colorimetry

Colorimetry, at its purest and most basic level, is quite simple. The technology was developed to answer one question, 'Does this test colour match this reference colour?'. Basic colorimetry can provide nothing more than the answer to that question. To obtain further information is to extend the technology from the measurement of colour into the measurement of colour appearance. Topics like colour difference, colour constancy or lack of constancy, corresponding colour and answers to the question 'How does this image or image element appear?' require additional assumptions and technologies beyond the scope of colorimetry. If there is one principle that needs to be kept in mind while becoming proficient in colour engineering, it is that 'Colorimetry does not describe what a person sees!'. Colorimetry is fully enveloped by the technology of colour matching. The details in colorimetry are found in how the colour match is created and reported.

2.1.3 Making the transition from what we 'see' to quantifying how we 'match' a colour

It can sometimes be difficult for the novice colour engineer to come to be completely comfortable with this limited definition of colorimetry. After all, as pointed out in many textbooks, colour is what you see and the human visual system is highly adapted to the visual perception of colour. Indeed this is true, but words are often pregnant with deeper meanings and implications. So it is with the concept of perception. The perception of colour involves many factors, most of which require the interaction of neural processes and physical phenomena that are poorly understood at best. Sensation involves fewer processes and occurs at much earlier stages in the visual system. The assessment of a colour match involves the sensation of identity or difference in the two sources of colour. Colorimetry, then, is still a visual measurement, even though we may utilize a colorimeter to make the measurements more precise. In the following paragraphs we will look at the various ways to construct a colorimeter and how that colorimeter can be used to answer the question of what reference colour is matched by the test colour. From an understanding of visual colorimetry, the development of an international standard of colorimetry can be understood and finally applied to automatic, electronic colorimeters.

2.2 VISUAL COLORIMETRY

2.2.1 A method to uniquely map the colour of lights and objects

Visual colorimetry is the most direct and accurate method for objective quantification of colour. It is also the most difficult. Visual colorimetry requires the colourist to mix lights of different colours until a match between the test colour and the mixture is obtained. There are several common components found in all colorimeters. These include a source of light, a source of primary or mixing colours, and viewing optics. Figure 2.1 shows a block diagram of a visual colorimeter. The principle of visual colorimetry is very simple and very familiar to most people today. It is the same principle that is used to set the mood on the stage in a theatre, generate colour signals on television screens, computer displays and slide or motion picture film projectors. This approach to colorimetry is also appropriate to teach primary school children to mix tempera paints, or to dye our clothing or hair, and is familiar to artists and technologists who mix colorants to match the colours of natural objects in a paint medium. The physics of how the primaries interact to form the final coloured stimulus is different, but the process and result are completely analogous.

The earliest commercial colorimeters were thus visual colorimeters. Donaldson [1] in the UK produced several different models of visual colorimeter. The light source was a stable, ribbon-filament incandescent lamp which illuminated both halves of a bipartite

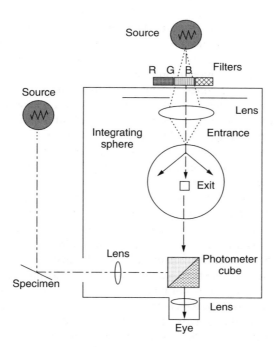

Figure 2.1 Typical visual colorimeter with three primary filters.

visual field. A transparent, coloured specimen could be placed in one half of the visual field and located between the coloured primaries opened full and the viewing optics. In the other half, the mixture of the coloured primaries is adjusted by opening or closing shutters over the primaries, allowing more or less light to pass through the primary filters. Opening the shutters together made the image brighter while closing them together made the image dimmer. The shutters on the test field could also be adjusted so that the light seen through the viewing optics, usually a telescope imaging the two light-mixing chambers side by side into the observer's eye, could be matched for both colour and brightness. When the observer was satisfied with the quality of the match, the positions of all the shutters would be recorded, fixing exactly the state of that colour mixture on that colorimeter. Those numbers could be communicated to anyone else, in the world, with the same model of Donaldson colorimeter and the match could be visualized by setting the shutters to the same positions. But if the second laboratory did not have the same model of colorimeter, the shutter settings would be of little value. Visual colorimetry as practised in this way was thus a rather limited tool, good for evaluation of repeated specimens of the same colour but not for communication of colours to engineers in outside laboratories.

2.2.2 Development of the CIE method of visual colorimetry

In the late 1920s two researchers in the London area began studies to quantify the methods of visual colorimetry. One was John Guild at the National Physical Laboratory and the other was a graduate student at Imperial College, W. David Wright. Being interested in both the basic science of colorimetry and the needs of commerce, Guild [2] built a colorimeter that was similar to those of Donaldson, with Red, Green and Blue filters as primaries but with finer resolution and the best optical characterization available to a national standardizing laboratory. Wright [3, 4], being a physics graduate student, was more interested in making the most thorough determination of these colour mixture functions and built his visual colorimeter using prism monochromators for the primaries; he seems to have first used the term 'Trichromator' for such an instrument. Guild was able to convince seven people to go through the difficult task of making colour matches to 30 or so narrow bands of wavelengths. The result was a spectral curve of colour mixtures representing the amounts of each of Guild's Red, Green or Blue primaries that would be mixed to match a given wavelength of light. Figure 2.2 shows a typical set of colour mixture curves. Wright was a bit more successful at recruiting observers and had 10 people make matches on his Trichromator. Wright's colorimeter then produced curves that described the amounts of his primary wavelengths (460 nm, 530 nm, 650 nm) required to match the spectral colours. Even before Guild had published, Wright took the two sets of data and compared them. They found that the colour mixtures curves in terms of primaries, RGB, of both experiments were very similar and so, using linear algebra, they transformed the two sets of results to a common set of monochromatic primaries and then averaged the results. Finally, they normalized the results using a white point determined by NPL and made the middle wavelength or green mixture function identical to the CIE 1929 standard of photometry. Details of how they were able to achieve this can be found in the review by Fairman *et al.* [5]. The resulting set of colour matching functions were transformed from real mixing primaries to an imaginary set of primaries with theoretical

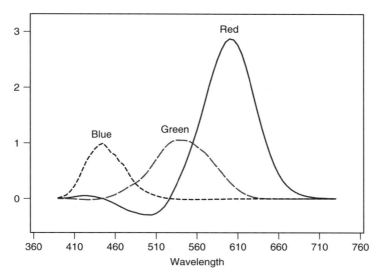

Figure 2.2 Colour mixture curves for matches to spectral lights using Red, Green and Blue primaries.

colours outside the gamut of spectral colours, and were adopted by the CIE in 1931. These colour matching functions, now known as spectral tristimulus values, have become an international Standard Observer for small visual fields (less than 4°). This is often referred to as the 2° Observer. The transformations resulted in colour mixing curves that were all positive, since the primaries were taken from points outside the gamut of real colours, as can be seen in Figure 2.3. The CIE named the primaries X, Y and Z; the colour matching functions, since they were based on the average of 17 observers, were given the symbols $\bar{x}$, $\bar{y}$ and $\bar{z}$, which in modern notation would now be termed $X(\lambda)$, $Y(\lambda)$ and $Z(\lambda)$, hence the name spectral tristimulus values. Since the whole system was linear by definition, this would allow any visual colorimeter readings to be converted to CIE equivalent readings and thus standardize the whole process. Having the middle primary equal to the photometric function allows one to estimate both chromaticness and luminance with one reading.

The whole process was repeated in the late 1950s by Stiles and Burch [6] in the UK and by Speranskya [7] in the USSR, but using field sizes of 10°. Unfortunately, no large field luminance or brightness function has ever been standardized, so they could not provide a reference to large field photometry. Even so, many people, most recently Trezona [8], have observed that the large field $Y_{10}(\lambda)$ function correlates well with luminance factor in surface colours. The data that were collected for the development of the large field colour matching functions are now considered to be the most accurate and reliable ever obtained. They have recently been confirmed by Stockman *et al.* [9, 10] along with the most reliable estimates of cone fundamentals or spectral response curves. This new information is being assembled by CIE TC 1-36 [11] into a new chromaticity diagram based on these physiologically significant coordinates.

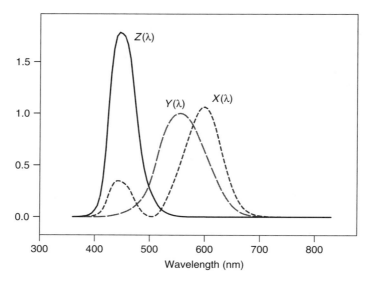

Figure 2.3 CIE 1931 Standard Observer's color matching functions.

Just as the observer had to be characterized and standardized, so too the sources of light used in colorimetry had to be standardized. The CIE is primarily an organization of illumination engineers and not colour scientists. Over the past 80 years, the CIE has standardized two spectral distributions of light. One was named Illuminant A and represents the light output of a ribbon filament (as opposed to a coiled wire) incandescent lamp operated at a voltage such that its chromaticity is equivalent to that of a theoretical black body heated to a temperature of 2856 kelvin. The second standard illuminant is known as D65 and represents an average of the light incident on the surface of the earth from an overcast sky when facing north in the northern hemisphere with an equivalent or correlated colour temperature of about 6500 kelvin. The CIE has defined an entire continuum of phases of daylight ranging in correlated colour temperature from about 4000 kelvin up to about 40 000 kelvin. The latter represents the clear, blue, north sky without the sun occluded.

Note that in the descriptions above, the term 'illuminant' has been used to define the table of numbers that represents some spectral distribution of radiation. This is also a CIE recommendation. The term 'illuminant' is used to describe a tabular or graphical display of the spectral distribution of radiant flux, while the term 'source' is reserved to describe the physical device or material that actually produces the radiant flux. In the case of Standard Illuminant A, the corresponding Source A is the ribbon filament incandescent lamp. In the case of Standard Illuminant D65 there is no source available. So the CIE has also developed a method [12] for testing the quality of daylight simulators – sources that approximate the spectral radiant distribution of D65.

While the CIE has standardized only two illuminants they have recommended many others. Of particular interest to colour engineers involved in imaging is CIE Recommended Illuminant D50 which is a standard in the graphic arts for viewing hardcopy images. This illuminant is similar to D65 but has a colour temperature near to 5000 kelvin and a more neutral spectral response making it more suitable for visual evaluation of tone scales and colour balance. An older illuminant, no longer recommended for colorimetry by the CIE, is Illuminant C, which represents daylight without the ultraviolet component that is present in D65 but has a correlated colour temperature of 6774 kelvin. It thus approximates daylight through a window, such as an artist's or graphic designer's skylight, and is the illuminant used in the definition of the Munsell [13] colour order system. The source associated with Illuminant C is constructed by placing a series of liquid-filled cells in front of a standard Source A. The chemicals required are now difficult to obtain but there are glass filters available that are capable of making an adequate simulation of illuminant C using an ordinary tungsten lamp. With the recommended illuminants and either an associated source or a source simulator and the standard observer functions, visual colorimetry can be applied to measurement problems in colour engineering. Figure 2.4 shows the spectral radiant distributions of CIE Illuminants A, C, D50 and D65.

2.2.3 Applications of visual colorimetry

Given any set of visual colorimeter readings (RGB) and the white point of the colorimeter, it is possible to calculate standard CIE values and compare them to those of someone else who has done the same experiment. But collecting the data is still very

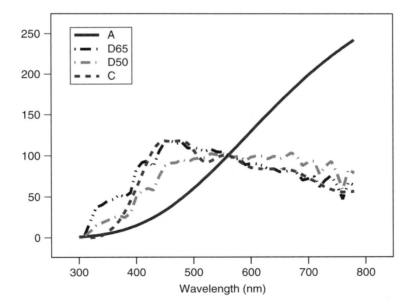

Figure 2.4   CIE Standard Illuminants A and D65, Recommended Illuminant D50 and Deprecated Illuminant C.

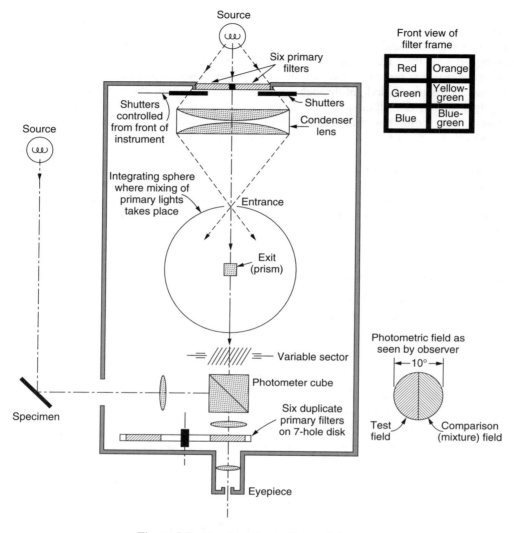

Figure 2.5   Donaldson's six-filter colorimeter.

tedious and the gamut is constrained by the luminance and chromaticity of the three primaries.

Donaldson [14] attempted to address each of these issues with the development of his 'Six Filter Colorimeter'. This instrument had the usual R, G, B filters but it also had orange, cyan and magenta filters. These three additional filters greatly increased the gamut but also greatly increased the complexity of the matching process. To aid the observer, Donaldson then added an equivalent set of six filters in the viewing optics. By placing a red filter in the viewing optics, only red light in the test field and in the match field would be viewed. The observer then could adjust the intensity of the red primary independently of the other five primaries. This was a far easier, unidimensional task and greatly improved both the speed and precision of matching. As shown in Figure 2.5, surface colours could be matched using the Donaldson six-filter colorimeter, but matches were limited to sources based on incandescent lamps.

2.2.4 Disadvantages of visual colorimetry

The most significant disadvantage of visual colorimetry is the tedium involved in the matching process. Even after much experience, each match is a struggle – overcoming the natural inclination for subtractive mixture matches and forcing oneself to think in the additive mixture paradigm. The result of the struggle is that even with the best of tools, the precision varies from day to day, from lab to lab and from person to person. The learning curve for understanding the controls and how they affect what is observed can also be quite steep, and it may take weeks or months before one is fully competent to deal with the struggle. Clearly, while the method is very accurate, its lack of consistency and precision adversely affects its usefulness in colour engineering. What is needed is a method that can give up some of the accuracy of the visual process for a significant improvement in the precision of the measurements.

2.3 ANALOGUE SIMULATION OF VISUAL COLORIMETRY

2.3.1 Replacing the human eye with an optoelectronic sensor

One of the surest ways to develop precision in a visual measurement is to substitute some form of electric eye or light meter for the human observer. Utilizing a method similar to Donaldson's compensation filter method in which the light from the specimen or test field and that from the match field were limited to one primary at a time, the photoelectric sensor would read the intensity through that filter and an equivalence between the filter readings and the visual values was observed.

2.3.2 Substituting coloured filters to approximate the CIE colour-matching functions

Optical instrument engineers soon took up the challenge to develop a photoelectric filter colorimeter. Illuminating engineers and photographers had already started using photoelectric photometers or light meters which were constructed by placing a well-designed green filter in front of a photocell and then reading the voltage across the cell as an indication of the light levels. To build a filter colorimeter all that was needed was to create filters to match the other two CIE colour-matching functions. This, it turns out, is much more difficult than it appears. Even with modern optical design tools, it is not possible to create a single piece of glass that matches the unusual shapes of the colour-matching functions. But approximations can be found and, when used to compare a product standard to a production sample, the errors are similar and subtract out, leaving difference components that are as exact as visual colorimetry and far more precise. Filter-based colorimetric systems have been constructed from three simple filters, four simple filters, three filter stacks and even mosaic arrays of small slivers of different filters. An amber filter cannot form as good a fit to the short-wavelength peak of the $X(\lambda)$ function as it does to the long-wavelength peak, so the short-wavelength bump is usually approximated by using a scaled-down portion of the blue filter or by adding a separate indigo filter, thus forming a four-filter colorimeter. Just as in the visual colorimeter, there is a source of light, a set of primary filters and a detector of the light that passes from the reference colour and from the test or trial colour. The green and blue filters are scaled using normal linear algebra and the amber filter must be combined with either the blue or indigo filters such that

$$X = a \cdot A + b_2 \cdot B \text{ or } X = a \cdot A + i \cdot I$$

$$Y = g \cdot G \tag{2.1}$$

$$Z = b \cdot B$$

Figure 2.6 shows the spectral approximation to the tristimulus values using three separate coloured glass filters. The differences between the solid curves, representing the filter colorimeter, and the dashed curves, representing the CIE colour-matching functions, are quite substantial. Note that the peak of the short-wavelength bump of the $X(\lambda)$ function does not align with that of the $Z(\lambda)$ function. Figure 2.7 shows two colorimeters, each using four filters. In one, the coloured glass filters are made from separate stacks of filters (known as serial filters) so that the combined transmission is an approximation to the colour-matching functions. In the other, the coloured glasses are mounted side-by-side (known as parallel or mosaic filters). The latter will usually give a better fit to the spectral tristimulus functions and have a higher throughput but are more difficult to construct

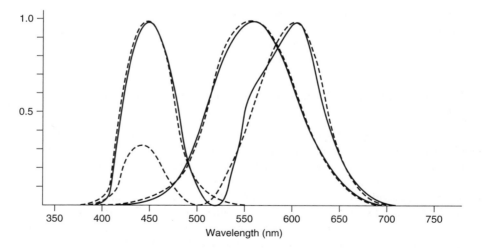

Figure 2.6 Filter fit for a three-filter analogue photoelectric colorimeter.

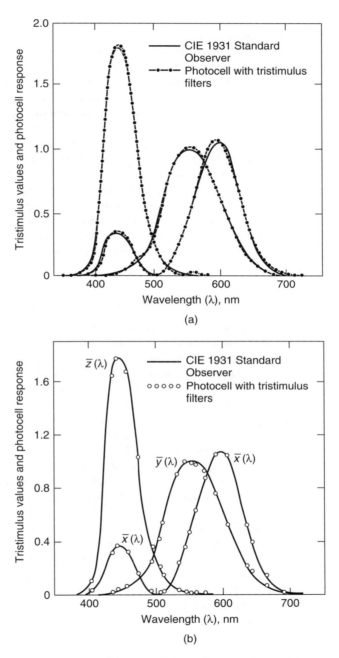

Figure 2.7 (a) Filter fits using four serial filters; (b) filter fits using four parallel filters.

and much more expensive to manufacture. A third method of constructing filters, only recently having been used commercially, involves state-of-the-art thin film deposition and monitoring to form an interference stack with a precisely defined but wide passband. Such filters are extremely difficult to make without stringent extreme process control on the deposition equipment. Still, reports in the literature by Engelhardt and Seitz [15] indicate that it is possible to manufacture exact copies of the tristimulus functions, including the double peak of the $X(\lambda)$ function, using modern thin film, interference filter technology, close monitoring of each deposition layer and adaptive changes in the filter design. Instruments designed for the measurement of the colour difference of objects will also include the spectral variations of the instrument source and the detector responsivity in the analogue computations, so that the final filter functions produce a set of approximate tristimulus values for a specific CIE illuminant and observer. Thus, these instruments are incapable of assessing the level of metamerism. Instruments designed for the measurement of the colour of self-luminous objects, such as lamps, CRT or flat-panel displays and LEDs, will have only the spectral variations of the detector system in the analogue computations, so that filter function becomes the ratio of the tristimulus function to the detector sensitivity function.

2.3.3 Assessing the 'goodness of fit' of a set of colorimeter filters

The CIE has developed a method for assessing the goodness of fit of photometer filters [16] and is working on the development of a metric of fit for filter colorimeters. At this point it is too early to say exactly how that metric will develop, but it is felt that it will be along the lines of the photometer conformance metric, where the average difference between the relative spectral responsivity of the optical sensors will be compared to the relative spectral distribution of the spectral tristimulus values or colour-matching functions. Currently, the only reliable way to assess the conformance of a filter colorimeter is to measure a series of standard colours, either object colours or standard lights in combination with stable, standard coloured filters. The object colours or filtered sources will have been characterized using spectral measurements and then converted to tristimulus values following methods described in the next section. Ceramic tiles, such as the Ceramic Colour Standards II [17], or plastic or paint panels characterized by standardizing or industrial consulting laboratories, or even chips from a colour atlas such as the Munsell Atlas [18] or the Swedish Natural Color System [19], can be used as diagnostic standards for filter colorimeters. If the measurement geometry between the spectral-based readings and the colorimeter-based readings are the same, the differences in the tristimulus values from the two measurements will be proportional to the filter fit differences. This is an important caveat, since most colour order systems are measured and maintained using integrating sphere instruments, while nearly all filter colorimeters are bidirectional instruments. Measurements are not transferable between geometries.

2.3.4 Schematic description of analogue filter colorimeters

Filter colorimeters are produced to assess the colour and colour-differences of both self-luminous objects, such as lamps, panel lights, LEDs and coloured visual displays, and

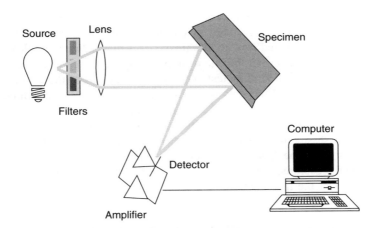

Figure 2.8 The Tristimulus Filter Colorimeter – an analogue approximation of the visual colorimeter.

reflecting or transmitting surfaces or objects. Figure 2.8 shows the schematic of a typical commercial filter colorimeter. It is readily apparent that the optoelectronic instrument has many components and structures in common with the visual instruments shown in Figure 2.1. The RGB filters have been replaced with the AGB (Amber, Green, Blue) filters that can form a better fit to the peak of the $X(\lambda)$, $Y(\lambda)$, $Z(\lambda)$ (or $\bar{x}$, $\bar{y}$, $\bar{z}$) colour-matching function. Such instruments will trade off the accuracy of a visual colorimeter for the improvements in precision of the solid state photo sensor, usually a photodiode. If greater accuracy is required, one can improve the accuracy by performing local standardizations to product standards or transfer standards with nearly identical colorimetric properties as the test materials. There is an ASTM standard [20] practice for improving the accuracy of filter colorimeters used to measure the colour of visual display units that utilizes the known properties of a similar light source. When used in the difference mode, if the reference colour and the test colour are not metameric, the errors in the colorimeter filter functions will be nearly identical and subtract out in the difference calculations. The use of local standards in object colour evaluation has been performed for decades and is often termed the 'hitching post' method of colour measurement. It is a very effective way of compensating for any number of instrumental errors. It is still in use today in the paper industry and is recommended by ISO/TC 6 [21] as the only method to provide both accurate and reproducible measurements of the whiteness and brightness of papers around the world.

2.3.5 Disadvantages of analogue filter colorimeters

While filter colorimeters can be used quite effectively in critical applications as colour-difference meters or in less critical applications as absolute colorimeters, there are some drawbacks to the instruments. First, there is a high cost to obtain accuracy. Whether the accuracy is obtained by developing mosaic filters or by maintaining a set of reference standards, in every colour that must be measured, this becomes a major logistical problem and someone has to pay for the logistics of selecting and cutting filter pieces or developing and characterizing the certified reference standards. Second, as pointed out above, most filter colorimeters are optimized for one illuminant. This makes them less flexible in terms of the assessment of metamerism, but more importantly in terms of the accuracy of the readings if the current production or test specimen deviates from the exact composition of the production standard. When this happens, there is no way to determine that it has occurred, from the readings taken with a filter colorimeter, and it will be a mathematical certainty that at some point in the life of that product a human observer will strongly disagree with the assessment of the colorimeter. The best-case scenario for that eventuality is for that person to be the Quality Assurance manager. The worst-case scenario is for that person to be the customer. What can be done to minimize the impact of this scenario? That is the subject of the next section.

2.4 DIGITAL SIMULATION OF VISUAL COLORIMETRY

2.4.1 Replacing the analogue filters with an abridged spectrometer

One of the simplest ways to create an abridged spectrometer is to replace the broadband coloured filters with a sequential series of narrowband interference filters. This was the method of choice for constructing colour-measuring instruments in the late 1960s through the mid 1970s. Several companies around the world created and refined such instruments. Usually, the instruments included a set of analogue tristimulus filters as well. The earliest instruments had only 10 filters representing a selected set of wavelengths or ordinates which could then be very simply summed to generate a fairly accurate set of tristimulus values [22]. The selected ordinate system takes a small number of wavelengths, optimally selected to provide an estimate of the area under the ideal tristimulus filter, and 'digitizes' the spectral product of the object spectral reflectance or transmittance, the colour-matching functions and the illuminant under which the colour is to be evaluated. This was done in much the same way that strip chart recordings in chromatography were digitized and converted into accurate estimates of the area. While 10 selected ordinates can provide adequate accuracy in the generation of a digital approximation to the ideal CIE tristimulus values, the 10 points do not provide a very meaningful sampling of the spectral reflectance factor or spectral transmittance factor for use in analytical applications, such as colorant recipe prediction, colour strength or any other spectral-based index. The next generation of instruments began using filters set at 20 nm intervals from 400 nm to 700 nm, 16 points in all. The filters were mounted in a large metal wheel and the wheel was rotated so that the filters were placed in the measurement beam between the specimen and the detector one at a time. The detectors would require two to four seconds each to collect enough light to produce a reliable reading. A full spectral scan would take 40 to 90 seconds. This instrument could not be described as 'blazingly fast'.

The biggest competitor to the filter instruments was the scanning spectrophotometer. This instrument would use a monochromator as the spectral device and would slowly and continuously change the centroid wavelength of the spectrometer from just below 400 nm to just above 700 nm. By installing a digital encoder onto the wavelength drive motor, this instrument could be set up to 'accumulate' readings at selected ordinates or to print out readings at a small number, generally 16, of evenly spaced sampling intervals while drawing a continuous plot of the spectral radiance, reflectance or transmittance.

Today, scanning monochromator systems and wheel-mounted interference filters have been replaced by arrays of photodiodes which collect spectral data in parallel, through either a series of discrete interference filters or flat-field holographic grating spectrographs. Such devices are rugged and permanently aligned and can capture the spectra of flashing lamps. The few scanning systems that are still available today have greatly increased the rotational speed of the filter wheel or dispersing element so that readings never take more than a couple of seconds, thus keeping the instruments competitive with the newer parallel designs. The same instrument, when equipped with the proper optical geometry, can also return reflection density values from prints, and is thus termed a spectrodensitometer. The distinction between the spectrometer-based instrument and the traditional colorimeter or densitometer is the use of the digital approximation to the spectral products. The older term for a transmission or reflection spectrometer, *spectrophotometer*, is still used in both the literature and language of colorimetry and analytical chemistry but is slowly being replaced by the more exact terms used above. The suffix 'photometer' literally means light measuring, and the field of 'photometry' involves the luminous efficiency function, termed $V(\lambda)$ by the CIE, implying a vision-based task. The reason for this term arises from the earliest days of spectroscopy before the days of photoelectric sensors. The two beams of a spectrometer (sample and reference) would be combined in a telescopic prism known as a *photometer* cube, and a human observer could look at a bipartite (side-by-side) image of the two beams and adjust the intensity of the reference beam until a visible match to the sample beam was obtained. Those early instruments really were spectro-*photometers* and that name stuck with the instruments even after photoelectric and optoelectronic sensors replaced the human observer. In modern colour engineering, the instruments are designed specifically for colorimetry; even though they may be capable of returning the underlying spectral data, their primary application is that of colorimetry and so the term spectrocolorimeter or spectrodensitometer has become the terminology of choice.

2.4.2 Assessing the 'goodness of fit' of abridged spectrometers

Just as with the visual and filter colorimeter, there are a number of parameters that must be controlled for a spectrometer to be useful as a spectrocolorimeter. The operating characteristics of a spectrometer for use in analytical chemistry are quite different from those of a spectrocolorimeter. While some chemical spectrometers can be used for colorimetry, their cost and complexity exclude them from most applications, both types of instrument have many characteristics in common and only the specifications of those characteristics need to be changed.

The main operating parameters of a spectrometer are the wavelength scale, radiometric scale, spectral sampling interval and spectral pass band. The radiometric scale transforms the spectrometer reading from the electrical domain into an optical property. This property may be spectral transmittance, spectral reflectance, spectral reflectance factor, spectral radiance, spectral radiance factor or spectral irradiance. There are accepted, international standards for each of the scales described above. The instrument manufacturer will generally supply a realization of one or more of these scales but it is the responsibility of the user to develop a plan and procedures to verify the scale realization. It is also the responsibility of the user to confirm the short-term and long-term repeatability of the instrument. The ISCC maintains a list of available standards and a guide on how to use them [23].

For verifying the scale of transmittance, series of neutral density filters are available from both primary and secondary standards laboratories and also from filter glass suppliers. These filters will test both the scale of transmittance and the linearity of the detector/amplifier/digitizer system. For verifying the scale of reflectance or reflectance factor, there are series of neutral plaques, either ceramic tiles or plastic chips, that can be used to test the scale of reflectance factor, the linearity of the detector/amplifier/digitizer system and the optical zero of the scale. This latter parameter is usually not an issue in transmittance readings because the light beam is simply blocked or shuttered and the readings are taken with no light passing through the sample beam. Reflection optics can and often do produce homochromatic (similar wavelength) stray light that must be measured and subtracted from all subsequent readings. This correction should always be done at the time the white tile is used to standardize ('calibrate' in historical instrument terminology) the radiometric scale of the instrument.

2.4.3 Schematic description of digital spectrocolorimeters

Figure 2.9 shows the schematic diagram of a typical spectrocolorimeter – that is a spectrometer designed for colorimetry and not for analytical spectroscopy. Note that the design is very similar to the visual colorimeter and the filter colorimeter. The main difference is that the set of RGB filters has been replaced by a spectrometer. That spectrometer may be a prism, a grating, or a series of filters or light-emitting diodes. In each of the schematics that have been shown, the filters have been associated with the light source. But most modern colour-measuring instruments have moved the filters from the light source side of the specimen to the detector side of the specimen. This has two advantages. First and foremost, it will correctly assess the reflectance and fluorescence from a specimen with mild levels of fluorescence, which almost all papers and textiles possess. Second, it makes the colorimetric radiometer independent of the light source so that the same filter/detector system can be used to read the colour of emitted light displays and reflected or transmitted light from prints or films. This feature has been shown to be very useful in many engineering applications of colour management. While it has not yet been produced, the ideal combination would be an instrument in which a relay lens could be mounted at the normal reflectance aperture so that the imaging system of the reflectance spectrocolorimeter could be transformed into a tele-spectroradiometric colorimeter.

2.4.4 Advantages and disadvantages of digital spectrocolorimeters

The main advantage of the digital spectrocolorimeter is the more accurate characterization of the colour and colour-difference of metameric pairs. This is achieved by sampling and digitizing both the standard illuminant and/or observer and the object or display at discrete

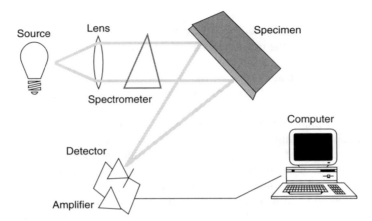

Figure 2.9   The Spectrocolorimeter – an analogue approximation of the visual colorimeter.

points along the visible spectrum and then numerically computing the tristimulus values. From the tristimulus values, any number of other colorimetric properties may be derived.

This advantage comes at a price in both complexity and size. It also requires that the spectral scale of the colorimeter be very accurately established and maintained. For strongly metameric specimens, it has been shown that errors in the wavelength scale of a few tenths of a nanometre translate into CIELAB colour differences of about the same magnitude. Likewise, errors in the radiometric scale of a few tenths of a percent will result in colour differences, between the readings, of a few tenths of a percent. If your product tolerance is to be 1.0 colour-difference unit then the instrument's contribution to that tolerance is required to be less than 20% or about 0.2 unit. That level of precision and reproducibility is difficult to obtain and maintain in a production environment. Maintaining a precision and reproducibility of 0.4 to 0.6 is much more typical in production control applications and is more than adequate for most imaging and electronic publishing situations.

2.5 SELECTING AND USING COLORIMETERS AND SPECTROCOLORIMETERS

2.5.1 Reading and understanding specifications and technical literature

Like any good engineering tool, a colorimeter should come with a set of specifications that describe how it will perform its function and what level of conformance to national or international standards the operator can expect from the instrument. Unfortunately, the history of the colour-measuring instrument business is littered with one-of-a-kind terminology and numerical values. Conformance to standards has never been a high priority for instrument makers. Since the early instruments were being built and installed in self-contained production facilities, size, durability and ease of use were always more important concerns. The operators were frequently technicians, tradesmen or craftsmen and not engineers. This section will try to list the more common terms and relate them to the operating conditions of the instruments as might be encountered in a colour engineering application.

The instrument specification page of the sales literature should clearly state what the general geometric conditions of the instrument are. These may include the use of an integrating sphere resulting in a diffuse illumination and a directional viewing or a pair of angles in a bidirectional configuration. The angle of view should be specified in terms of centroid and aperture angles. Today, almost all instruments with integrating sphere are di:8° or de:8° (see 2.6.2 for an explanation of terminology associated with instrument geometry) or capable of being switched between the two or reading in both conditions simultaneously. The other geometry for colour engineering is 45°c:0°, where the illumination is circumferential from several (6 to 8) points at 45° from the normal to the sampling aperture and the sensing is along the normal to the sampling aperture. The exact number and placement of the illumination points should be specified. It is rare for a manufacturer to provide more information to the user than this general synopsis. The documentary standards on the design of instruments are quite general and so the manufacturer merely claims conformance to those general standards.

The next parameter of interest in any colour-measuring instrument is its repeatability. This number represents a great many things but is usually given only in terms of short-term repeatability. It will be a very small number and identifies the drift of an instrument when making a few (20 to 30) readings of a white or pale grey ceramic tile as quickly as the instrument will allow. This measurement rarely requires more than two or three minutes to complete. In a well-designed instrument the number will be approximately equal to the resolution of the analogue to digital converter used to process the photocurrent. Values of this property above 0.1 should be considered as unacceptable for engineering applications but may be used for certain kinds of sorting or palette searches.

A second form of repeatability is often given, which represents the collection of the readings (30 or more) over an 8-hour period. The represents the drift or noise level across a single work shift. Again, this number should be small, but for instruments in which it is not, the standard operating conditions should indicate that re-standardization is required on a more frequent basis than once per shift. Longer-term repeatability, often called reproducibility, may also be of interest but is infrequently cited in instrument literature. Here one would obtain an estimate of the instrument drift over weeks or months. The longest study ever reported involved readings taken over a period of more than 3 years [24].

The next parameter found in most instrument specifications is inter-instrument agreement or instrument reproducibility. This property is generally stated in terms of a colour difference between two instruments or between a test instrument and a master instrument or the average of a group of production instruments (often called the production aim point). The confusion begins with how that colour difference is then reported. Sometimes it is reported as an average, sometimes as a maximum, a typical value, an RMS value, or the MCDM or Mean Colour Difference from the Mean. Some instrument data sheets have even broken the specifications down into agreement on neutral and pastel colours versus high chroma or dark colours. On occasion, a standard deviation or variance may

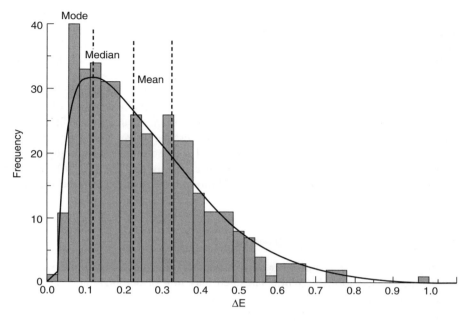

Figure 2.10   Statistical distribution of total colour difference (DE) showing parameters of most frequent (Mode), 50% above/below (Median) and arithmetic average (Mean). Based strictly on probabilities, a random reading is more likely to fall near the mode or median than near the mean.

be reported on the colour difference. Such a term is truly meaningless. Figure 2.10 shows the distribution of colour differences from a pair of well-standardized and maintained instruments. Note that it does not have the usual bell-shaped curve of the normal statistical distribution. Adding more and more readings to this data will never result in a normal distribution. The ability of an instrument to agree with its 'brothers' and with the instruments of one's customers or suppliers is an important property. The common parameter in all of the above definitions is the average or mean value. As can be seen from Figure 2.10, that is a rather conservative estimate of the actual performance of the instrument or instruments, and the more frequently observed or 'typical' readings will be closer to 0.0 than to the mean. Maximum is rather more elusive because it depends on the sample set. If the sample set is well defined then maximum may also convey some information about the length of the tail of the distribution. The critical issue then is not the magnitude of the average or maximum but how those numbers were derived. How many samples were used? What kind of samples were used? Answers to these questions are rarely given and yet they are critically important. Readings taken in a controlled environment on a small number of pristine (ceramic tile) materials with the two instruments side by side will generally give better agreement than readings taken against stored standard values in a production-floor environment using plastic, porcelain or paint references. The best way to glean useful information about inter-instrument agreement from a manufacturer's specification data is to look for the average number and see whether it is based on the 'BCRA'[25] ceramic tiles or other materials. Keeping Figure 2.10 in mind, it should be possible to convert any of the above definitions into the average estimate. If a standard deviation is given, it will be close to the RMS value. Both of these numbers will be observed to be smaller than the average or mean value.

The final parameter of interest on a specification sheet will be an accuracy assessment. Not all instrument data sheets carry this specification, since the assessment of accuracy is somewhat difficult to define in colorimetry. By convention, accuracy is defined to be the conformance to the correct or accepted value of the test parameter. In the few cases in the literature where national laboratories have reported round-robin or inter-comparison data on coloured materials, the results have not been encouraging [26, 27]. Still, as pointed out earlier, improvements in spectral scale calibration have resulted in a routine accuracy of 0.1 nm or better.

2.5.2 Verifying performance specifications

Once a high-performance colorimeter or spectrocolorimeter has been purchased, it must be maintained conscientiously. While salesmen may tout or demonstrate ruggedness and reliability of their products, keeping the instrument operating at factory specifications will require a significant amount of effort on the part of the colour engineer. One of the activities that must be performed on a routine basis is the verification of the instrument's performance to its field specifications. Generally, performance in the field is given a slightly downgraded set of values compared to the new, factory specifications listed in the sales literature.

2.5.3 Standards of colour and colour-difference

Verifying performance in the field requires the use of real-world or typical materials rather than pristine standard materials. Like Berns [28], this text recommends the use of production materials as field verification standards to supplement rather than replace ideal materials such as the Ceramic Colour Standards, Series II [21]. For some materials that may be easy to do since the products themselves are intended to have stable, consistent colour for long periods of time. For more fugitive materials, new specimens will need to be produced before the old ones change and the new standards characterized relative to the older ones. In some cases, there are certified reference materials available or routine inter-comparison programs [29] that can be utilized which incorporate standards produced from similar materials to those used in the products. Again, the ISCC Technical Report [19] contains information about standard materials, material standards or standards-issuing laboratories. For graphic arts materials, technical organizations such as the GCA [30] or GATF [31] can be a source of standard materials, including standards of colour, density and various printing properties such as dot gain, contrast and trap. A simple pair of standards, such as a medium green and a medium magenta, can be used for weekly analysis and the results stored and plotted against the initial reading to give a time-series process chart.

It can also be recommended to have a check-white standard and a check-black standard. The check-white is used to verify the scale transfer and the check-black to verify the status of the specular port or the optical zero level. Having a history of readings on a secondary white standard also creates an insurance policy against the loss of an instrument's working standard. Until a new working standard can be obtained from the manufacturer, the check-white standard can be used to standardize the instrument, and the standard's values directly traced back to the original, but now missing, working standard. There will be no loss of long-term reproducibility, thus maintaining the consistency and quality of the products being tested with the instrument.

2.5.4 Sources of error and uncertainty in the measurement of reflectance, transmittance and radiance

There is an ISO standard [32] on how to create and use certified reference standards for colour measurement in the graphic arts. There is also an international document that describes how and why uncertainty is different from experimental errors and what must be done to determine the uncertainty of a measurement. That document is titled the *Guide to the Expression of Uncertainty in Measurement* [33] and is referred to as 'The GUM'. It is entirely general and is not specific to colorimetry or imaging. A similar publication is the WECC [34]. The CIE has held symposia in which radiometric scientists and engineers debated, discussed and finally agreed on how to use and interpret the statements and procedures in these documents. The fundamental measurement of light and colour does not have the same engineering significance as the metre, the kilogram or the volt. Fortunately, there have been several recent papers on how to perform the uncertainty analysis for colour measurements, including those of Burns and Berns [35], Fairchild and Reniff [36] and Gardner [37], which provide a practical description of the methods and requirements of an uncertainty analysis. One of the most important points brought out in these documents is that the uncertainty in a colour measurement is a characteristic of the measurement and not of the instrument alone.

Systematic errors and uncertainty appear in many places in a measurement sequence. The major sources are systematic uncertainty in the values assigned to the working standard, errors in the alignment or scale transfer of the wavelength scale, and inconsistent bandwidths across the spectral range of measurements. The seminal paper by Robertson [38] gives a method for easily assessing the severity of these kinds of errors. It is beyond the scope of this book to discuss all the issues related to the assessment of uncertainty in colour engineering measurements. It is sufficient to warn the reader that the magnitude of these uncertainties is not inconsequential in a programme or project dependent on absolute colorimetry. The many chemical and process industries that have utilized relative colour measurements for years are being surprised when they try to apply colorimetry to global colour communications and find the results far poorer than they have been accustomed to observing. Technologists in image reproduction and the graphic arts have struggled with this for some time but have also been quick to adopt the tolerances and specifications of relative colorimetry to their more nearly absolute requirements. At the time of writing, the total uncertainties for instrumental colour measurements, at a coverage factor $k = 2$, are approximately equal to that of the average human observer and slightly worse than that of a well-trained and highly motivated master colour technologist.

2.6 GEOMETRIC REQUIREMENTS FOR COLOUR MEASUREMENTS

2.6.1 Colour measurements from self-luminous objects

Self-luminous coloured objects present some unique measurement problems. Any measurement of a self-luminous object must be, by the nature of the tests, an absolute measurement. Any absolute light measurement introduces a large array of geometric problems. There is a CIE Technical Committee [39] currently writing recommendations on the self-luminous object measurement issue. All other descriptions and standards on colour measurement of self-luminous objects come from the illumination engineering literature. Extended sources are always more difficult to characterize than small, point sources. VDU sources are non-uniform, temporally changing, polarization dependent, affected by the gravitation of the earth and moon, and subject to flux variances on the order of the square of the differences in absolute distance between the surface of the source and the surface of the detector. For extended sources, the measurement of choice is generally radiance or spectral radiance. Radiance is the light flux per unit area per unit solid angle in a given direction from the extended source. Either imaging optics or precision apertures are used to define and restrict the field of view. This is the geometry of the common telespectroradiometer used to measure the colour of displays from across the room. Irradiance, spectral irradiance or radiant exitance from a small or point source is defined in terms of the light flux per unit area, integrated over all solid angles of view. This usually requires some form of flux integrating device such as an integrating sphere or a domed diffuser. This is the geometry of the typical CRT calibrator device which attaches to the face plate of the monitor. Figure 2.11 illustrates the light flux in a radiance and an irradiance measurement.

  Geometric parameters that will affect the results of any colour measurement, instrumental or visual, include the angle between the normal to the source and the optical axis of the instrument, the solid angle of the receiver of the instrument, the distance between the receiver and the source, the kind and amount of flux integration (relay lens, pigmented plastic, flashed opal glass, packed white powder, integrating sphere or tunnel). Each of these parameters must be documented and logged along with the light readings for every measurement. Even for instruments in which the normal operating mode is to have each of these fixed, it should be standard procedure to record them along with the colorimetric values so that there is never any doubt as to the geometric state of the measurements. As a result, flux-integrating devices measuring irradiance are easier to construct but require exact positioning. A device that attaches the light sensor directly to the face of a display solves that problem easily. Radiance is a more difficult measurement to make but is more representative of the way displays are viewed. So some engineering trade-off is always accepted in the colour measurement of a display.

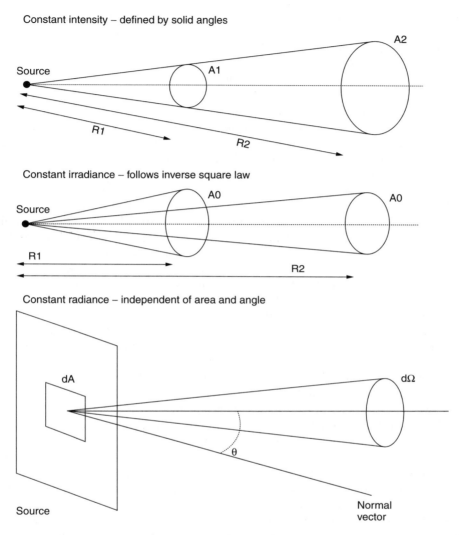

Figure 2.11 Geometry of intensity, irradiance and radiance in terms of flux per unit area and per unit solid angle.

2.6.2 Colour measurements from reflecting or transmitting objects

The geometric requirements for colour measurement of transmitting and reflecting objects are no less stringent, though they have been given little attention over the past two decades. One reason for this is that most standardizing laboratories concern themselves with ideal or near-ideal materials. In these materials, the errors and uncertainties associated with geometry are not of much concern. Only recently, with the rising need for global standardization, has it become apparent that many practical materials, even those materials used as verification or diagnostic standards, have significant interactions with the optical measuring geometry of the instrument. The most comprehensive treatment of the notation and description of the geometry of instruments for the measurement of objective colour is that of ASTM E 1767 [40]. The system of geometric notation used there was originally developed by McCamy [41] and utilized first in the ISO standards on densitometry [42]. This notation differs from that used in many CIE and historical documents on colorimetry. Earlier documents focused on the concepts of illuminating/viewing conditions, while E 1767 is based on a specimen-centred coordinate system that captures the essence of the influx and efflux beams. That notation is 'influx:efflux'.

There are two widely accepted and recommended geometries for reflected colour and two for transmitted colour. These recommended geometries are currently documented in Publication CIE 15.2, *Colorimetry* [43]. The transmission geometries are by far the simplest definition. One is a regular transmittance geometry with the influx and efflux along the normal ($0°$). The cone angles of the two beams are to be matched with half-angles of $5°$. There are no tolerances on the optical angles, even though it is well known that surface reflection directly back through the optics creates a systematic measurement error. Specimens with large surface reflectances, such as metal film on quartz plate filters, must be positioned so that the surface reflectance does not re-enter the optical path. This is usually accomplished by rotating the specimen so that the normal to its surface is no longer aligned with the optical axis of the instrument. The other geometry is a diffuse transmittance geometry. This geometry is implemented either with an integrating sphere as the efflux receiver or with an opal glass plate as the influx diffuser, as in a transmission densitometer. Figure 2.12 shows the full set of CIE recommended geometries for transmittance and diffuse reflectance.

The two recommended diffuse reflection geometries are bidirectional ($45°:0°$) and hemispherical diffuse (d:n). In the CIE publication on colorimetry no further definitions are given, Some general tolerances are applied, such that any angle less than $10°$ can be considered equivalent to normal viewing and the illumination (influx) cone half-angles should be $5°$ or less. This latter restriction was added to the CIE recommendations when the second edition of *Colorimetry* (15.2) was issued and was based partly on the reports of Johnson and Stephenson [44], Rich [45] and Roos [46, 47] which first reported on geometry-based measurement errors. More recently, several researchers, involved in a CIE technical committee on instrument geometry, have reported on similar findings and have better documented the exact nature of the geometric dependence of these errors. These reports, published in the proceedings of the 3rd Oxford Conference of the CORM – UV Spectrophotometry Group on Advances in Spectroscopy, include work started by John Verrill [48] of the NPL and Gorow Baba [49] of the Murakami Color Research Laboratory in Japan. The results of these studies will serve to help standards-writing bodies to develop better definitions and tolerances on the geometric requirements for colour-measuring instruments. That will, in turn, provide the necessary requirements for instrument makers to build instruments with improved inter-instrument agreement and then, coupled with an anticipated reduction in the uncertainty of the measurement of object colour, will provide the foundation for consistent global colour communications.

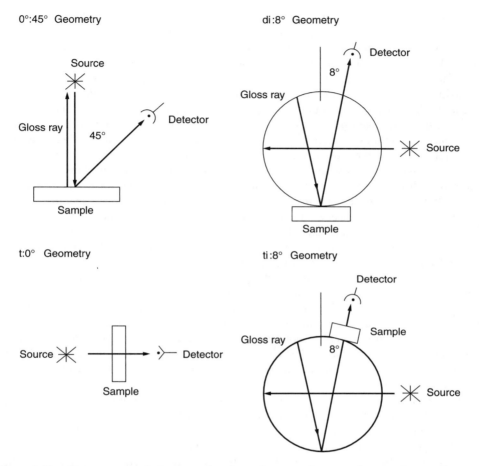

Figure 2.12 CIE recommended geometries for reflectance factor and transmittance factor of coloured objects.

There is a law in optical physics known as the Helmholtz law of reciprocity. It is based on a conservation of energy argument and says that in a completely linear optical system, the light source and detector can be exchanged without changing the results of a measurement. This means that a 45°:0° instrument can be replaced with a 0°:45° instrument without affecting the measurements. For most materials this statement has been shown to be true. The equivalent reverse geometry for an integrating sphere is 8°:d, indicating that the influx is incident at 8° from the specimen normal and the efflux is integrated over the 2π steradians above the specimen surface. In an integrating sphere instrument the surface reflectance is mirror-like or specular and can be included or excluded from the readings. This is indicated by the addition of the letter i or e to the d, so that a directional-hemispherical instrument that includes the specular reflectance is described as

8°:di and one that excludes the specular reflectance as de:8°. Further, a 45°:0° instrument might use only one beam, along a specific azimuthal angle at 45° to the normal angle, or it might use a small number of such beams creating a discrete ring of illumination points known as circumferential illumination, and is given the shorthand notation 'c'; while another approach might be to spread the influx uniformly and continuously around the azimuth, and is given the name 'annular' and the shorthand notation 'a'. A circumferential influx instrument would thus be generally described as '45°c:0°' and an annular efflux instrument would be described as '0°:45°a'. Historically, there has been no general preference among these variations around the recommended geometries. There have been a few reports in the literature claiming to have identified a preferential geometry and thus a failure of the reciprocity law. In every case, close examination has shown some small but critical difference between the 'forward' and 'reverse' geometries so that exact reversal had not been achieved. There are some justifications for choosing a specific geometry based more on historical beliefs than on documented evidence. In the development of the ISO 5 series it was claimed that a 45°:0° geometry agreed best with the visual evaluation of optical density. When colorimetry first began to be applied to graphic arts, it was again asserted that this correlation to visual perception should be recognized. This was done in spite of the fact that every major study of visual scaling of colour used hemispherical geometry for the instrumental readings and that, even to this day, all colour order systems are set up and maintained using hemispherical diffuse instruments. It is beyond the scope of this chapter to make explicit recommendations on the 'best' geometry for colour measurements, except to note that it would probably be a good idea to apply the principle of simulation when either choosing an instrument or designing a visual experiment so that the influx:efflux geometries are consistent between instrumental evaluation and visual evaluation.

One last aspect of measurement geometry must be presented in this closing paragraph. It is included here not because it represents a small or insignificant error but because it is observed only under certain conditions of measurement with certain kinds of specimens. This last problem is known by various names – edge loss errors, translucent blurring effect, lateral diffusion error – but describes the measurement error that arises when light from the influx beam escapes from the measurement system and is not included in the efflux beam, or conversely when light not from the influx beam intrudes into the measurement system and is included in the efflux beam. The effect was first reported by Atkins and Billmeyer [50] with respect to translucent plastic sheeting, then studied fairly carefully by Hsia [51] at the NIST for standard materials and finally and most thoroughly by Spooner [52] in a series of papers that both describe and quantify the practical aspects of this effect for the graphic arts and in particular for electronic, digital printing. This can be a very serious error and most modern instruments provide a buffer zone between the area of the sampling aperture that is filled with light by the influx and the area of the sampling aperture from which light is collected by the efflux optics. As long as there is a 1 – 2 mm difference in the diameter of these areas, the errors will be tolerable. The problem arises when one wishes to read very small test targets. The buffer zone does not scale with the measurement sizes but is a function of the distance a light ray can travel inside the material. This then sets a practical lower limit on the minimum sampling aperture size of about 3 mm for an efflux area of 1 mm.

2.7 CONCLUSIONS AND EXPECTATIONS

2.7.1 Current CIE activities in colour and colour-difference measurements

The CIE continues to make efforts to document current technologies and make recommendations on the correct measurement procedures and the definition of what constitutes an 'accurate' measurement. Division 1 on Vision and Color is in the process of issuing a third edition of the publication *Colorimetry* and has recently published a standard based on its former recommendations concerning tests of conformance of daylight simulators. The new standard covers D50, D65 and two other daylight illuminants. This standard fits nicely with the work of a committee that has surveyed the world to identify and rank the quality of daylight simulators in lighting cabinets for visual inspection of coloured objects. Recently, a new colour tolerance equation has been recommended, known as CIEDE2000,[53] that is intended to replace the previous recommendation CIE94[54] as well as the more widely used CMC($l:c$)[55] equation. Division 2 has several technical committees looking into measurement areas of interest to colour engineers. One committee is close to completing a proposed method for assessing the quality of the fit of filter colorimeters; another is defining the geometries and tolerances for highly reproducible colour measurements. Finally, very soon there should be a new CIE recommendation on what operating parameters of a spectroradiometer must be carefully controlled or monitored to obtain radiometric readings suitable for use in modelling the colour of a CRT or flat panel display.

2.7.2 Quality management systems and colour measurements

One of the most common questions heard at colour conferences these days concerns quality management systems, such as ISO 9000, and colour measurement traceability. Most quality systems require that any measurement that can affect the quality of the product be certified to a authorized scale. For devices such as voltmeters, temperature gauges and distance guides this is not a problem, since the definition of each of these scales is produced from first principles and transfer devices can be easily produced and carried into the field. Not so with many optical sensors. Colorimetry is a measurement based on a theoretical standard, either the perfect reflecting diffuser or the lumen. Each of these is defined in terms of a fundamental measurement that requires several corrections for experimental deviations from the ideal. As a result the uncertainty associated with scale determination and transfer results in scales that can vary from time to time or from lab to lab by more than a visually distinguishable amount. The result is a measurement scale which may be absolutely traceable and yet inconsistent with the intent of the product. These errors are not purely random and can be assessed and corrected by careful comparison of visual colour with instrumental colour. But these will be unique corrections, peculiar to a given product or material. They also make the colour measurement relative rather than absolute. What is important is that this correlation be determined and then maintained strictly. Consistency becomes far more important than absolute traceability – even though traceability is the first step towards global consistency. It is like surveying without the aid of a GPS.

At some point, a stake is driven into the ground and all readings are relative to that stake. So it is with quality management systems and colorimetry.

As presented in this chapter, colorimetry is a measurement science that began as visual sensation and progressed through many technologies to end with a measurement science that must correlate with visual sensation. Reaching that goal has not been easy. Many obstacles have been encountered and surpassed. The last few problems are in sight and researchers from around the world are chipping away at them. It is easy to speculate that within a decade of this writing, absolute colour communications will be a reality. Even now, in many applications, the long-term stability of colour measurement is better than the stability of material or product standards and much more consistent than the processes used to create test specimens. This has prompted many industries to stop making replicate material standards and begin submitting digital files of reflectance, radiance or colorimetric coordinates as the definition of the target colour. The result has been a decrease in disagreements and rejected products and an increase in the efficiency and throughput of the production process.

REFERENCES

1. Donaldson, R. (1954) Spectrophotometry of fluorescent pigments. *British Journal of Applied Physics*, **5**, 210–214.
2. Guild, J. (1931) The colorimetric properties of the spectrum. *Philosophical Transactions of the Royal Society of London, Ser A*, **230**, 149–187.
3. Wright, W. D. (1928–29) A re-determination of the trichromatic coefficients of the spectral colours. *Transactions of Optical Society of London*, **30**, 141–164.
4. Wright, W. D. (1929–30) A re-determination of the mixture curves of the spectrum. *Transactions of the Optical Society of London*, **31**, 201–211.
5. Fairman, H. S., Brill, M. H. and Hemmendinger, H. (1997–98) How the CIE 1931 color-matching functions were derived from the Wright-Guild data. *Color Research & Application*, **22**, 11–23 (1997), 259 (1998).
6. Stiles, W. S. and Burch, J. M. (1959) N.P.L. colour-matching investigation: Final Report (1958). *Optica Acta*, **6**, 1–26.
7. Speranskya, N. I. (1959) Determination of spectrum color co-ordinates for twenty-seven normal observers. *Optics and Spectroscopy*, **7**, 424–435.
8. Trezona, P. (2001) Derivation of the 1964 CIE 10° XYZ Colour-Matching Functions and Their Applicability in Photometry. *Color Research & Application*, **26**, 67–75.
9. Stockman, A., Sharpe, L. T. and Fach, C. C. (1999) The spectral sensitivity of the human short-wavelength cones. *Vision Research*, **39**, 2901–2927.
10. Stockman, A. and Sharpe, L. T. (2000) The spectral sensitivities of the middle- and long-wavelength-sensitive cones derived from measurements in observers of know genotype. *Vision Research*, **40**, 1711–1737.
11. CIE Technical Committee 1-36, *Chromaticity diagram with physiological significant axes*.
12. CIE (1981) Publication CIE 51, *A method for assessing the quality of daylight simulators for colorimetry*, Commission Internationale de l'Éclairage, Vienna, Austria.
13. *Munsell Book of Color*, Munsell Color division, GretagMacbeth, New Windsor, NY 12553, USA.
14. Donaldson, R. (1947) A colorimeter with six matching stimuli. *Proceedings of the Physical Society of London*, **59**, 544–550.
15. Engelhardt, K. and Seitz, P. (1993) Optimum color filters for CCD digital cameras. *Applied Optics*, **32**(16), 3015–3023.

16. CIE (1987) Publication CIE 69, *Methods of characterizing Illuminance meters and luminance meters: Performance, characteristics, and specifications*. Vienna, Austria: Commission Internationale de l'Éclairage.
17. *Ceramic Colour Standards II* are produced by CERAM Research, Queens Road, Penkhull, Stoke-on-Trent ST4 7LQ, UK.
18. *Munsell Atlas* is a product of the Munsell Color division, GretagMacbeth, New Windsor, NY 12553, USA.
19. *NCS Color Atlas* is a product of the Swedish Colour Research Institute.
20. ASTM E1455–97, *Standard Practice for Obtaining Colorimetric Data from a Visual Display Unit Using Tristimulus Colorimetry*. ASTM, 100 Barr Harbor Drive, West Conshohocken, PA 19428–2959, USA.
21. ISO/TC 6–Paper, board and pulps, WG 3–Optical Properties. International Organization for Standardization, Geneva, Switzerland.
22. MacAdam, D. L. (1970) *Sources of Color Science*. Cambridge, MA: MIT Press.
23. ISCC Technical Report 89-1 *Guide to Material Standards and Their Use in Calibration*. Inter-Society Color Council, 11491 Sunset Hill Road, Reston, VA 20190, USA (iscc@compuserve.com).
24. Rich, D., Battle, D., Malkin, F., Williamson, C. and Ingleson, A. (1995) Evaluation of the long-term repeatability of reflectance spectrophotometers. In *Spectrophotometry, Luminescence and Color*, ed. Burgess, C. and Jones, D. G., pp. 137–154. Elsevier, New York.
25. BCRA Tiles are more correctly known as the Ceramic Colour Standards, Series II and are produced by CERAM Research, Queens Road, Penkhull, Stoke-on-Trent ST4 7LQ, UK.
26. Verrill, J. F. (1993) Intercomparison of colour measurements synthesis report. *Report EUR 14982 EN*, European Commission, Brussels.
27. Verrill, J. F., Clarke, P. J and O'Halloran, J. (1997) Study of colorimetric errors on industrial instruments. *NPL Report COEM 2*. Project 7, Reference MPU 8/36.3.
28. Berns, R. (2000) *Billmeyer and Saltzman's Principles of Color Technology*, 3rd edn., p. 98, John Wiley & Sons.
29. *Color and appearance collaborative reference program for color and color-difference*, Collaborative Testing Services, Inc., McLean, VA, USA.
30. GCA, Graphic Communications Association, 100 Daingerfield Road, Alexandria, VA 22314-2888, USA.
31. GATF, Graphic Arts Technical Association, 200 Deer Run Road, Sewickley, PA 15143-2600, USA.
32. ISO/DIS 15790 (2000) *Graphic technology and photography – Reflection and transmission metrology – Certified reference materials – Documentation and procedures for use, including determination of combined standard uncertainty*. International Organization for Standardization, Geneva, Switzerland.
33. ISO (1993) *Guide to the Expression of Uncertainty in Measurement*. International Organization for Standardization, Geneva, Switzerland. This Guide was prepared by ISO Technical Advisory Group 4 (TAG 4), Working Group 3 (WG 3). ISO/TAG 4 has as its sponsors the BIPM, IEC, IFCC (International Federation of Clinical Chemistry), ISO, IUPAC (International Union of Pure and Applied Chemistry), IUPAP (International Union of Pure and Applied Physics), and OIML.
34. WECC Document 19, (1990) *Guidelines for the expression of uncertainty of measurement in calibrations*.
35. Burns, P. D. and Berns, R. S. (1997) Error propagation analysis for color measurement and imaging. *Color Research & Application*, **22**, 280–289.
36. Fairchild, M. D. and Reniff, L. (1991) Propagation of random errors in spectrophotometric colorimetry. *Color Research & Application*, **16**, 361–367.
37. Gardner, J. L. (2000) Uncertainty estimation in colour measurement. *Color Research & Application*, **25**, 349–355.

38. Robertson, A. R. (1987) Diagnostic performance evaluation of spectrophotometers. In *Advances in Standards and Methodologies in Spectrophotometry*, ed. C. Burgess and K. D. Mielenz, Elsevier, New York, pp. 277–286.

39. CIE Technical Committee 2-42, *Colorimetric measurements for visual displays*, Christine Wall, Chairperson (NPL, Teddington, UK).

40. ASTM E1767, *Standard Practice for Specifying the Geometry of Observations and Measurements to Characterize the Appearance of Materials*. American Society for Testing and Materials, West Conshohocken, PA, USA.

41. McCamy, C. S. (1966) Concepts, terminology, and notation for optical modulation. *Photographic Science and Engineering*, **10**, 314–325.

42. ISO 5/1, *Photography Density Measurements, Part 1: Terms, Symbols and Notation*. International Organization for Standardization, Geneva, Switzerland.

43. CIE (1986) Publication CIE 15.2, *Colorimetry*, 2nd edn. Commission Internationale de l'Éclairage, Vienna, Austria.

44. Johnson, N. L. and Stephenson, H. F. (1983) The influence of geometric tolerances on 45°/0° and 0°/45° colorimetric measurements. *Proceedings of the CIE 20th Session*, Amsterdam, CIE Publication No. 56, D202/1–D202/4.

45. Rich, D. C. (1988) The effect of measuring geometry on color matching. *Color Research & Application*, **13**, 113–118.

46. Roos, A. (1988) Anomalies in integrating sphere measurements on structured samples. *Applied Optics*, **27**, 3828–3832.

47. Roos, A. (1988) Interpretation of integrating sphere signal output for non-Lambertian samples. *Applied Optics*, **27**, 3833–3837.

48. Hanson, A. R. and Verrill, J. F. (1999) Study of the dependency of spectral radiance factor and colorimetric values on geometric tolerances. *Analytica Chemica Acta*, **380**, 363–367.

49. Baba, G. and Suzuki, K. (1999) Gonio-spectrophotometric analysis of white and chromatic reference materials. *Analytica Chemica Acta*, **380**, 173–182.

50. Atkins, J. T. and Billmeyer, F. W. Jr. (1966) Edge-Loss Errors in Reflectance and Transmittance Measurement of Translucent Materials. *Materials Research and Standardization*, **6**, 564.

51. Hsia, J. J. (1976) The Translucent Blurring Effect – Method of Evaluation and Estimation. *NBS Technical Note 594-12*, US Department of Commerce.

52. Spooner, D. L. (1991) Translucent blurring errors in small area reflectance spectrophotometer & densitometer measurements. *TAGA Proceedings*, 130–143.

53. CIE Technical Report 142, (2001) *Improvement to Industrial Colour Difference Equation*, Central Bureau of the CIE, Vienna, Austria.

54. CIE Technical Report 116, (1995) *Industrial Colour-Difference Evaluation*, Central Bureau of the CIE, Vienna, Austria.

55. Clark, F. J. J., McDonald, R. and Rigg, B. (1984) Modification to the JPC 79 colour-difference formula. *Journal of the Society of Dyers and Colorists*, **100**, 128–132.

3

Colorimetry and colour difference

Phil Green

3.1 INTRODUCTION

In this chapter we present a brief summary of colorimetry, whose main purpose is to identify the colorimetric quantities and related computations upon which later sections rely. Readers who wish to know more about the principles of colour vision and colorimetry are invited to study the other chapters in this section and other texts where such concepts are presented in greater detail [7, 35, 22, 23, 28, 47, 29, 66, 75, 76].

This is followed by consideration of colour difference metrics. When evaluating the performance of different colour reproduction systems, it is often desirable to be able to assess the magnitude of difference between the reproduced colour and some reference, whether the original or some other reproduction. In industrial colour reproduction such assessments are of great importance, as the acceptance or rejection of a quantity of material will often depend on whether the colour is within some agreed tolerance.

Some of the advanced colour difference formulae are presented, together with a brief summary of the methods used to derive them.

3.2 COLORIMETRY

The ultimate aim of colorimetry is to model the human perception of colour. This implies the following objectives:

Colour Engineering, edited by P.J. Green and L.W. MacDonald.
© 2002 John Wiley & Sons Ltd.

1. Each unique colour should have a unique colorimetric coordinate: i.e. two colours that have the same colorimetric values should have the same appearance, and two colours with different coordinates should have a different appearance.
2. Colorimetric values should be represented in a coordinate system whose dimensions are perceptually orthogonal and represent some correlate of the principal perceptual attributes of colour.
3. The magnitude of difference between the numerical values representing two colours should be proportional to the perceived difference between them.

The first objective was achieved by the adoption of the CIE 1931 Standard Observer colour matching functions. Experimental data were used to model the response of an average person with normal colour vision to a set of theoretical primaries XYZ, which are themselves linear transforms of a particular set of monochromatic RGB primaries. These colour matching functions are shown in Figure 3.1.

Multiplying the colour matching functions with the spectral power of the illuminant and the reflectance or transmittance of a sample at finite wavelength intervals gives rise to a unique set of *XYZ* tristimulus values for that sample:

$$X = k \sum_{l}^{u} \overline{x}(\lambda) S(\lambda) R(\lambda)$$

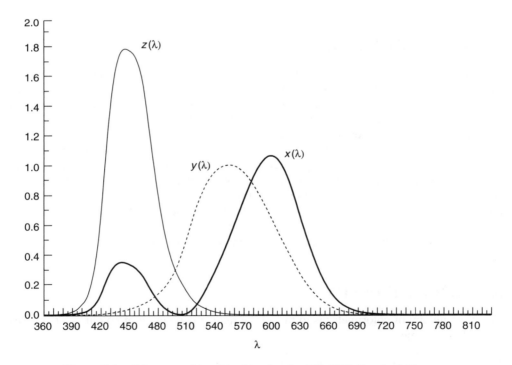

Figure 3.1 Colour matching functions for the CIE 1931 Standard Observer.

$$Y = k \sum_{l}^{u} \overline{y}(\lambda) S(\lambda) R(\lambda) \qquad (3.1)$$

$$Z = k \sum_{l}^{u} \overline{z}(\lambda) S(\lambda) R(\lambda)$$

where $S(\lambda)$ is the spectral power distribution of the illuminant, $R(\lambda)$ is the spectral reflectance or transmittance factor of the sample, $\overline{x}$, $\overline{y}$ and $\overline{z}$ are the colour matching functions, u and l are the upper and lower bounds of the wavelength range in $S(\lambda)$, $R(\lambda)$ and $\overline{x}$, $\overline{y}$, $\overline{z}$, and k is a normalising constant. $S(\lambda)$ can be expressed as either factors or percentages. The constant k is given by:

$$k = \frac{100}{\sum_{l}^{u} \overline{y}(\lambda) S(\lambda)} \qquad (3.2)$$

The primaries XYZ were chosen in such a way that negative values in the colour matching functions are avoided and the Y tristimulus value is proportional to perceived luminance.

It is important to note that a set of tristimulus values are valid only for the illuminant used in the computation. Where it is necessary to predict the tristimulus values that would arise under a different viewing condition, a more complex model of colour appearance is required, as described in Chapter 4.

In CIE Publication 15.2 [12] (the main recommendation on CIE colorimetry and colorimetric computations, shortly to be updated by Publication 15.3), the wavelength range u:l in equations (3.1) above is recommended to be the entire visible spectrum (360–830 nm) at 1-nanometre intervals. However, it suggests that for practical purposes 5 nm intervals over the range 380–760 nm are used. Colour matching function and illuminant spectral power data for 5 nm intervals are given in Publication 15.2.

Of the standard and recommended illuminants for which data are published by the CIE, it is most common to use either D50 or D65 in colour imaging, the former particularly for graphic arts applications and the latter for CRT colorimetry.

The possibility of other wavelength ranges and intervals is permitted in colorimetric computations, and is indeed essential owing to the range and precision of the available instruments. A consequence is that a single set of reflectance data can give rise to different tristimulus values depending on the method of computation. Previous studies have suggested that the effect of using different wavelength intervals is itself relatively small [62] but that the combined effect of different wavelength ranges and intervals and different methods of interpolation and abridgement of the matching functions can lead to large errors [38]. In cases where instrumental data for the full wavelength range are not available, the CIE recommends that the wavelength range be truncated if the consequent errors are negligibly small; otherwise the missing data should be either interpolated from the known data or found by replicating the nearest measured value.

It should be noted from Figure 3.1 that the values of the colour matching functions are very low at the extremes of the visible spectrum (as a result of the low sensitivity of the human visual system at these extremes) and as a result errors arising from the absence of measurement data in these regions will not necessarily be significant.

The CIE matching functions for 1 nm intervals are linearly interpolated from the 5 nm data, and similarly 10 nm interval data are linearly abridged from the 5 nm data. However, the functions implied by the 5 nm CMF data can be better approximated by a non-linear calculation [29,38,61], and alternative data based on a Lagrangian polynomial have been published by the ASTM [5,6].

These possible variations in the methods of calculating XYZ tristimulus values from spectral power data are addressed in ISO 13655 [32], which mandates a wavelength range of 340–780 nm and intervals of 10 nm, and specifies that where the highest wavelength of the reflectance or transmittance data is less than 780 nm, or the lowest wavelength greater than 340 nm, the data are extended. The method described is equivalent to replicating the data corresponding to the nearest wavelengths for which data exist. Further, ISO 13655 specifies the use of D50 as the illuminant and provides weighting factors based on the ASTM standard [6], with CMF data premultiplied by the D50 illuminant spectral power distribution at 10 nm intervals, in order to minimise any variation that may arise from differences in the precision of the data. In this case the first of equations (3.1) becomes:

$$X = \sum_{\lambda=340}^{\lambda=780} W_X(\lambda) R(\lambda)$$

where W_X is the weighting function for the X tristimulus value; and similarly for Y and Z. Since the weighting function values below 360 nm and above 770 nm are zero, reflectance or transmittance data are in fact needed only for the range 360–770 nm.

Other issues of accuracy and uncertainty in the calculation of XYZ tristimulus values have been extensively studied [66, 19].

3.2.1 Calculation of chromaticity

Chromaticity coordinates are calculated by:

$$x = \frac{X}{X + Y + Z}$$

$$y = \frac{Y}{X + Y + Z} \tag{3.3}$$

$$z = \frac{Z}{X + Y + Z}$$

It can be seen from equations (3.3) that $x + y + z = 1$. The redundancy implied makes it possible to describe the chromaticity of any colour by using just two coordinates, and it is usual to specify colours either in terms of their x, y coordinates, or by including the Y tristimulus value (i.e. xyY).

3.2.2 Calculation of CIE 1976 $L^* a^* b^*$

Although tristimulus and chromaticity meet the requirement of first objective identified above, they have a high degree of perceptual non-uniformity and the quantities do not directly represent any useful perceptual attribute. Hence in order to meet our second objective it is necessary to apply a further transformation to these quantities. This is achieved by the CIE 1976 Uniform Colour Spaces, which includes the CIELAB colour space:

$$L^* = 116 \left(\frac{Y}{Y_n} \right)^{1/3} - 16$$

$$a^* = 500 \left(\left(\frac{X}{X_n} \right)^{1/3} - \left(\frac{Y}{Y_n} \right)^{1/3} \right) \tag{3.4}$$

$$b^* = 200 \left(\left(\frac{Y}{Y_n} \right)^{1/3} - \left(\frac{Z}{Z_n} \right)^{1/3} \right)$$

where X_n, Y_n, Z_n are the tristimulus values of a reference white. This reference white used can be a source, a perfect diffuser, or the white point of the transmissive or reflective media. If the media white is taken as the reference white, the resulting values are often referred to as 'media-relative'. ISO 13655 [32] states that the tristimulus values of the reference white shall be those of the illuminant D50. Table 3.1 gives the tristimulus values of several illuminants.

In the case of very low values of X, Y or Z, equations (3.4) must be modified. In CIE Publication 15.2 [12], equations (3.4) become:

$$L^* = 116(Y/Y_n)^{1/3} - 16, \quad Y/Y_n > 0.008856$$
$$L^* = 903.3(Y/Y_n), \quad Y/Y_n \leqslant 0.008856$$

Table 3.1 Tristimulus values and chromaticity coordinates of some CIE Standard Illuminants, calculated using the CIE 1931 Observer at 10 nm intervals (Source: ASTM [6])

	X	Y	Z	x	y
D50	96.422	100	82.521	0.3457	0.3585
D55	95.682	100	92.149	0.3324	0.3474
D65	95.047	100	108.883	0.3127	0.329
A	109.85	100	35.585	0.4476	0.4074
C	98.074	100	118.232	0.3101	0.3161

and

$$a^* = 500[f(X/X_n) - f(Y/Y_n)] \tag{3.5}$$

$$b^* = 200[f(Y/Y_n) - f(Z/Z_n)]$$

where:

$$
\begin{aligned}
f(X/X_n) &= (X/X_n)^{1/3}, & X/X_n &> 0.008856 \\
f(X/X_n) &= 7.787(X/X_n) + 16/116, & X/X_n &\leqslant 0.008856 \\
f(Y/Y_n) &= (Y/Y_n)^{1/3}, & Y/Y_n &> 0.008856 \\
f(Y/Y_n) &= 7.787(Y/Y_n) + 16/116, & Y/Y_n &\leqslant 0.008856 \\
f(Z/Z_n) &= (Z/Z_n)^{1/3}, & Z/Z_n &> 0.008856 \\
f(Z/Z_n) &= 7.787(Z/Z_n) + 16/116, & Z/Z_n &\leqslant 0.008856
\end{aligned}
\tag{3.6}
$$

Assuming a value of 100 for Y_n, the knee point of the L^* function is at approximately 0.9 and 8.0 for Y and L^* respectively.

In ISO 13655 [32], the calculation of lightness is slightly different:

$$L^* = 116[f(Y/Y_n)] - 16$$

where $f(Y/Y_n)$ is calculated as in equations (3.6) above. Although the two methods of calculating lightness for low values of Y have a different form, the results are not significantly different.

The a^*, b^* coordinates correspond approximately to the dimensions of redness–greenness and yellowness–blueness respectively, and are orthogonal to the L^* dimension. Hence a colour whose coordinates $a^* = b^* = 0$ is considered achromatic regardless of its L^* lightness.

Transforming the Cartesian L^*, a^*, b^* coordinates to polar coordinates results in the quantities h_{ab} (hue angle) and C_{ab}^* (chroma). These can be considered as correlates for the perceptual quantities of hue and chroma.

$$C^* = (a^2 + b^2)^{1/2} \tag{3.7}$$

$$h_{ab} = \tan^{-1}\left(\frac{b^*}{a^*}\right) \tag{3.8}$$

Note that in computing hue angle it is necessary to ensure that the resulting angle is correctly ranged from 0° to 360°. This is usually done by using the two-quadrant *atan2* function, and adding 360 to the result in cases where b^* is negative. As computed in equation (3.8), h_{ab} (hue angle) is expressed in radians, and must be converted to degrees by multiplying by $180/\pi$.

3.2.3 Inversion of colorimetric transforms

Equations (3.7) and (3.8) can readily be inverted to return from polar to Cartesian coordinates:

$$a^* = C^* \cos(h_{ab})$$

$$b^* = C^* \sin(h_{ab})$$

(3.9)

Note that the conversion from radians to degrees described above must be reversed before applying equation (3.9).

If the tristimulus values of the reference white are known, it is also possible to invert equations (3.4)–(3.6).

$$Y = Y_n \left(\frac{L^* + 16}{116} \right)^3$$

(3.10)

$$X = X_n \left(\frac{a^*}{500} + \left(\frac{L^* + 16}{116} \right) \right)^3$$

(3.11)

$$Z = Z_n \left(\left(\frac{L^* + 16}{116} \right) - \frac{b^*}{200} \right)^3$$

(3.12)

Note that where the ratios (X/X_n), (Y/Y_n), $(Z/Z_n) \leqslant 0.008856$, it is necessary to adjust for the non-linearity in equation 3.6, so that equations 3.10–3.12 become:

$$X = X_n f(X/X_n)^3, \quad f(X/X_n) > 0.008856^{\frac{1}{3}}$$

$$X = X_n((f(X/X_n) - 16/116)/7.787), \quad f(X/X_n) \leqslant 0.008856^{\frac{1}{3}}$$

(3.13)

$$Y = Y_n f(Y/Y_n)^3, \quad f(Y/Y_n) > 0.008856^{\frac{1}{3}}$$

$$Y = Y_n((f(Y/Y_n) - 16/116)/7.787), \quad f(Y/Y_n) \leqslant 0.008856^{\frac{1}{3}}$$

(3.14)

$$Z = Z_n f(Z/Z_n)^3, \quad f(Z/Z_n) > 0.008856^{\frac{1}{3}}$$

$$Z = Z_n((f(Z/Z_n) - 16/116)/7.787), \quad f(Z/Z_n) \leqslant 0.008856^{\frac{1}{3}}$$

(3.15)

where:

$$f(Y/Y_n) = \frac{L^* + 16}{116}$$

$$f(X/X_n) = \frac{a^*}{500} + f(Y/Y_n)$$

$$f(Z/Z_n) = f(Y/Y_n) - \frac{-b^*}{200}$$

It can be seen that equations (3.1) are not analytically invertible and hence it is not possible to recover the original spectral power distribution from the tristimulus values. Where there is a need to calculate new *XYZ* values for a different illuminant, a chromatic adaptation transform should be used, as described in Chapter 4.

3.2.4 Implementation issues

Most colour image file formats adopt a *uint8* (8-bit unsigned integer) encoding, although index and *uint16* encodings are also used and there is provision in the TIFF 6.0 file format for encoding floating-point data.

Device-dependent encodings either scale device scalars into a 0–255 range, or adopt a slightly smaller range in order to allow for the possibility of encoding negative values into the 'headroom' thus created.

CIELAB encodings have greater potential for ambiguity. Although an upper bounding value of 100 is appropriate for data expressed in percentages (as is the case with CMYK), L^* values arising from actual measurements can in some circumstances validly exceed 100, particularly when fluorescence is present in the sample. However, in most cases L^* is encoded with a range of 0–100 and scaled into the range 0–255 in the same way as described above. In the TIFF 6.0 format, CIELAB a^* and b^* values have a range of −127 to 127, and actual values outside this range are unlikely.

The TIFF 6.0 format suggests a signed 8-bit integer for CIELAB encodings, although in practice TIFF files are more often encoded as *uint8* values, with an offset of 128 to range the values appropriately. The Profile Connection Space (PCS) encoding of CIELAB described in the current ICC profile format [30] has a similar format. In it, the range of L^* values encoded is 0–100, and that of a^* and b^* is −128 to 127. Again there is an offset of 128. A 16-bit encoding is also defined, and conversion between 8-bit and 16-bit encodings is performed by multiplying or dividing by 255.

The published TIFF 6.0 specification does not include an encoding for *XYZ* data. However, such encodings are possible through the flexibility of the TIFF format in allowing the generation of new tag definitions.

Although 8-bit formats have adequate precision for transporting colour data, when performing computations on such data there is a choice between using integer and floating-point operations. While the former tend to be significantly faster, there will be a loss of precision which in some circumstances can have the potential to cause objectionable artefacts.

3.3 COLOUR DIFFERENCE

If a colour space is perceptually uniform and its dimensions orthogonal, then the difference between two colours can be represented as the Euclidean distance between their coordinates. The 1976 Uniform Colour Spaces CIELAB and CIELUV were intended to provide just such a perceptually uniform colour space through a transformation of tristimulus *XYZ*, and in doing so to overcome the very significant non-uniformity of tristimulus space that had been discussed by MacAdam and others from the early 1940s [45].

Colour difference in CIELAB is thus defined as follows:

$$\Delta E_{ab}^* = (\Delta L^{*2} + \Delta a^{*2} + \Delta b^{*2})^{1/2} \tag{3.16}$$

where ΔL^*, Δa^* and Δb^* are the arithmetic differences between the coordinates of the sample and reference colours. For example,

$$\Delta L^* = L_r^* - L_s^* \tag{3.17}$$

where the subscripts r and s refer to the reference and sample respectively.

In CIELUV, colour difference is defined similarly:

$$\Delta E_{uv}^* = (\Delta L^{*2} + \Delta u^{*2} + \Delta v^{*2})^{1/2} \tag{3.18}$$

The use of CIELUV in colour difference is less common in colour imaging and so the remainder of this chapter will focus on CIELAB and colour difference metrics derived from it.

It is desirable to have the ability to define components of colour difference in the polar coordinates of chroma and hue angle. However, since h_{ab} is expressed in degrees, Δh_{ab} cannot be combined with ΔL^* and ΔC^* to form an expression analogous to equation (3.18) above. If a dimension orthogonal to L^* and C_{ab}^* is assumed to exist, and is given the label H_{ab}, then:

$$\Delta E_{ab}^* = (\Delta L^{*2} + \Delta C_{ab}^{*2} + \Delta H_{ab}^2)^{1/2} \tag{3.19}$$

and thus ΔH_{ab} can be calculated:

$$\Delta H_{ab} = (\Delta E_{ab}^{*2} - \Delta L^{*2} - \Delta C_{ab}^{*2})^{1/2} \tag{3.20}$$

Care should be taken with the use of ΔH_{ab} as equation (3.20) results in it being unsigned, unlike the other components of colour difference which are usually found by subtraction. Where the application requires that ΔH_{ab} is signed, it is considered positive if indicating an increase in Δh_{ab} (i.e. a counter-clockwise direction) and negative if indicating a decrease in Δh_{ab}.

An alternative calculation of hue difference is given by:

$$\Delta H_{ab}^* = 2\sqrt{C_{ab,r}^* C_{ab,s}^*} \sin\left(\frac{\Delta h_{ab}}{2}\right) \tag{3.21}$$

where the subscripts r and s again refer to the reference and sample. This latter method is generally preferred.

If a colour space has perfect perceptual uniformity, the locus of colours which are not perceptibly different from a given sample forms a sphere around that colour's coordinates in the space, and such spheres would have constant radius in all regions of colour space.

However, there remains considerable perceptual non-uniformity in both CIELAB and CIELUV, and in practice the locus of discrimination for a given colour can be better represented as an ellipsoid. Although the magnitudes of the semi-axes of these ellipsoids are more uniform than those in the xy chromaticity diagram, there is still significant variation in their sizes. The main trend is an increase in semi-axis size with greater distance from the achromatic axis, particularly of the major semi-axis.

Figure 3.2 illustrates the non-uniformity of CIELAB. Each ellipse (or ellipsoid if considered in three dimensions) indicates the region of colour space around a target colour which is not discriminated from the target colour.

3.3.1 Problems with using UCS colour difference

CIELAB colour difference continues to be widely used, but is inadequate in certain applications such as those where a single value is required to define a tolerance throughout colour space. The major problems that are encountered in using a 1976 UCS metric to define colour difference are as follows.

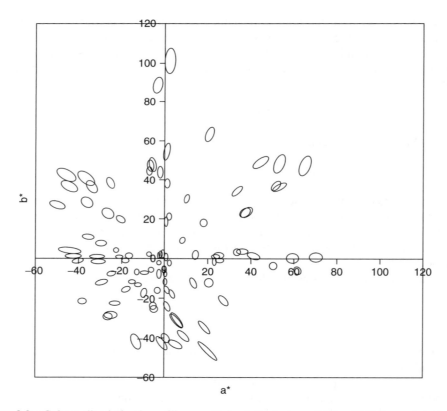

Figure 3.2 Colour discrimination ellipses obtained by Luo and Rigg [43] (reproduced from Figure 1 of 'The development of the CIE 2000 colour difference formula, CIEDE 2000', *Col. Res. App.*, **26**: 340–350, by permission of M. Ronnier Luo).

Uniformity of the components of colour difference

- *Chroma.* It can be seen in Figure 3.2 that in regions of the colour space distant from the achromatic axis, the magnitude of ΔE_{ab}^* is far larger than is justified by the perceptual difference between the two samples. Similarly, in a pair of neutral samples of the same lightness, the perceptual difference will tend to be considerably greater than is suggested by the numerical values of ΔC_{ab}^* and ΔE_{ab}^*. The size and tilt of the ellipses suggest that C_{ab}^* (chroma) is considerably less perceptually uniform than h_{ab} (hue angle).
- *Hue.* The threshold of perceptible colour difference varies with the hue angle of the samples being compared. For certain hues, particularly yellows, the numerical value of ΔE_{ab}^* at this threshold has been found to be larger than for other hues. The lack of hue uniformity in CIELAB, particularly in the blue region, can also reduce the reliability of a UCS difference metric. This arises because the Euclidean distance between two coordinates is the length of the straight line connecting them, whereas the line of constant perceived hue between them may in fact be curved [37].
- *Lightness.* Although the performance of ΔL^* in predicting perceptual differences in lightness has often been assumed to be satisfactory, analyses of several experimental data sets [16, 51, 11] have shown that the magnitude of ΔL^* varies with different values of L^*.

Viewing conditions

The detector of a colour measurement instrument such as a spectrophotometer simply records the stimuli presented to it, without regard to the conditions under which the samples are viewed.

A colour difference metric (even a perfectly uniform one) can be strictly correct for only one condition of viewing. As adaptation effects take place as the level or colour of the illumination changes, or simultaneous contrast effects are introduced, the uniformity of the space must deteriorate. For many practical purposes such problems are not significant, since the changes due to adaptation are small compared to the non-uniformity of the metrics themselves, but in some applications adaptation effects are sufficient to give rise to considerable difficulties in quantifying colour differences.

Surface characteristics

Media characteristics such as gloss and texture are known to induce changes in the way that small colour shifts as are perceived. For example, in textured samples the importance of changes in lightness increases relative to hue and chroma [72]. It is also likely that in most viewing conditions flare will have a greater effect in glossy samples, making colour shifts less apparent [21]. However, there is still work to be done to understand and model these effects.

3.3.2 Acceptability of colour differences

Establishing metrics of perceived colour difference is clearly important but is not the final stage in defining reproduction tolerances. It is also important to understand (and where possible model) the factors that affect tolerances of acceptability. These are not trivially related to perceived difference, but include a range of contextual factors. The magnitude of acceptable tolerances will vary according to the application; for example, colour matching in packaging or advertising is much more critical than in 'editorial' content in periodicals.

It is often found that the ΔL^*, ΔC_{ab}^* and ΔH_{ab} components of colour difference are given different weight when considering acceptability than when judging the magnitude of perceptible difference. Acceptability tolerances are also dependent on the location in colour space, the direction of colour shift, and (in colour images) scene content. For example, a colour difference that results in the final colour crossing a 'name boundary' (e.g. from yellow to green) will be more objectionable than a shift of equal magnitude that remains within a colour category (e.g. where a yellow becomes more greenish but is still identifiably yellow). It is also frequently acceptable for a colour to shift in one direction in colour space but not in another; for example, it is often acceptable to reproduce flesh tones with a redder hue but not with a bluer hue.

3.4 OVERCOMING THE LIMITATIONS OF UCS COLOUR DIFFERENCE WITH ADVANCED COLOUR DIFFERENCE METRICS

The problems identified above can be addressed in two ways. Either the colour space itself can be modified in a way that makes its perceptual uniformity adequate; or its limitations can be overcome by weighting colour difference equations in a way which corrects for the principal sources of non-uniformity.

In future it is possible that colour differences will be based on a colour appearance metric, rather than simple measures of the stimulus [2]. However, more work is needed to demonstrate the ability of appearance metrics to define colour difference in a uniform way. In the meantime, considerable effort has gone into deriving and testing equations that correct for the residual non-uniformity in the 1976 colour spaces.

Colour difference metrics for industrial use arose from the first uniform colour spaces. Prior to the adoption of the 1976 uniform colour spaces CIELAB and CIELUV, a colour difference equation had been defined in ANLAB and had found some industrial applications. An equation based on ANLAB, but with modifications for local distortions in colour space, continues to be used by the well-known retailer Marks and Spencer. Most advanced colour difference equations in current use (including all those described below) are based on a transformation of coordinates in the CIELAB colour space.

3.4.1 CMC

In 1984 the CMC $(l:c)$ equation [15] was adopted by the Colour Measurement Committee of the Society of Dyers and Colourists (SDC), following previous use of ANLAB-based metrics. It was derived from a considerable body of experimental data concerning the acceptability of colour matches in the textile industry and appears to give substantially better uniformity than ΔE_{ab}^* for small differences in colour.

$$\Delta E_{CMC} = \left[\left(\frac{\Delta L^*}{lS_L} \right)^2 + \left(\frac{\Delta C_{ab}^*}{cS_C} \right)^2 + \left(\frac{\Delta H_{ab}^*}{S_H} \right)^2 \right]^{\frac{1}{2}} \qquad (3.22)$$

where

$$S_L = 0.040975L^*/(1 + 0.01765L^*) \qquad L^* \geqslant 16$$
$$S_L = 0.511 \qquad\qquad\qquad\qquad\qquad L^* < 16$$

and

$$S_C = 0.0638C_{ab}^*/(1 + 0.0131C_{ab}^*) + 0.638$$

$$S_H = S_C(fT + 1 - f)$$

where

$$f = [(C_{ab}^*)^4/((C_{ab}^*)^4 + 1099)]^{0.5}$$

and

$$T = 0.36 + |0.4\cos(h_{ab} + 35)| \text{ for } 164° < h_{ab} > 345°$$

or

$$T = 0.56 + |0.2\cos(h_{ab} + 168)| \text{ otherwise}$$

The lightness and chroma weightings l and c are set to a ratio of 1:1 relative to hue for the prediction of perceptibility; while for acceptability prediction the weightings may be set higher than 1.

The weightings S_L, S_C and S_H adjust the sizes of the semi-axes of the ellipsoids defining the tolerance volume in CIELAB. The magnitude of ΔE_{CMC} for a constant numerical difference in a^* and b^* is shown in Figure 3.3. This shows that the same CIELAB colour difference could lead to ΔE_{CMC} values varying by an appreciable factor, and that very large numerical values of ΔE_{CMC} occur around the neutral axis.

3.4.2 BFD

A number of experimental data sets which included both perceptibility and acceptability results were combined to derive the BFD(l:c) equation in 1987 [40]:

$$\Delta E(BFD) = \left\{ \left[\frac{\Delta L(BFD)}{l} \right]^2 + \left[\frac{\Delta C_{ab}^*}{cD_C} \right]^2 + \left[\frac{\Delta H_{ab}^*}{D_H} \right]^2 + R_T \left(\frac{\Delta C_{ab}^*}{D_C} \right) \left(\frac{\Delta H_{ab}^*}{D_H} \right) \right\}^{\frac{1}{2}}$$
$$(3.23)$$

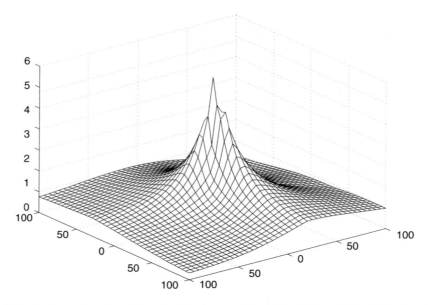

Figure 3.3 ΔE_{CMC} plotted on the $a^* - b^*$ plane at $L^* = 50$, $\Delta a^* = \Delta b^* = 3.55$, showing the greater weighting given to colours close to the achromatic axis.

where:

$$D_C = 0.035 C_m^* / (1 + 0.00365 C_m^*) + 0.521$$

$$D_H = D_C (GT' + 1 - G)$$

$$G = \{(C_m^*)^4 / [(C_m^*)^4 + 14000]\}^{\frac{1}{2}}$$

$$T' = 0.627 + 0.055 \cos(h_m - 254) - 0.04 \cos(2h_m - 136) + 0.07 \cos(3h_m - 32)$$

$$- 0.049 \cos(4h_m + 114) - 0.015 \cos(5h_m - 103)$$

$$R_T = R_H R_C$$

$$R_H = 0.26 + 0.055 \cos(h_m - 308) - 0.379 \cos(2h_m - 160) - 0.636 \cos(3h_m - 254)$$

$$+ 0.226 \cos(4h_m + 140) - 0.194 \cos(5h_m - 280)$$

$$R_C = \{(C_m)^6 / [(C_m^*)^6 + 7 \times 10^7]\}^{\frac{1}{2}}$$

where C_m^* and h_m are the mean of the CIELAB C_{ab}^* and h_{ab} values for the reference and sample, and h_{ab} values are expressed in radians.

The BFD equation introduces a $\Delta H/\Delta C$ hue–chroma interaction, which improved the fit with the experimental data sets. The R_T term controls the rotation of the tolerance ellipses with maximum effect in high-chroma blue samples.

3.4.3 CIE94

A Technical Committee of the CIE (TC 1-29), set up to study industrial colour difference evaluation, published its recommendations as CIE Technical Report in 1995 [13]. The new equation was known as CIE94.

The report proposed a new colour difference equation whose structure is similar to the CMC(l:c) equation, but whose weighting functions are largely based on the RIT/DuPont tolerance data derived from experiments with automotive paints [4, 9]. The definitions of S_C and S_H are simplified, and parametric factors k_L, k_C and k_H are introduced.

The role of the parametric factors is to enable the sensitivity of the equation to lightness, chroma and hue components of colour difference to be adjusted to allow for evaluation tasks which differ from the basis conditions for which CIE94 was derived. These parametric factors are listed in Table 3.2.

The CIE94 equation is:

$$\Delta E_{94}^* = \left(\left(\frac{\Delta L^{*2}}{k_L S_L}\right) + \left(\frac{\Delta C_{ab}^{*2}}{k_C S_C}\right) + \left(\frac{\Delta H^{*2}}{k_H S_H}\right)\right)^{\frac{1}{2}} \tag{3.24}$$

where

$$S_L = 1, \; S_C = 1 + 0.045 C_{ab}^*, \; S_H = 1 + 0.015 C_{ab}^*$$

and $k_L = k_C = k_H = 1$ for a set of reference sample, viewing and illuminating conditions, but can be set differently according to the context of use.

The effect of the CIE94 equation on the magnitude of colour difference in different regions of colour space can be seen in Figures 3.4 and 3.5.

The Technical Committee proposed that the equation should continue to evolve as new experimental data became available. It was envisaged that the equation will be optimised for assessment tasks which depart from the basis conditions, by deriving modified values for the parametric weighting factors and k_L, k_C, k_H. This is of course independent of the need to set overall tolerance thresholds according to customer requirements and manufacturing variances. It was also anticipated that the basic structure of the equation and its weighting functions would undergo revision in the light of experience.

Table 3.2 Parametric factors standardised in research on colour difference

Illumination	D65
Illuminance	1000lux
Background	uniform neutral grey, $L^* = 50$
Sample size	angular subtense greater than $4°$
Sample separation	Direct contact
Sample structure	Visually homogeneous
Magnitude of colour difference	$\Delta E_{ab} \leqslant 5.0$

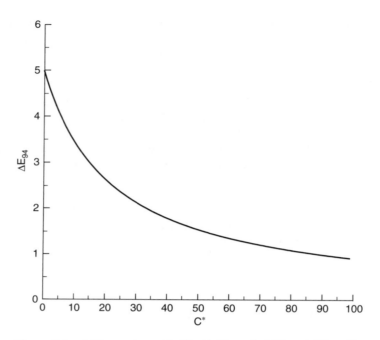

Figure 3.4 ΔC_{94}^* varying with C_{ab}^* ($\Delta C_{ab}^* = 5$, $\Delta L^* = \Delta H_{ab}^* = 0$).

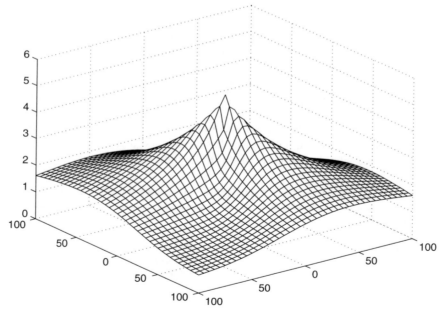

Figure 3.5 ΔE_{94}^* plotted on the $a^* - b^*$ plane at $L^* = 50$, $\Delta a^* = \Delta b^* = 3.55$.

3.4.4 CIEDE2000

CIE Technical Committee 1-47 (Hue and Lightness Dependent Correction to Industrial Colour Difference Evaluation) proposed the first major revision to CIE94 in 2001 [43], where a hue–chroma interaction term similar to that of the BFD equation was introduced and adjustments made to the weighting functions S_L and S_H. This revision was based on experimental data accumulated through a number of different studies, and its structure is similar to that of BFD(l:c).

$$\Delta E_{00} = \left(\left(\frac{\Delta L'}{k_L S_L} \right)^2 + \left(\frac{\Delta C'_{ab}}{k_C S_C} \right)^2 + \left(\frac{\Delta H'_{ab}}{k_H S_H} \right)^2 + R_T \left(\frac{\Delta C'_{ab}}{k_C S_C} \right) \left(\frac{\Delta H'_{ab}}{k_H S_H} \right) \right)^{\frac{1}{2}} \quad (3.25)$$

In CIE$_{94}$, the assumption that L^* adequately predicts perceived difference in lightness was reflected in the setting of S_L to unity. CIEDE2000 marks a departure from $S_L = 1$, based on observations [7] that the optimum weighting function for lightness varies according to the value of L^*, and in particular that ΔL^* overpredicts the magnitude of perceived lightness difference in samples at the light and dark ends of the lightness range.

The new S_H function was derived [44] from five experimental data sets with largely hue-only differences, where it gave a better fit than the weighting functions for hue in other equations.

CIEDE2000 colour difference is calculated as follows [43]:

1. Calculate a', C' and h' from CIELAB L^*, a^*, b^* and C^*:

$$
\begin{aligned}
L' &= L^* \\
a' &= (1 + G)a^* \\
b' &= b^* \\
C'_{ab} &= \sqrt{a'^2 + b'^2} \\
h'_{ab} &= \tan^{-1}(b'/a')
\end{aligned}
\quad (3.26)
$$

where

$$G = 0.5 \left(1 - \left(\frac{\overline{C^*_{ab}}^7}{\overline{C^*_{ab}}^7 + 25^7} \right)^{\frac{1}{2}} \right)$$

where $\overline{C^*_{ab}}$ is the arithmetic mean of the C^*_{ab} values for a pair of samples.

2. Calculate $\Delta L'$, $\Delta C'$ and $\Delta H'$:

$$
\begin{aligned}
\Delta L' &= L'_b - L'_s \\
\Delta C'_{ab} &= C'_{ab,b} - C'_{ab,s} \\
\Delta H'_{ab} &= 2\sqrt{C'_{ab,b} C'_{ab,s}} \sin \left(\frac{\Delta h'_{ab}}{2} \right)
\end{aligned}
\quad (3.27)
$$

where

$$\Delta h'_{ab} = h'_{ab,b} - h'_{ab,s}$$

3. Calculate the weighting functions S_L, S_C and S_H:

$$S_L = 1 + \frac{0.015 \left(\overline{L'} - 50\right)^2}{\left(20 + \left(\overline{L'} - 50\right)^2\right)^{\frac{1}{2}}}$$
$$S_C = 1 + 0.045 \overline{C'_{ab}}$$
$$S_H = 1 + 0.015 \overline{C'_{ab}} T$$

(3.28)

where

$$T = 1 - 0.17 \cos\left(\overline{h'_{ab}} - 30°\right) + 0.24 \cos\left(2\overline{h'_{ab}}\right) + 0.32 \cos\left(3\overline{h'_{ab}} + 6°\right)$$
$$- 0.20 \cos\left(4\overline{h'_{ab}} - 63°\right)$$

where $\overline{L'}$ and $\overline{C'_{ab}}$ are the arithmetic means of the L' and C'_{ab} values for a pair of samples; and $\overline{h'_{ab}}$ is the arithmetic mean of the h_{ab} hue angles of the two samples:

$$\overline{h'_{ab}} = (h'_{ab,s} + h'_{ab,b})/2, \qquad \overline{h'_{ab}} \leqslant 180$$
$$(h'_{ab,s} + h'_{ab,b})/2 - 180, \ \overline{h'_{ab}} > 180$$

4. Calculate the hue–chroma interaction factor R_T:

$$R_T = - \sin(2\Delta\theta) R_C$$

(3.29)

where

$$\Delta\theta = 30 \exp\left\{-\left[\left(\overline{h'_{ab}} - 275°\right)/25\right]^2\right\}$$

and

$$R_C = 2 \left(\frac{\overline{C'_{ab}}^{-7}}{\overline{C'_{ab}}^{-7} + 25^7}\right)^{\frac{1}{2}}$$

5. Select appropriate values for the parametric constants k_L, k_C and k_H ($k_L = k_C = k_H = 1$ for reference conditions, with variations from these values where necessary).
6. Calculate CIE ΔE_{00} by using the results of steps 2–5 in equation (3.25).

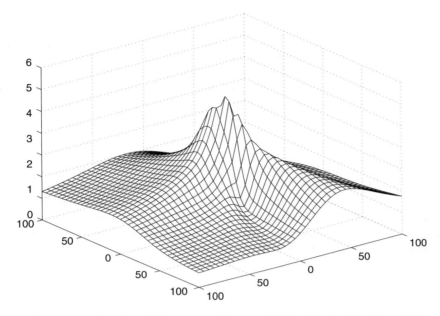

Figure 3.6 ΔE_{00} plotted on the $a^* - b^*$ plane at $L^* = 50$, $\Delta a^* = \Delta b^* = 3.55$.

Figure 3.6 shows the variation in ΔE_{00} for a constant value of CIELAB. The numerical value of ΔE_{00} reaches a similar maximum to ΔE_{94} and ΔE_{CMC} at the neutral axis, but the function falls off less sharply and has greater hue angle dependency.

3.5 ADVANCED COLOUR DIFFERENCE METRICS IN COLOUR IMAGING

The colour difference metrics described above have been extensively tested with large experimental data sets [3, 4, 11, 14–17, 25, 27, 41–43, 46, 50, 51, 60, 69, 74]. However, there are some limitations in their applicability to colour imaging, and further work is needed in order to understand how best to apply them and in particular to optimise the parametric factors k_L, k_C and k_H. (Parametric effects are summarised in Ref. [14])

3.5.1 Basis conditions

CIE$_{94}$ and CIEDE2000 were derived for a set of basis conditions described above. Many of these conditions do not represent the circumstances under which colour difference assessments take place in colour imaging. The major departures are listed below.

Illuminant

Although D65 is the illuminant most commonly associated with colorimetry and colour difference, and in several industries is used as the basis of standard viewing sources, in other industries such as the graphic arts D50 has been used for a number of years and is specified in both ISO 13655 [32] and ISO 3664 [33] for colorimetry and viewing respectively.

Illumination

While the guidelines specify 1000 lux, on the basis that lower levels of illumination would lead to reduced colour discrimination, for critical appraisal in the graphic arts a level of 2000 lux is specified in ISO 3664. The higher level of illumination is likely to alter the size and shape of the discrimination ellipse [10] and possibly affect thresholds of acceptability, while the resulting increase in flare is also likely to have an effect on some judgements, particularly in glossy samples [34].

Sample separation

In the reference conditions, edge contact of samples is specified which in practical assessment tasks enables the greatest discrimination when comparing samples. However, in many imaging applications the physical separation of colours is dependent on page structure. The media white and other coloured areas on the page are likely to provide an immediate surround to the region being evaluated, in addition to other regions that are in the field of view.

Sample size and image structure

The basis conditions relate to samples that subtend an angle of 4° or more to the observer, and are homogeneous without apparent pattern.

Homogeneity is unattainable in many hard copy reproduction systems, where the halftoning pattern or in some cases the raster structure of the marking technology is likely to be visible to some degree. In commercial printing processes, there is also a tendency for local inking variations to occur.

Although a proportion of typical page content comprises uniform areas of colour, complex images play an important role in many colour reproduction tasks. By their nature, colour images are highly inhomogeneous and are composed of small elements that are frequently below the resolving power of the eye at normal viewing distances.

3.5.2 Colour difference in complex images

Most work on colour difference has considered uniform colour samples. CIE Technical Committee TC 8-02 was established in 1998 to develop recommended methods of deriving colour differences for colour images.

The main parameters that will affect the magnitude of perceived difference in addition to the stimulus itself are the media and its associated mode of viewing (i.e. emissive, reflective or transmissive), and the viewing conditions (source and level of illuminance, surround and background). These parameters are largely accounted for in appearance models such as CIECAM97s. However, further factors which contribute to the perceived difference between two colours within a complex image have been identified [64], such as simultaneous contrast, assimilation, and the relationship between the elements of image content (including their 3D organisation, subjective contours, apparent lighting and transparency).

If an appropriate appearance model is chosen and the factors that affect perceived difference in colour images are well understood, it will in principle be possible to derive measures of difference for both inter-image and intra-image comparison tasks. There is, however, much work to do to achieve this, and in particular a need for more experimental data on how the factors described above affect perceived differences.

Since in a complex image there will be a large number of pixels for which a colour difference can be found, it is important to develop appropriate summary statistics for reporting the differences of interest in a given application. This is particularly problematic as there is insufficient evidence on the relationship between simple mean ΔE_{ab} measures (in any difference metric) and the perceived differences between images. In one study [65] a mean perceptibility tolerance of $3\Delta E_{ab}$ was found; mean acceptability tolerances were found to be $6\Delta E_{ab}$, but a simple linear scaling from perceptibility did not model the results well. Suggested summary statistics [64] include the mean ΔE_{ab} and its standard deviation, but also (recognising that the distribution of differences is unlikely to be Gaussian or to have a mean of zero) the coefficient of skewness and ΔE_{ab} at selected quantiles, together with the mean and standard deviation of the colour difference components ΔL^*, Δa^* and Δb^* and the correlations between them. This latter approach could also be extended to ΔL^*, ΔC^* and ΔH^* components.

3.5.3 Acceptability and perceptibility

Although the perceptibility of colour difference can reasonably be assumed to be constant among observers of normal colour vision, the same assumption cannot be made in relation to the acceptability of such differences. Both inter-observer and intra-observer variation will tend to be much higher than for perceptibility judgements, and it has to be accepted that there will inevitably be a degree of subjectivity in any one decision.

Relatively few studies have investigated acceptability in colour difference, particularly in colour imaging media. Where mean acceptability thresholds have been determined they are typically twice as large as those for perceptibility. However, it is probably unwise to assume that weighting functions derived from perceptibility data can be linearly scaled to define acceptability thresholds. In one study of samples produced by four-colour offset lithography, it was found that CIE₉₄ performed no better than ΔE_{ab}^* in predicting acceptability weightings and thresholds [21]. CIEDE2000 performed little better in predicting this data set, and it was suggested that weighting functions derived from perceptibility data may not be able to make better predictions of acceptability thresholds unless the parametric factors are modified according to the region of colour space (thus becoming functions rather than constants).

In a study of images displayed on a CRT [60], acceptability tolerances were found to be approximately twice those for perceptibility. Another study [65] concluded that scene content affects perceptibility but not acceptability.

3.5.4 Large vs. small differences

Colour difference metrics are intended to apply to colour difference magnitudes in the range of $0-5\Delta E^*_{ab}$ [12, 36], and most of the experimental data from which they were derived are limited to differences of this order. However, in colour imaging considerably larger differences are often experienced, and in some regions of colour space it is possible to find differences of $12\Delta E^*_{ab}$ or even more being judged acceptable.

The advanced colour difference metrics are non-Euclidean, since the way in which the formulae are defined leads to a difference in ΔE^*_{ab} according to which of the two samples is taken as the reference. For some high-chroma sample pairs which are judged acceptable, this can lead to a difference of about $0.5\Delta E^*_{ab}$ depending upon which sample is taken as the reference. Recently more work has been done on larger colour differences [21, 24, 26, 54, 68].

3.6 DERIVING COLOUR DIFFERENCE TOLERANCES

A great deal of work has been carried out to date to reach our present level of understanding of colour difference. A number of previous studies have investigated the discrimination tolerance around selected colours and quantified the resulting discrimination ellipsoids or acceptability thresholds [57, 70, 39]. The advanced colour difference equations described above were largely derived from efforts to combine the data sets arising from such experiments, in order to establish metrics with wide application.

Further work is needed in order to understand how to apply the existing knowledge of colour difference to colour imaging and CIE TC 8-02 is actively addressing this question at present. The purpose of this section is to provide an overview of some the experimental methods used to derive colour tolerances.

There are three main phases in an experiment to derive colour difference tolerances. First, colour samples are prepared in the media being studied; secondly, a psychophysical experiment is performed in which the colour difference of each sample is assessed relative to a reference colour by a panel of observers; and finally the visual differences between the samples are compared with the corresponding colorimetric differences to compute the required tolerances.

Further analysis may also take place, for example to derive weighting functions that fit the experimental data, to test how well a colour difference metric predicts the experimental data, or to compare different experimental data sets in order to explore the effect of varying parametric conditions.

3.6.1 Sample preparation

In an experiment, one or more colour centres are selected and samples generated on the media being studied. Seventeen colour centres have been recommended for study [73]. These colour centres are coordinates in different regions of CIELAB colour space. Among the media investigated in previous studies are glossy acrylic paints, textiles such as wool samples, colour prints and CRT-generated colours [52].

For each colour centre, multiple samples are prepared with small incremental colour variations. The aim is to create a distribution of samples around the centroid that vary in lightness, chroma and hue. Each sample is measured and its colour space coordinates recorded.

Ease of creating suitable samples and their stability over time are important factors in selecting the media to study. CRT-generated samples avoid some of the problems with 'hard copy' media and in some situations can be used to simulate colour assessment tasks in other media [52].

3.6.2 Psychophysical experiments

Samples are judged by a panel of observers in order to derive a visual scale of difference for each colour centre that can be compared with the differences found by measurement.

The judgement task can be a pair comparison (in which each sample is compared relative to a reference sample and a judgement made as to whether the difference between them is perceptible), magnitude estimation (in which the observer estimates the degree of difference in lightness, chroma or hue between each sample and the reference colour), or category judgement (in which the observer judges each sample relative to the reference and assigns it to one of a number of predetermined categories). Pair comparison and category judgement techniques can be applied to judgements of acceptability as well as perceptibility.

In studies seeking to determine perceptibility thresholds, a visual reference is included in the experimental task. Such references typically include an 'anchor pair' of achromatic samples with an L^* lightness of approximately 50, differing from each other by around $1\Delta E_{ab}^*$ [58]; or a grey scale of achromatic samples with a range of L^* values [39, 71].

Observers may experience some difficulty in scaling the difference between such achromatic references to large colour differences or to hue differences in chromatic samples. Where acceptability tolerances are sought, it is often better to anchor the judgement by the observer's experience of acceptable reproduction, and the colour-matching experience of the observer can then affect the repeatability and inter-observer reliability of the data [21].

Psychophysical experiments yield ordinal data about the underlying magnitudes of perceived difference, but it is possible to transform such data to interval or ratio scales [20, 59, 63, 67]. The aim is usually to derive a ratio scale of visual difference, ΔV, from the judgements made.

3.6.3 Observer variability and experience

Because the physiological and affective differences between observers will have an impact on the judgements made, deriving tolerance data from a small number of observers is highly unreliable [56]. A minimum of 20 observers is required to provide a good basis for analysis.

It is desirable to investigate the repeatability of the judgements made in any colour assessment experiment. This can be done by analysing the variation between the judgements made by individual observers (inter-observer variation) [70] and between repeated judgements made by the same observer (intra-observer variation). Inter-observer variation has been found to be higher than intra-observer variation, typically by a factor of 2 (for both homogeneous samples and complex images). One study that focused on this question [1] reported mean inter-observer variability for homogeneous samples of $2.5 \Delta E_{ab}^*$, with small differences in different types of media.

The colour-matching experience of the observers in psychophysical experiments is an important factor in the judgements they make. In one study where observers were divided into those with production experience of colour matching and those without [21], the former group were found to achieve greater scores on the Farnsworth–Munsell 100 Hue test [18] of colour discrimination, have lower mean acceptability thresholds, and make more judgements before experiencing task fatigue. The covariance of their visual scale judgements with the colorimetric difference between samples was also greater. These features are presumably the result of extensive experience in critical colour appraisal in colour reproduction tasks in which tolerances have to be set sufficiently small to meet the expectations of the majority of clients.

Experimenters should be aware of the need to minimise task difficulty and avoid observer fatigue or loss of motivation, as the judgements made under such conditions are likely to be less reliable.

3.6.4 Calculating colour tolerances from experimental data

The calculation of three types of colour tolerance is briefly described here: the ellipsoids that represent the locus of discrimination for a given colour centre; the parametric factors that can be used to account for media or viewing conditions that are different from the basis conditions described earlier; and acceptability thresholds. For more detail on these methods the reader is referred to the sources given in the references.

3.6.5 Calculation of discrimination ellipsoids and tolerance distributions

The semi-axes of a tolerance ellipsoid are notionally oriented along the dimensions of lightness, hue and chroma, although, as can be seen in Figure 3.2, in practice some rotation occurs. An ellipsoid in a three-dimensional space with axes x, y and z can be represented by a set of six coefficients:

$$a_{11}x^2 + a_{22}y^2 + a_{33}z^2 + 2a_{12}xy + 2a_{13}yz + 2a_{23}xz - j = 0 \qquad (3.30)$$

It can be seen that equation (3.26) is of quadratic form in three variables. It can thus be rewritten in matrix form as:

$$\mathbf{x}^t \mathbf{A} \mathbf{x} - j = 0 \qquad (3.31)$$

where $\mathbf{x}$ is the vector $[x, y, z]$, $\mathbf{x}^t$ is its transpose, $\mathbf{A}$ is a symmetric 3×3 matrix and j is a scalar. Coefficients a_{11}, a_{22} and a_{33} form the diagonal of the matrix $\mathbf{A}$, and a_{12}, a_{13} and a_{23} form the upper triangular part of $\mathbf{A}$ (and are repeated in the lower triangular part).

The axes of the colour space of interest can be substituted in equation (3.27) above. For colour difference in CIELAB the equation becomes:

$$\Delta E = [\Delta L^*, \Delta a^*, \Delta b^*]\mathbf{A}[\Delta L^*, \Delta a^*, \Delta b^*] \tag{3.32}$$

or alternatively

$$\Delta E = [\Delta L^*, \Delta C^*, \Delta H^*]\mathbf{A}[\Delta L^*, \Delta C^*, \Delta H^*]$$

Several methods of deriving the matrix $\mathbf{A}$ of ellipsoid coefficients from a set of experimentally derived visual scale values have been described [9, 31, 39, 48, 49, 55, 58]. A common feature of many methods is the minimisation of the function

$$S^2 = \sum (\Delta V_i - \Delta E_i)^2 \tag{3.33}$$

where ΔV are the experimentally determined visual scale differences, ΔE are the corresponding differences determined by measurement, and the subscript i refers to the individual sample pairs.

There will inevitably be a degree of uncertainty around the precise location of the boundaries of a tolerance ellipsoid for a given population of observers, and as a result the tolerance volume may be better expressed as the median tolerance of the population. It has been shown that probit analysis and logistic analysis provide appropriate methods of determining median tolerances [9], and the associated uncertainty can be defined as confidence intervals or as fiducial limits [4, 9].

3.6.6 Calculation of parametric constants in weightings functions

Parametric factors for hue, chroma and lightness that optimise the visual data can be calculated using a logistic regression method described by Berns [8].

1. The visual scale difference ΔV is calculated. It is assumed that the relationship between ΔV and the stimulus difference is non-linear, and can be approximated by a suitable function such as the logit function:

$$\Delta V = \log_e[f/(1 - f)] \tag{3.34}$$

where f is the frequency of rejection. Because logit values cannot be calculated where the frequency of rejection is 1 or 0 (i.e. where a sample is accepted or rejected by all observers), such samples are normally excluded. The visual scale values are transformed to an all-positive scale by an additive constant (e.g. 0.5).

2. The values of the weighting functions S_L, S_C and S_H are calculated for each sample, with the reference sample for each colour centre used as the standard in the calculation. A least-squares regression equation can then be defined:

$$(\Delta V)_i^2 = \beta_L(\Delta L^*/S_L)_i^2 + \beta_C(\Delta C_{ab}^*/S_C)_i^2 + \beta_H(\Delta H_{ab}^*/S_H)_i^2 \qquad (3.35)$$

where i refers to the individual observations, and β_L, β_C and β_H are the coefficients to estimate in the regression. The coefficients β_L, β_C and β_H are then converted to $l{:}c{:}h$ weightings by

$$l = (1/\beta_L)^{\frac{1}{2}}$$

and similarly for c and h. Finally these weightings are divided by h in order to normalise the $l{:}c$ weights relative to $h = 1$.

3.6.7 Calculation of acceptability thresholds

Optimum acceptability thresholds can be determined for each colour centre by finding the colour difference at which the number of instrumental wrong decisions is minimised [8, 21]. This is the threshold which minimises the number of times samples are judged to be acceptable (or unacceptable) when the metric difference is greater (or less) than the threshold.

The pass/fail judgements are first sorted according to the magnitude of the colour difference in the selected metric. Each adjacent sample in turn can then be considered as a threshold and the number of resulting instrumental wrong decisions counted. The sample pair for which the number of wrong decisions was smallest is taken as marking the transition between pass and fail, and the mean of the two samples gives the threshold.

3.7 CONCLUSION

Colour difference has a wide range of applications in colour engineering. The CIE 1976 Uniform Colour Spaces L^*, a^*, b^* and L^*, u^*, v^* provide the basis for a colour difference metric calculated as the Euclidean distance between coordinates within the space. These metrics are sufficient for many applications, but have some limitations. The advanced colour difference formulae described in this chapter have greater perceptual uniformity, particularly in the prediction of just-perceptible colour differences. The recent adoption of CIEDE2000 incorporates a significant body of research on colour tolerances, and it is hoped that its increased complexity will be justified by an improvement in perceptual uniformity and its adoption by the colorant industries.

There remain questions that are still to be answered in understanding and modelling the perception of colour difference. In particular there is insufficient understanding of the applicability of colour difference metrics to complex images. This is an interesting and continuing field of research.

REFERENCES

1. Alfvin, R. L. and Fairchild, M. D. (1997) Observer variability in metameric color matches. *Col. Res. App.*, **22**, 174–188.
2. Alman, D. H. (1993) CIE Technical Committee 1-29 Industrial colour difference evaluation progress report. *Col. Res. App.*, **18**, 137–139.
3. Alman, D. H. (1999) *Improvement to industrial colour-difference evaluation* (draft). Report to CIE TC 1-47.
4. Alman, D. H., Berns, R. S., Snyder, G. D. and Larsen, W. A. (1989) Performance testing of color difference metrics using a color appearance data set. *Col. Res. App.*, **14**, 139–151.
5. ASTM (2000) *Standards on color and appearance measurement*, 6th edn. ASTMS, West Conshohocken, PA.
6. ASTM E308-95 (1995) *Standard method for computing the colours of objects by using the CIE system*. ASTMS, West Conshohocken, PA.
7. Berns, R. S. (2000) *Billmeyer and Saltzman's Principles of Colour Technology*. Wiley, New York.
8. Berns, R. S. (1996) Deriving instrumental tolerances from pass-fail and colorimetric data. *Col. Res. App.*, **21**, 459–472.
9. Berns, R. S., Alman, D. H., Reniff, L., Snyder, G. D. and Balonen-Rosen, M. R. (1991) Visual determination of suprathreshold color-difference tolerances using probit analysis. *Col. Res. App.*, **16**, 297–316.
10. Carreno, F. and Zoido, J. M. (2001) The influence of luminance on colour difference thresholds. *Col. Res. App.*, **26**, 362–368.
11. Chou, W., Lin, H., Luo, M. R., Westland, S., Rigg, B. and Nobbs, J. (2001) The performance of lightness difference formulae. *Coloration Technology*, **117**, 19–29.
12. CIE Technical Report (1986) CIE Publication 15.2, *Colorimetry*, 2nd edn. Central Bureau of the CIE, Vienna, Austria.
13. CIE Technical Report (1995) *Industrial Colour Difference Evaluation*. CIE Publication No. 116. Central Bureau of the CIE, Vienna, Austria.
14. CIE Technical Report (1993) *Parametric Effects in Colour-Difference Evaluation*. CIE Publication No.101. Central Bureau of the CIE, Vienna, Austria.
15. Clarke, F. J., McDonald, R. and Rigg, R. (1984) *Modification to the JPC79 colour-difference formula*. J. Soc. Dyers Col., **100**, 117–148.
16. Coates, E., Fong, K. Y. and Rigg, B. (1981) Uniform lightness scales. *J. Soc. Dyers Col.*, **97**, 179–183.
17. Cui, G., Luo, M. R. and Li, W. (2000) Colour-difference evaluation using CRT colours. *Proc. Conf. Colour and Visual Scales*, National Physical Laboratory, UK.
18. Farnsworth, D. F. (1957) *The Farnsworth-Munsell 100-Hue Test for the Examination of Colour Discrimination – Manual*. Munsell Color Company, Baltimore, MD.
19. Gardner, J. L. (2000) Uncertainty estimation in colour measurement. *Col. Res. App.*, **25**, 349–355.
20. Gescheider, G. A. (1997) *Psychophysics*, 3rd edn. Mahwah, NJ: Lawrence Earlbaum Associates.
21. Green, P. J. and Johnson, A. J. Issues of measurement and assessment in hard copy colour reproduction. *Proc. SPIE Conf. Col. Imaging: Device-Independent Color, Color Hardcopy and Graphic Arts*, **5**.
22. Grum, F. and Bartleson, C. J. (1980) *Optical Radiation Measurement: 2. Colour Measurement*. Academic Press.
23. Grum, F. and Becherer, R. (1981) *Optical Radiation Measurement: 1. Radiometry*. Academic Press.
24. Guan, S. S. and Luo, M. R. (1999) A colour-difference formula for assessing large colour differences. *Col. Res. App.*, **24**, 344–355.
25. Guan, S. S. and Luo, M. R. (1999) Investigation of parametric effects using small colour differences. *Col. Res. App.*, **24**, 331–343.

26. Guan, S. S. and Luo, M. R. (1999) Investigation of parametric effects using large colour differences. *Col. Res. App.*, **24**, 356–368.
27. Heggie, D., Wardman, R. H. and Luo, M. R. (1996) A comparison of the colour differences computed using the CIE94, CMC(*l:c*) and BFD(*l:c*) formulae. *Journal of the Society of Dyers and Colourists*, **112**, 264–269.
28. Hunt, R. W. G. (1995) *The Reproduction of Colour*, 5th edn. Fountain Press.
29. Hunt, R. W. G. (1998) *Measuring Colour*, 3nd edn. Fountain Press.
30. ICC (2001) *File Format for Color Profiles: specification* ICC.1:2001–12.
31. Indow, T. and Morrison, M. L. (1991) Construction of discrimination ellipsoids for surface colors by the method of constant stimuli. *Col. Res. App.*, **16**, 42–56.
32. ISO 13655:1996 *Graphic technology – Spectral measurement and colorimetric computation for graphic arts images*.
33. ISO 3664 *Viewing conditions – Prints, transparencies and substrates for graphic arts technology and photography*.
34. Johnson, A. J. and Green, P. J. (2001) The CIEDE2000 colour difference formula and its performance with a graphic arts data set. *Proc. 28th Conf. IARIGAI*.
35. Kaiser, P. K. and Boynton, R. M. (1996) *Human Color Vision*, 2nd edn. Holt, Rinehart and Winston, New York.
36. Kuehni, R. G. (1982) Advances in color difference formulas. *Col. Res. App.*, **7**, 19–23.
37. Kuehni, R. G. (1998) Hue uniformity and the CIELAB space and color difference formula. *Col. Res. App.*, **23**, 314–322.
38. Li, C. J., Luo, M. R. and Huang, H. (1999) Interpolation and abridgement of spectrophotometric data. *Int. Col. Management Forum*, University of Derby, pp. 51–60.
39. Luo, M. R. and Rigg, B. (1986) Chromaticity-discrimination ellipses for surface colours. *Col. Res. App.*, **11**, 25–42.
40. Luo, M. R. and Rigg, B. (1987) Colour difference formula BFD Part 1 – Development of the formula. *J. Soc. Dye Col.*, **103**, 86–94.
41. Luo, M. R. and Rigg, B. (1987) Colour difference formula BFD Part 2 – Performance of the formula. *J. Soc. Dye Col.*, **103**, 126–132.
42. Luo, M. R. (1996) The LLAB model for colour appearance and colour difference evaluation. *Proc. SPIE Conf. Color Imaging: Device-Independent Color, Color Hard Copy, and Graphic Arts*, **2658**, 261–269.
43. Luo, M. R., Cui, G. and Rigg, B. (2001) The development of the CIE 2000 colour difference formula. CIEDE2000. *Col. Res. App.*, **26**, 340–350.
44. Luo, M. R., Cui, G. and Rigg, B. (2000) Derivation of a rotation function for the new CIE colour difference formula. *Proc. Conf. Colour and Visual Scales*, National Physical Laboratory, Teddington, UK.
45. MacAdam, D. L. (1943) Specification of small chromaticity differences. *J. Opt. Soc. Am.*, **33**, 18–26.
46. Mahy, M., Van Eycken, L. and Osterlink, A. (1994) Evaluation of uniform colour spaces developed after the adoption of CIELAB and CIELUV. *Col. Res. App.*, **19**, 105–121.
47. McDonald, R. M. (ed) (1997) *Colour Physics for Industry*, Society of Dyers and Colourists.
48. McDonald, R. (1980) Industrial pass/fail colour matching. Part II – Methods of fitting tolerance ellipsoids. *J. Soc. Dyers Col.*, **96**, 418–433.
49. Melgosa, M., Hita, E., Poza, A. J., Alman, D. H. and Berns, R. S. (1997) Suprathreshold color-difference ellipsoids for surface colors. *Col. Res. App.*, **22**, 148–155.
50. Melgosa, M., Hita, E. and Perez, M. M. (1995) Sensitivity differences in chroma, hue and lightness from several classical threshold datasets. *Col. Res. App.*, **20**, 220–225.
51. Melgosa, M., Hita, E., Poza, A. J. and Perez, M. M. (1996) The weighting function for lightness in the CIE94 color-difference model. *Col. Res. App.*, **21**, 347–352.
52. Montag, E. D. and Berns, R. S. (1999) Visual determination of hue suprathreshold colour-difference tolerances using CRT-generated stimuli. *Col. Res. App.*, **24**, 164–176.
53. Paul, A. (1998) Colour fluctuations in printing. *J. Prepress & Printing Tech.*, **2**, 18–28.

54. Pointer, M. R. and Attridge, G. G. (1997) Some aspects of the visual scaling of large colour differences. *Col. Res. App.*, **22**, 298–307.
55. Qiao, Y., Berns, R. S., Reniff, L. and Montag, E. (1998) Visual determination of hue suprathreshold color-difference tolerances. *Col. Res. App.*, **23**, 302–313.
56. Rich, D. and Jalijali, J. (1995) Effects of observer metamerism in the determination of human color-matching functions. *Col. Res. App.*, **20**, 29–35.
57. Rich, D. C. and Billmeyer, F. W. (1983) Small and moderate color difference IV Color difference perceptibility. *Col. Res. App.*, **8**, 31–39.
58. Robertson, A. (1978) CIE guidelines for co-ordinated research on colour difference evaluation. *Col. Res. App.*, **3**, 149–151.
59. Shepard, R. N. (1966) Metric structures in ordinal data. *J. Mathematical Psychology*, **3**, 287–315.
60. Song, T. and Luo, M. R. (2000) Testing color-difference formulae on complex images using a CRT monitor. *Proc. IS&T/SID 8th Color Imaging Conf.*, 44–48.
61. Stearns, E. I. (1981) The determination of weights for use in calculating tristimulus values. *Col. Res. App.*, **6**, 210–212.
62. Stearns, E. I. (1987) The influence of spectral bandpass on accuracy of tristimulus data. *Col. Res. App.*, **12**, 282–284.
63. Stevens, S. S. (1971) Issues in psychophysical measurement. *Psychological Review*, **78**, 426–450.
64. Stokes, M. (1999) *Methods to derive colour differences for images*. CIE Technical Report (draft 0.3).
65. Stokes, M., Fairchild, M. D. and Berns, R. S. (1992) Colorimetrically quantified visual tolerances for pictorial images. *Proc. TAGA Conf.*, 757–778.
66. Thornton, W. A. (1998) Towards a more accurate and extensible colorimetry. Part IV Visual experiments with bright fields and both 10° and 1.3° field sizes. *Col. Res. App.*, **23**, 92–103.
67. Thurstone, L. (1927) A law of comparative judgement. *Psycholog Review*, **34**, 273–286.
68. Uroz, J., Luo, M. R. and Morovic, J. (2000) Colour difference perceptibility for large-size printed images. *Proc. Conf. Colour Imaging Science*, University of Derby, UK, pp. 138–151.
69. Witt, K. (1999) Geometric relations between scales of small colour differences. *Col. Res. App.*, **24**, 78–92.
70. Witt, K. (1987) Three-dimensional threshold of color difference perceptibility in painted samples: variability of observers in four CIE color regions. *Col. Res. App.*, **12**, 128–134.
71. Witt, K. (1990) Parametric effects on surface color-difference evaluation at threshold. *Col. Res. App.*, **15**, 185–195.
72. Witt, K. (1994) Modified CIELAB formula tested using a textile pass-fail data set. *Col. Res. App.*, **19**, 273–276.
73. Witt, K. (1995) CIE Guidelines for co-ordinated future work on industrial colour-difference evaluation. *Col. Res. App.*, **20**, 399–403.
74. Witt, K. (1995) Linearity and additivity of small color differences. *Col. Res. App.*, **20**, 36–43.
75. Wright, W. D. (1973) *The Measurement of Colour*, 4th edn. Adam Hilger.
76. Wyszecki, G. and Stiles, W. S. (1982) *Colour Science: Concepts and Methods, Quantitative Data and Formulae*, 2nd edn. Wiley, New York.

4

The CIE 1997 colour appearance model: CIECAM97s

M. Ronnier Luo

4.1 INTRODUCTION

The colour appearance of an object, or an image, changes according to different viewing conditions such as different media, different light sources, different background colours, and different luminance levels. This phenomenon causes severe problems in industrial colour control. For example, in the surface colour industries colourists require to know the degree of colour change across a wide range of illumination conditions. Lighting engineers need to evaluate the colour rendering property between a test and a reference illuminant. Colour reproduction engineers desire to reproduce faithfully the original presented on a different medium, known as cross-media reproduction. The media involved might include original scenes, transparencies, monitors, photographs or reflection prints. A dress normally seen in daylight might be photographed and appear in a catalogue, possibly under tungsten room lighting generally with a white surround, and also on a VDU screen in a dimly lit room. It is important that the dress looks the same under all conditions. Traditionally, obtaining the same appearance requires visual judgement by experienced workers. This process is subjective and expensive. Hence, there has long been a great demand by industrialists for the ability to accurately quantify changes in colour appearance so as to minimise observer dependencies.

Colour Engineering, edited by P.J. Green and L.W. MacDonald.
© 2002 John Wiley & Sons Ltd.

There are many colorimetric measures that have been recommended by the CIE *(Commission Internationale de l'Eclairage)* [1]. Those such as tristimulus values, chromaticity coordinates, dominant wavelength and excitation purity were used to indicate whether two stimuli match each other. On the other hand, the measures used to quantify colour differences such as CIELAB and CIELUV were developed so that equal scale intervals represent approximately equally perceived differences in the attributes considered. However, all the above measures are limited to being used under a set of fixed viewing conditions: the two stimuli under question should be presented using identical media and be viewed under the same daylight viewing conditions. These measures would not work at all in the above example on the reproduction of colour appearance from a dress viewed under daylight to that printed in a catalogue or displayed on a VDU screen under tungsten light. If, say, the tristimulus values are kept constant, the different conditions mean that the colour seen will change appreciably. Hence, neither of these two classes of measures provides a satisfactory means for solving the above problems. The solution is to devise an international standard colour appearance model, which is capable of accurately predicting the colour appearance of colours under a wide range of viewing conditions.

A general consensus has been formed that a five-stage transform is required to achieve successful cross-media colour reproduction, as shown in Figure 4.1. The transform includes four major components: a colour appearance model (CAM) including a forward and a reverse mode, a device characterisation model for converting between the colour primaries of particular imaging devices and the CIE specification, a device profile for defining a translation from a device characterisation under a set of viewing conditions to a standard colour space under a reference set of viewing conditions, and a gamut mapping model for mapping out-of-gamut colours from the input to output devices while preserving overall appearance. All components except the colour appearance model are described elsewhere in this book. Figure 4.1 clearly shows that the forward (Step 2) and reverse (Step 4) modes of the colour appearance model play very important roles. The model also provides a uniform colour space in which to perform colour gamut mapping.

Various colour appearance models have been developed over the years. At the CIE Expert Symposium'96 on Color Standards for Image Technology, held in Vienna in 1996 [7], there was great demand by the industrialists present to ask the CIE to recommend a particular colour appearance model for industrial application. Four colour appearance models were considered to be most promising: Hunt [2, 3], Nayatani [4], RLAB [5] and LLAB [6]. Agreement was achieved that Hunt and Luo should examine the existing colour appearance models, try to combine the best features of these models into a high performance model for general use, and test its performance against available experimental data. It was also agreed that the model should be available in a comprehensive version, and in a relatively simple version for use in limited conditions. At its meeting held in Kyoto in 1997, CIE Technical Committee TC1-34 (Testing colour appearance models) agreed to adopt a simple version, which is named CIECAM97s [8]. The 's' stands for the 'simple'. The inclusion of the year 97 in the designation is intended to indicate the interim nature of the model, depending upon the availability of better models expected to emerge in the future. The full CIECAM97s model, including its forward and reverse modes together with test data, is given in Appendix 1. It is recommended that

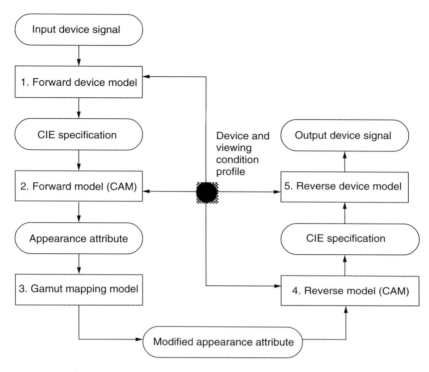

Figure 4.1   The data flow used in the five-stage transform.

the model should be applied for colour image processing to achieve successful colour reproduction across different media such as prints, monitors and transparencies. It should be pointed out that the adjective 'simple' is relative. The model is simple in the sense that various features which should be included in a full model are not considered. However, even a casual glance at Appendix 1 shows that the model is quite complicated.

4.2 CIE 1997 COLOUR APPEARANCE MODEL: CIECAM97s

Although there are many colour appearance models, their basic structures, input parameters, predicted attributes and colour appearance phenomenon are quite similar. CIECAM97s, the current CIE recommendation, is taken to typify colour appearance models in general and is described in this section.

4.2.1 Structure of the model

Figure 4.2 shows the structure of CIECAM97s. It comprises three parts: a cone response transform, a chromatic adaptation transform, a dynamic response function and colour spaces formed by different combinations of colour appearance attributes.

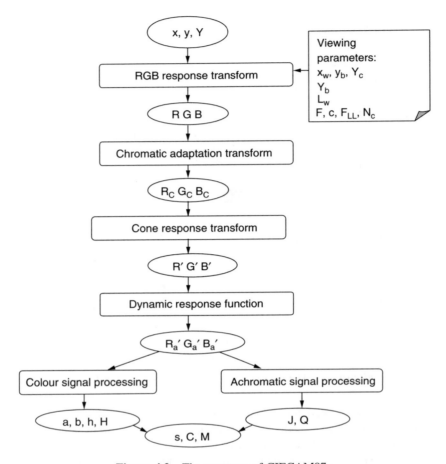

Figure 4.2   The structure of CIECAM97s.

The computation procedures for the forward and reverse modes are given in Appendix 1. The input parameters include x, y and Y (colorimetric values of the sample under the test illuminant), x_w, y_w and Y_w (colorimetric values of the reference white under the test illuminant), Y_b (the luminance factor of the background) and L_w (luminance of the reference white in cd/m^2 unit), together with those parameters defining the surround conditions: F, c, F_{LL} and N_c (see Appendix 1).

As shown in Figure 4.2, the initial step is to convert the colorimetric values (x, y, Y) to $R\,G\,B$ signals via a matrix transform. This is followed by transforming them to R_C, G_C and B_C signals via a chromatic adaptation transform, which is capable of predicting the corresponding colour from the test to reference illuminant (S_E, equal energy illuminant with $x = y = 0.3333$). Note that a pair of corresponding colours appears the same when viewed under test and reference illuminants. The CMC 1997 chromatic adaptation transform (CMCCAT97) is used in CIECAM97s [9].

The R_C, G_C and B_C signals are then transformed to cone responses. The dynamic response function in the form of a hyperbolic equation is then used to predict the extent of changes of responses due to different luminance levels. The predicted adapted cone responses R'_a, G'_a and B'_a are further used to calculate colour difference signals: a (red-ness–greenness) and b (yellowness–blueness), and A and A_W, achromatic signals for sample and reference white respectively. Subsequently, the a and b values are used to calculate h (hue angle) and H (hue composition). The A and A_W values are used to compute J (lightness) and Q (brightness). Finally, the s (saturation), C (chroma) and M (colourfulness) values are obtained (see next section for definitions of these attributes).

According to different applications, different colour spaces can be constructed by using different combinations of perceived attributes, i.e. typically $J\ C\ h$, $Q\ M\ h$, $J\ C\ H$, or $Q\ M\ H$. The former two spaces based upon hue angle, h, are approximately uniform colour spaces for calculating colour differences, and their cartesian coordinates can be calculated as $a_C = C\cos(h)$, $b_C = C\sin(h)$, or $a_M = M\cos(h)$ and $b_M = M\sin(h)$. The latter two colour spaces are used to describe colour appearance in terms of the hue composition expressed by the percentages of the four psychophysical hues (red, yellow, green and blue), e.g. an orange may consist of 60% of red and 40% of yellow.

4.2.2 Colour attributes

The colour appearance attributes predicted by CIECAM97s are described in this section. Their definitions can also be found in the CIE International Lighting Vocabulary [10].

Brightness

This is a visual perception according to which an area appears to exhibit more or less light. This is an open-ended scale with a zero origin defining the black.

The brightness of a sample is affected by the luminance of the light source used. A surface colour illuminated by a higher luminance would appear brighter than the same surface illuminated by a lower luminance.

Lightness

This is the brightness of an area judged relative to the brightness of a similarly illuminated reference white.

The lightness scale runs from black, 0, to white, 100. The lightness of the background used can cause a change of the lightness of the sample. This is called the *lightness contrast effect* (see next section). For example, a colour appears lighter against a dark background than when seen against a light background.

Colourfulness

Colourfulness is that attribute of a visual sensation according to which an area appears to exhibit more or less chromatic content.

This is an open-ended scale with a zero origin defining the neutral colours. Similar to the brightness attribute, the colourfulness of a sample is also affected by luminance. An object or image illuminated by a higher luminance would appear more colourful than when illuminated by a lower luminance. This is known as the Hunt effect (see next section).

Chroma

This is the colourfulness of an area judged as a proportion of the brightness of a similarly illuminated reference white. This is an open-ended scale with a zero origin representing neutral colours.

Saturation

This is the colourfulness of an area judged in proportion to its brightness. This scale again runs from zero, representing neutral colours, with an open end.

It can be considered as the ratio of colourfulness to brightness, or chroma to lightness, like that defined in the CIELUV space (s_{uv}) [1]. When adding a coloured pigment to a white paint, the resultant mixture increases the saturation by decreasing the brightness and increasing colourfulness.

Hue

Hue is the attribute of a visual sensation according to which an area appears to be similar to one, or to proportions of two, of the perceived colours red, yellow, green and blue.

Each colour appearance model predicts hue with two measures: hue angle ranging from 0° to 360°, and hue composition ranging from 0, through 100, 200, 300 to 400 corresponding to the psychological hues of red, yellow, green, blue and back to red. These four hues are the psychological hues, which cannot be described in terms of any combinations of the other colour names. All other hues can be described as a mixture of them. For example, purple colours should be described as mixtures of red and blue in the appropriate proportions.

4.2.3 Colour appearance phenomena predicted by the model

As mentioned earlier, CIE colorimetry considers only whether two stimuli match each other. The two stimuli in question must be viewed with identical surround, background, size, shape, texture, illumination, viewing geometry, etc. If any these constraints are violated, the colour match may no longer hold. Some important colour appearance phenomena are introduced in this section, which the basic colorimetry fails to predict.

Chromatic adaptation

One of the most important colour appearance phenomena is chromatic adaptation. Human eyes have a remarkable capability to maintain the colour appearance of an object after a period of adaptation across a wide range of illuminants regardless of large variations of spectral power distributions between different illuminants. Thus a piece of white paper appears white under daylight, and still appears white under tungsten light, despite the fact that the composition of the light reaching the eye is vastly different. Under daylight there is roughly the same amount of light at each wavelength. Under tungsten light, there is much more light at the longer wavelengths and much less at the shorter wavelengths.

Figure 4.3 illustrates 52 pairs of corresponding colours predicted by CIECAM97s (or its chromatic adaptation transform, CMCCAT97) from Illuminant A (open circle) to S_E (open end) plotted on the CIE $u'v'$ chromaticity diagram. The open circle colours were selected to have L* equal to 50 according to CIELAB under Illuminant A. These were then transformed by the model to the corresponding colours under illuminant S_E. The ends of each vector represent a pair of corresponding colours under two illuminants. The input parameters (see Appendix 1) are 1000 lux for the illuminance of white, average surround, and $Y_b = 20$.

The results show that there is a systematic pattern: for colours below v' of 0.48 or with u' larger than 0.44 under illuminant A, these vectors shift to the blue direction under the illuminant S_E. For colours not within the above region, their appearance changes following a counter-clockwise direction, i.e. red shifts to yellow, yellow to green, green to cyan as the illuminant changes from A to S_E. Note that some of these changes in $u'v'$ are about 0.1. Under constant conditions a change of about 0.003 corresponds to a noticeable difference.

Hunt effect

Hunt [11] found that the colourfulness of an object increases due to the increase of luminance. A typical outdoor scene appears much more colourful in bright sunlight than it does on a dull day. This is known as the Hunt effect. Figure 4.4 illustrates this effect as predicted by the CIECAM97s model. Five colours were selected from CIELAB having

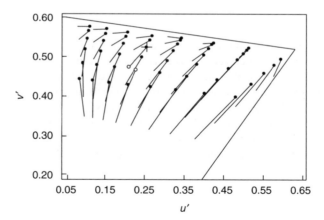

Figure 4.3 The corresponding colours predicted by CIECAM97s from illuminant A (closed circles of vectors) to illuminant S_E (opposite ends of vectors) plotted in CIE $u'v'$ chromaticity diagram. The cross and open circle represent illuminants A and S_E respectively.

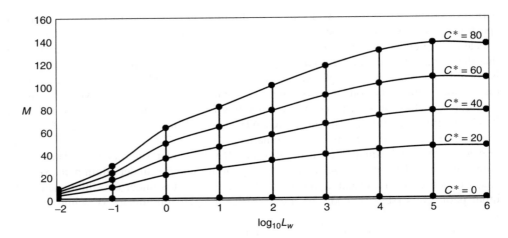

Figure 4.4 The Hunt effect predicted by the CIECAM97s model.

a constant L^* of 50 and a hue angle of 20 (red) with C^* varying from 0 (a neutral colour) to 80 (a high chroma colour). These colours were predicted by CIECAM97s under nine luminance levels, i.e. luminances of reference white from 0.01 (10^{-2}) to 1 000 000 (10^6) cd/m^2.

The results clearly demonstrate the Hunt effect, i.e. each sample represented by each curve increases its colourfulness (M) (except for the neutral colours) when the luminances of the reference white (L_W) increase until reaching about 100 000 cd/m^2. In addition, the colourfulness contrast is increased from dark to bright illuminance levels (compare the lengths of the vertical lines between dark and bright levels).

Stevens effect

Stevens and Stevens [12] found an increase in brightness (or lightness) contrast with an increasing luminance. Five neutral samples having L^* values of 0.01, 20, 40, 60 and 80 were selected to demonstrate the Stevens effect as predicted by CIECAM97s.

Figure 4.5 illustrates the Stevens effect by plotting brightness (Q) against $\log_{10} L_W$, i.e. an increase of brightness contrast with an increase of the luminance. The degree of increment in brightness is marked for the lighter samples ($L^* > 20$) but the effect disappears for the darkest sample ($L^* = 0.01$). This leads to an increase of brightness contrast, i.e. the lighter samples appear much brighter and the darker samples appear darker.

Surround effect

Bartleson and Breneman [13] found that the perceived contrast in colourfulness and brightness increased with increasing luminance level from dark surround (projection viewing), dim surround (CRT viewing) to average surround (reflection viewing). This is an important

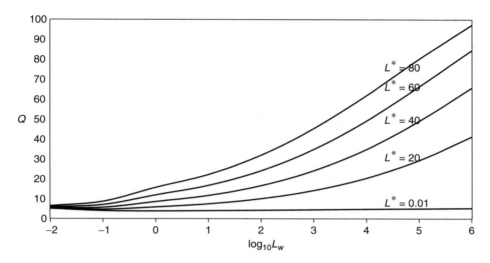

Figure 4.5   The Stevens effect predicted by the CIECAM97s model.

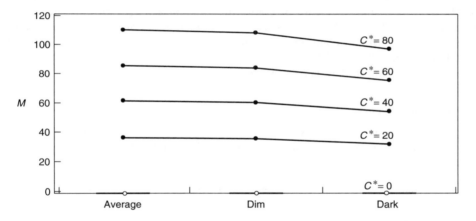

Figure 4.6   The surround effect predicted by the colourfulness (M) scale of CIECAM97s.

colour appearance phenomenon to be modelled, especially for the imaging and graphic arts industries, in which it is often required to compare different media under quite different viewing conditions.

Two figures are used to illustrate the surround effect: the colourfulness (M) and lightness (J) predicted by CIECAM97s under average, dim and dark surrounds. These are plotted in Figures 4.6 and 4.7 respectively. Figure 4.6 shows the colourfulness (M), for different surrounds, of samples with CIELAB C^* values of 0, 20, 40, 60 and 80 with constant L^* of 50 and h of 20 (red). (The other parameters were set to $N_c = 0.95$ rather than 1.1 as suggested by CIECAM97s (see Section 4.4), illuminance of 1000

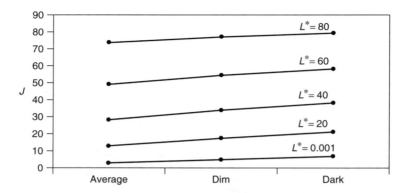

Figure 4.7   The surround effect predicted by the lightness (J) scale of the CIECAM97s.

lux and Y_b of 20.) Figure 4.7 shows the lightness (J), for different surrounds, of samples with CIELAB L^* of 0.001 to 80 neutral colours with the model parameters as for Figure 4.6.

Figure 4.6 shows that for each of the five test colours having C^* values of 20, 40, 60 and 80, there is a slight decrease of colourfulness from average, dim to dark surround conditions except for C^* of zero. This leads to a reduction of colourfulness contrast from average to dark surround conditions. Figure 4.7 shows that for each of the five neutral test colours having L^* of 0.001, 20, 40, 60 and 80, there is an increase of lightness from average, through dim to dark surround conditions.

Lightness contrast effect

The perceived lightness increases when colours are viewed against a darker background. This is called the lightness contrast effect. This is illustrated in Figure 4.8 by plotting the

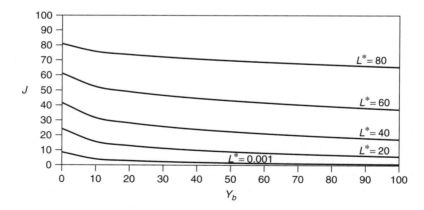

Figure 4.8   The lightness contrast effect predicted by the CIECAM97s model.

lightness (J) predicted by CIECAM97s against the luminance factor of the backgrounds (Y_b) for five neutral test colours having L^* of 0.001, 20, 40, 60 and 80.

It can be seen in Figure 4.8 that for all test colours, their lightness reduces when the background becomes lighter.

Effects not considered in CIECAM97s

Some other effects that are more complicated and are not considered in the CIECAM97s model include the *Helmholtz–Kohlrausch, simultaneous contrast* and *Helson–Judd* effects.

The *Helmholtz–Kohlrausch effect* refers to the increase of brightness (or lightness) for colour having the same luminance but higher chroma (or colourfulness). The *simultaneous contrast effect* refers to the change of colour appearance due to different backgrounds. This effect has been widely studied. It is well known that a change in the background colour may have a large impact on the lightness and hue percepts. There is some effect on colourfulness (or chroma), but this is not as large as that for the former two attributes. The lightness contrast effect has been discussed in the previous section.

The *Helson–Judd effect* [14] is that when a grey scale is illuminated by a light source, the lighter neutral stimuli will exhibit a certain amount of the hue of the light source and the darker stimuli will show its complementary hue. Thus for tungsten light, which is much yellower than daylight, the light stimuli appear yellowish, and the darker stimuli appear bluish. This effect was only found under haploscopic experimental conditions, i.e. different backgrounds were used for each eye (see below).

4.3 TESTING COLOUR APPEARANCE MODELS USING COLOUR APPEARANCE DATA SETS

4.3.1 Colour appearance data sets

Colour appearance models are developed partly on colour vision theories and partly empirically to fit some experimental results. Hence, the CIE Technical Committee 1-34 was formed to gather experimental data sets in order to test models' performance. The best model can then be recommended as an international standard model for industrial applications. The data collected by the CIE are given in Table 4.1. These data sets were selected from many data sets and are considered to be more reliable and representative than the others. These were accumulated using three experimental techniques: haploscopic matching, memory matching and magnitude estimation, which will be described in the next section. The results of testing colour appearance models' performance will then be given.

4.3.2 Psychophysical methods

Haploscopic matching

Haploscopic matching is the most widely used experimental technique. This technique requires specially designed viewing apparatus which presents a different adapting stimulus to each of the observer's two eyes. His or her task is to adjust the stimulus at one eye to

Table 4.1 List of important colour appearance data sets for each technique

Experiment	Year	Reference
Haploscopic matching		
CSAJ	1991	15
McCann *et al.*	1976	16
Breneman	1987	17
Memory matching		
Helson *et al.*	1952	18
Lam & Rigg	1985	19
Braun & Fairchild	1996	20
Magnitude estimation		
Luo *et al.* (Reflection and monitor stimuli)	1991	21
Luo *et al.* (Cut-sheet and 35 mm stimuli)	1993	22
Kuo & Luo (Reflection – Large textile)	1995	23

match that at the other eye. The task is relatively simple and the results in general have higher precision than the other two techniques. However, its validity is dependent on the assumption that the adaptation of one eye does not affect the sensitivity of the other eye. The technique imposes unnatural viewing conditions with constrained eye movement. In addition, when two eyes are presented with two stimuli under different adapting fields, observers tend to show bias towards one field rather than the other. This is known as *binocular rivalry*. The haploscopic technique can be improved to overcome some of the above shortcomings by using a complex field with free eye movement.

Memory matching

Memory matching is carried out under normal viewing conditions using both eyes and without the interposition of any optical devices. This technique provides a steady-state of adaptation with free eye movement. Observers are first trained to describe colours using the three colour appearance attributes of a colour descriptive system such as Munsell or the Natural Colour System. This means that they are able to describe with reasonable accuracy and precision the colour of any object in these terms under any viewing conditions. However, this is not a widely used technique and has drawbacks such as a substantial training period being required, complicated procedures for data analysis, lower precision than that of the haploscopic technique, limited capacity for retaining information, and memory distortion.

Magnitude estimation

Magnitude estimation has been used increasingly in recent years. Observers are asked to scale colour appearance attributes such as lightness, colourfulness and hue under fully adapted viewing conditions. Its only disadvantage is in having lower precision than that of the haploscopic technique. Many advantages are associated with this technique, such as normal viewing conditions using both eyes, steady state of adaptation, results that are described in terms of perceived attributes which can be directly compared with the colour appearance models' predictions, and a shorter training period than that of memory matching. The author and his coworkers have conducted a large set of experiments using this technique. The data is named the LUTCHI (produced at Loughborough University of Technology Computer Human Interface Research Centre) data set [21–23]. The results were used to develop the Hunt, LLAB and CIECAM97s colour appearance models.

4.3.3 Testing colour appearance models' performance

The data in Table 4.1 may be divided into two categories according to the nature of the data presented: those specifying corresponding colours (mainly obtained using the haploscopic matching and memory matching methods), and those representing average results for the magnitude estimation of the lightness, colourfulness and hue percepts.

Quantitative methods have also been developed to compare the performance between different colour appearance models in fitting these data sets. The method for testing models using the corresponding colour data sets is illustrated in Figure 4.9. The points E and F represent the experimental results for a pair of corresponding colours plotted in any suitable colour space, e.g. CIELAB, CIE $u'v'$. These two specimens (corresponding colours) are perceived to be the same colour when one is viewed under the reference viewing conditions and the other under the test viewing conditions, i.e. the colours look the same but the measured coordinates are different. A particular colour appearance model is then used to calculate the corresponding colour G under the reference viewing conditions (from the measurements for F under the test viewing conditions). The point F under the test viewing condition is then predicted by a particular model to point G under the reference viewing condition. The points E and G are coincident if there is perfect agreement between the visual and predicted results. The vector EG therefore represents the direction and magnitude of the error of prediction of the model for this pair of colours. The mean or root mean square (RMS) of the distances of vectors EG from a

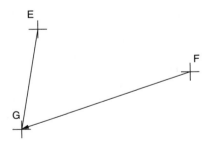

Figure 4.9   Method for estimating the error of prediction (vector EG) by a colour appearance model. Points E and F represent a pair of corresponding colours, and point G is the prediction from F.

number of corresponding-colour pairs has been used to measure the overall performance
of each transform. These are expressed in terms of colour difference (ΔE) calculated
by a suitable colour difference formula. For a perfect agreement between the predicted
and visual results, these measures should be zero. A larger value means worse performance.

The performance of the Hunt96, Nayatani97, RLAB96, LLAB96 and CIECAM97s
models is summarised in Table 4.2 in terms of CMC ΔE. These results are taken from
reference [24].

For testing models using the colour appearance percept data, the measure of coefficient
of variation (CV) was used as given in the following equation:

$$\text{CV} = 100 \left[\sum (V_i - P_i)/n \right] \bigg/ \left(\sum V_i/n \right) \qquad (4.1)$$

where V_i and P_i are the visual result and prediction from a model for sample i, and n is
the number of samples. For perfect agreement between the two sets of data, CV should
be zero. A CV of 30 indicates a 30% variation between the visual and predicted results.

Table 4.2 shows the models' performance in terms of $\Delta E_{\text{CMC}(1:1)}$. There appear to be
more data sets given in Table 4.2 than in Table 4.1 because those in Table 4.2 include
subsets of those in Table 4.1, i.e. three sets of CSAJ data for investigating chromatic
adaptation (−C), Helson−Judd (−H) and Stevens (−S) effects; three sets of Luo *et al.*
data for transforming from illuminants A, D50 and White Fluorescent (WF) to illuminant
D65; two sets of Kuo & Luo data for transforming from illuminants A and TL84 to
illuminant D65; and two sets of Breneman data for investigating chromatic adaptation

Table 4.2 Testing colour appearance models' performance (in $\Delta E_{\text{CMC}(1:1)}$) using corresponding
colour data sets. Best results for each data set are underlined

Data set	No. of phases	No. of samples	Hunt96	Nayatani97	RLAB96	LLAB96	CIECAM97s
CSAJ-C	1	87	3.7	6.1	5.4	3.7	3.7
CSAJ-H	1	20	3.5	4.3	6.6	3.5	3.5
CSAJ-S	1	19	4.1	6.9	4.1	4.1	4.1
Helson	1	59	3.8	7.5	5.2	3.8	3.8
Lam & Rigg	1	58	3.5	5.7	5.1	3.5	3.5
Luo *et al.* (A)	1	43	4.0	5.3	4.1	4.0	4.0
Luo *et al.* (D50)	1	44	4.4	4.1	3.5	4.4	4.4
Luo *et al.* (WF)	1	41	4.7	4.4	7.0	4.7	4.7
Kuo & Luo (A)	1	40	3.4	5.4	6.3	3.4	3.4
Kuo & Luo (TL84)	1	41	2.8	2.8	4.0	2.8	2.8
Breneman-C	9	108	5.7	4.0	5.3	5.7	5.7
Breneman-L	3	36	4.6	5.7	6.6	4.6	4.6
Braun & Fairchild	4	66	3.9	6.0	5.6	3.9	3.9
McCann	4	68	10.6	10.5	8.7	10.6	10.6
Weighted mean			4.8	5.6	5.6	4.8	4.8

(−C) and luminance adaptation (−L). Note that although many data sets were used to test models, the weighted mean was calculated by taking into account the number of stimuli in each data set.

The most promising five models at that time were compared: Hunt96 [25], Nayatani97 [4], RLAB96 [5], LLAB96 [25] and CIECAM97s [8]. For each data set, the best model (or the least ΔE value) is underlined to ease the comparison. The results clearly show that the CIECAM97s, Hunt96 and LLAB models outperform the Nayatani97 and RLAB96 models.

The magnitude estimation data sets in Table 4.1 were also used to test the above five colour appearance models. The CIECAM97s model gave an overall best performance for predicting the lightness and hue, and ranked second for predicting the colourfulness results. However, CIECAM97s is much simpler than Hunt96, which performed best for the colourfulness results. Hence, the performance of CIECAM97s shown in Tables 4.2 and 4.3 is considered to be highly satisfactory, and it was agreed by CIE TC1-34 members in 1997 to recommend the CIECAM97s model for general industrial applications.

It is worth mentioning that CIECAM97s was developed to follow the 12 principles set by Hunt in 1996 [7]. The majority of these have been successfully achieved. These principles are:

1. To cover as many (comprehensive) functions as possible in order for the model to be used in a variety of applications.
2. To cover a wide range of stimulus intensities by setting a maximum in the dynamic range. (A hyperbolic function was adopted to achieve this; see Step 5 of Appendix 1.)
3. To include and exclude the rod vision for stimuli viewed under very low scotopic levels. (The CIECAM97s model excludes this function, but will be included in CAM97c, a proposed comprehensive model [8].)
4. To cover a wide range of viewing conditions, including backgrounds of different luminance factors, and simplified media (surround) viewing conditions: average (such as prints, coatings, textiles, etc.), dim (such as broadcast television) and dark (projected images). (These effects were described in the previous section.)
5. The spectral sensitivities of the cones should be a linear transform of the CIE 1931 or 1964 standard colorimetric observers, and the $V'(\lambda)$ curve should be used to approximate rod vision. (The latter is included in the CAM97c model, not in CIECAM97s.)
6. The model should include an incomplete adaptation factor, which allows for adaptation between complete adaptation and no adaptation. (The D factor in Step 2 of Appendix 1 is used for this purpose.)
7. The model should predict a wide range of percepts: hue angle, hue composition (or hue quadrature), brightness, lightness, colourfulness, chroma and saturation. All these percepts are important in certain applications. (All these percepts are included in the model.)
8. The model must be able to be reversed. This is particularly important in colour management systems for imaging applications. (The reverse mode is also given in Appendix 1.)
9. The model should be no more complicated than is necessary to meet the above requirement.

Table 4.3 Testing colour appearance models' performance (in $\Delta E_{\text{CMC}(1:1)}$) using colour appearance percept data sets

Data set	No. of phases	No. of samples	Hunt96	Nayatani97	RLAB96	LLAB96	CIECAM97s
Lightness							
Reflection (high-luminance)	6	105	12	25	19	12	11
Reflection (low-luminance)	6	105	11	29	23	10	10
Reflection (variable-luminance)	6	40	15	17	13	16	13
Reflection (large-textile)	3	270	8	8	7	7	8
Monitor	11	103–61	10	21	27	9	9
Cut-sheet transparency	10	98–94	12	20	11	10	11
35 mm projection	6	99–36	12	20	32	18	11
Weighted mean			12	21	20	12	11
Colourfulness							
Reflection (high-luminance)	6	105	17	49	28	19	19
Reflection (low-luminance)	6	105	18	35	31	19	20
Reflection (variable-luminance)	6	40	18	79	33	21	20
Reflection (large-textile)	3	270	18	39	38	25	18
Monitor	11	103–61	17	32	29	22	19
Cut-sheet transparency	10	98–94	16	38	26	19	18
35 mm projection	6	99–36	18	25	24	19	19
Weighted mean			17	41	29	20	19
Hue							
Reflection (high-luminance)	6	105	7	12	8	7	7
Reflection (low-luminance)	6	105	8	16	8	7	8
Reflection (variable-luminance)	6	40	6	15	7	6	6
Reflection (large-textile)	3	270	9	9	9	8	9
Monitor	11	103–61	7	15	11	7	7
Cut-sheet transparency	10	98–94	6	10	7	7	6
35 mm projection	6	99–36	7	15	10	8	8
Weighted mean			7	14	9	7	7

10. It may be necessary to have two models for dealing with all possible and limited (but most frequent) applications respectively. (This is the reason why two models were published: CIECAM97s and CAM97c. The latter has not been finalised or adopted by CIE.)

11. The model should perform better than or as well as the existing best colour appearance models in predicting the selected experimental data sets (see results in Tables 4.2 and 4.3).

12. The comprehensive model should be available for application to unrelated colours such as those seen in dark surrounds in isolation from other colours. In addition, the

model should be able to predict the simultaneous colour contrast effect. (These aims have not yet been achieved.)

In addition to the above principles, CIECAM97s was developed to incorporate the best features of all the five colour appearance models tested. For example, the basic structure is based upon the Hunt96 model with significant simplifications. It also includes the concept of the incomplete adaptation factor from the RLAB96 and Nayatani96 models and employs the chromatic adaptation transform used in LLAB96.

In conclusion, the CIECAM97s model has been recommended by CIE for industrial applications and it has been extensively tested [26, 27]. In general, it performs well in many areas dealing with digital colour imaging. There are still some shortcomings, which have been investigated by a new CIE Technical Committee (TC8-01 on Colour Appearance Models for Colour Management Applications). The next section gives the latest modifications to the CIECAM97s model.

4.4 RECENT MODIFICATIONS TO CIECAM97s

More recently, it was found that there are some shortcomings in CIECAM97s. Hence, modifications were made by Li *et al.* [28], Step 9 in the forward mode was modified by the following equations:

$$A = [2R'_a + G'_a + (1/20)B'_a - 3.05]N_{bb}$$

$$A_w = [2R'_{aw} + G'_{aw} + (1/20)B'_{aw} - 3.05]N_{bb}$$

The modification will make the black point with Y of zero correspond to a lightness (J) of zero. In addition, the N_c value for the dim surround condition was changed from 1.1 to 0.95. This makes the colour gamut for the dim surround condition smaller than that for the average surround condition.

4.5 CONCLUSION

A colour appearance model is an important element for cross-media colour management applications. Over the last few years, tremendous efforts have been spent by colour scientists from many countries. This led to the recommendation of CIECAM97s by CIE. Although there are some minor modifications intended to further simplify the model, the original model is theoretically sound and should be safely applied.

It is expected that there will be models derived from the CIECAM97s for certain industrial applications. These models should at least give a similar performance to that of CIECAM97s in the prediction of most sub-data sets.

REFERENCES

1. CIE Technical Report (1986) CIE Publication 15.2, *Colorimetry*, 2nd edn. Vienna, Austria: Central Bureau of the CIE.

2. Hunt, R. W. G. (1991) Revised colour-appearance model for related and unrelated Colours. *Color Res. Appl.*, **14**, 146–165.
3. Hunt, R. W. G. and Luo, M. R. (1994) An improved predictor of colourfulness in a model of colour vision. *Col. Res. & Appl.*, **19**, 23–26.
4. Nayatani, Y., Sobagaki, H., Hashimoto, K. and Yano, T. (1997) Field trials of a nonlinear color-appearance model. *Col. Res. Appl.*, **22**, 240–258.
5. Fairchild, M. D. (1996) Refinement of the RLAB color space. *Color Res. Appl.*, **21**, 338–346.
6. Luo, M. R., Lo, M.-C. and Kuo, W.-G. (1996) The LLAB(l:c) colour model. *Col. Res. Appl.*, **21**, 412.
7. CIE (1996) *Proceedings of the CIE Expert Symposium'96 Colour Standards for Image Technology*, CIE Publ. No. x010–1996. Vienna, Austria: Central Bureau of the CIE.
8. Luo, M. R. and Hunt, R. W. G. (1998) The structure of the CIE 1997 colour appearance model (CIECAM97s). *Col. Res. & Appl.*, **23**, 138–146.
9. Luo, M. R. and Hunt, R. W. G. (1998) A chromatic adaptation transform and a colour inconstancy index. *Col. Res. & Appl.*, **23**, 154–158.
10. CIE (1987) *CIE International Lighting Vocabulary*, CIE Publ. No. 17.4. Vienna, Austria: Central Bureau of the CIE.
11. Hunt, R. W. G. (1952) Light and dark adaptation and the reception of color. *J. Opt. Soc. Am.*, **42**, 190–199.
12. Stevens, J. C. and Stevens, S. S. (1963) Brightness function: effects of adaptation. *J. Opt. Soc. Am.*, **53**, 375–385.
13. Bartleson, C. J. and Breneman, E. J. (1967) Brightness perception in complex fields. *J. Opt. Soc. Am.*, **57**, 953–957.
14. Helson, H. (1938) Fundamental problems in color vision. I. The principle governing changes in hue, saturation, and lightness of non-selective samples in chromatic illumination. *J. Exp. Psych.*, **23**, 439–477.
15. Mori, L., Sobagaki, H., Komatsubara, H. and Ikeda, K. (1991) Field trials on CIE chromatic adaptation formula. *Proc. CIE 22nd Session*, 55–58.
16. McCann, J. J., McKee, S. P. and Taylor, T. H. (1976) Quantitative studies in Retinex theory: a comparison between theoretical predictions and observer responses to the 'color mandarin' experiments. *Vis. Res.*, **16**, 445–458.
17. Breneman, E. J. (1987) Corresponding chromaticities for different states of adaptation to complex visual fields. *J. opt. Soc. Amer.*, **4A**, 1115–1129.
18. Helson, H., Judd, D. B. and Warren, M. H. (1952) Object-color changes from daylight to incandescent filament illumination. *Illum. Eng.*, **47**, 221–233.
19. Lam, K. M. (1985) *Metamerism and colour constancy*. Ph.D. thesis, University of Bradford.
20. Braun, K. M. and Fairchild, M. D. (1996) Psychophysical generation of matching images for cross-media colour reproduction. *Proc. 4th Color Imaging Conference*, pp. 214–220, IS&T, Springfield, VA.
21. Luo, M. R., Clarke, A. A., Rhodes, P. A., Schappo, A., Scrivener, S. A. R. and Tait, C. (1991) Quantifying colour appearance. Part I. LUTCHI colour appearance data. *Col. Res. Appl.*, **16**, 166–180.
22. Luo, M. R., Gao, X. W., Rhodes, P. A., Xin, H. J., Clarke, A. A. and Scrivener, S. A. R. (1993) Quantifying colour appearance. Part IV. Transmissive media. *Col. Res. Appl.*, **18**, 191–209.
23. Kuo, W. G., Luo, M. R. and Bez, H. E. (1995) Various chromatic adaptation transforms tested using new colour appearance data in textiles. *Color Res. Appl.*, **20**, 313–327.
24. Luo, M. R. and Hunt, R. W. G. (1998) Testing colour appearance models using corresponding colour and magnitude estimation data sets. *Col. Res. Appl.*, **23**, 147–153.
25. Luo, M. R. and Morovic, J. (1996) Two unsolved issues in colour management – colour appearance and gamut mapping. *5th International Conference on High Technology*, Chiba, Japan, pp. 136–147.
26. Sueeprasan, S. and Luo, M. R. Investigation of Colour Appearance Models for Illumination Changes across Media, submitted to *Col. Res. Appl.*

27. Alessi, P. (2000) Private communication, April.
28. Li, C. J., Luo, M. R. and Hunt, R. W. G. (2000) A revision of the CIECAM97s Model. *Color Res. Appl.*, **25**, 260–266.

APPENDIX 1. COMPUTATION PROCEDURES FOR CIECAM97s

The CIECAM97s colour appearance model is given in this appendix (without the modifications mentioned in the 'Recent modifications' section). The viewing conditions assumed for the test sample by the model are illustrated in Figure 4.10. A test sample and a reference white is viewed against an achromatic background under a particular luminance of a test illuminant. Parameters for different surrounds are given in Table 4.4.

Forward CIECAM97s colour appearance model

Data input

x, y, Y	Colorimetric values of sample under test field
x_w, y_w, Y_w	Colorimetric values of adopted white under test field
x_{wr}, y_{wr}, Y_{wr}	Colorimetric values of reference white under reference field (equal energy illuminant, $x_{wr} = y_{wr} = 1/3$ and $Y_{wr} = 100$)
Y_b	Y of achromatic background (%)
L_w	Luminance of reference white in cd/m^2
L_A	Luminance of achromatic background in cd/m^2 (calculated by $L_W Y_b/100$).

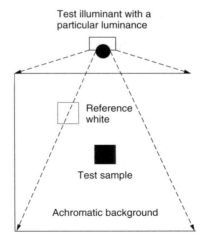

Figure 4.10   Condition of viewing in test field.

Table 4.4 Parameters of CIECAM97s

Surround parameters	F	c	F_{LL}	N_c
Average with sample $>4°$ viewing field	1.0	0.690	0.0	1.0
Average with sample $\leqslant 4°$ viewing field	1.0	0.690	1.0	1.0
Television and VDU displays in dim surrounds	0.9	0.590	1.0	0.95 (originally 1.1)
Projected photographs in dark surrounds	0.9	0.525	1.0	0.8
Large cut-sheet transparency	0.9	0.410	1.0	0.8

Computational procedures

Step 1 Calculate R, G and B values of the test sample (R G B), and similarly for the reference white under the test and reference fields respectively (R_w G_w B_w and R_{wr} G_{wr} B_{wr}).

$$\begin{bmatrix} R \\ G \\ B \end{bmatrix} = M_{BFD} \begin{bmatrix} X/Y \\ Y/Y \\ Z/Y \end{bmatrix}$$

where $M_{BFD} = \begin{bmatrix} 0.8951 & 0.2664 & -0.1614 \\ -0.7502 & 1.7135 & 0.0367 \\ 0.0389 & -0.0685 & 1.0296 \end{bmatrix}$

and $X = xS$, $Z = (1 - x - y)S$, where $S = Y/y$

Step 2 Calculate the corresponding R, G and B values of the test sample (R_c G_c B_c), and similarly for the reference white (R_{wc} G_{wc} B_{wc}).

$$R_c = [D(R_{wr}/R_w) + 1 - D]R$$

$$G_c = [D(G_{wr}/G_w) + 1 - D]G$$

For $B_c > 0$, $B_c = [D(B_{wr}/B_w^p) + 1 - D]B^p$

Otherwise, $B_c = -[D(B_{wr}/B_w^p) + 1 - D]|B|^p$

where $p = (B_w/B_{wr})^{0.0834}$
$D = F - F/[1 + 2(L_A^{1/4}) + (L_A^2)/300]$

Step 3 Calculate luminance level adaptation factor (F_L), chromatic background induction factor (N_{cb}) and brightness background induction factors (N_{bb}).

$$F_L = 0.2k^4(5L_A) + 0.1(1 - k^4)^2(5L_A)^{1/3}$$

where $k = 1/(5L_A + 1)$
$N_{cb} = N_{bb} = 0.725(1/n)^{0.2}$
where $n = Y_b/Y_w$

Step 4 Calculate the corresponding tristimulus values of the test sample $(R'\ G'\ B')$, and similarly for the reference white $(R'_w\ G'_w\ B'_w)$.

$$
\begin{bmatrix} R' \\ G' \\ B' \end{bmatrix} = M_H M_{BFD}^{-1} \begin{bmatrix} R_c Y \\ G_c Y \\ B_c Y \end{bmatrix}
$$

where $M_H = \begin{bmatrix} 0.38971 & 0.68898 & -0.07868 \\ -0.22981 & 1.18340 & 0.04641 \\ 0.00000 & 0.00000 & 1.00000 \end{bmatrix}$

and $M_{BFD}^{-1} = \begin{bmatrix} 0.9870 & -0.1471 & 0.1600 \\ 0.4323 & 0.5184 & 0.0493 \\ -0.0085 & 0.0400 & 0.9685 \end{bmatrix}$

Step 5 Calculate cone responses after adaptation of test sample $(R'_a\ G'_a\ B'_a)$, and similarly for the reference white $(R'_{aw}\ G'_{aw}\ B'_{aw})$.

$$
R'_a = \frac{40(F_L R'/100)^{0.73}}{[(F_L R'/100)^{0.73} + 2]} + 1
$$

$$
G'_a = \frac{40(F_L G'/100)^{0.73}}{[(F_L G'/100)^{0.73} + 2]} + 1
$$

$$
B'_a = \frac{40(F_L B'/100)^{0.73}}{[(F_L B'/100)^{0.73} + 2]} + 1
$$

If R'_a is less than zero use:

$$
R'_a = \frac{-40(-R'/100)^{0.73}}{(-R'/100)^{0.73} + 2} + 1
$$

and similarly for R'_{aw} and for the G'_{aw} and B'_{aw} equations.

Step 6 Calculate the red-green (a) and yellow-blue (b) opponent correlates.

$$
a = R'_a - 12G'_a/11 + B'_a/11
$$

$$
b = (1/9)(R'_a + G'_a - 2B'_a)
$$

Step 7 Calculate hue angle (h).

$$
h = \tan^{-1}(b/a)
$$

Step 8 Calculate eccentricity factor (e) and hue quadrature (H).

$$
H = H_1 + \frac{100(h - h_1)/e_1}{(h - h_1)/e_1 + (h_2 - h)/e_2}
$$

$$
e = e_1 + (e_2 - e_1)\frac{h - h_1}{h_2 - h_1}
$$

where H_1 is 0, 100, 200 or 300, according to whether red, yellow, green or blue, respectively, is the hue having the nearest lower value of h. The values of h and e for the four unique hues are:

	Red	Yellow	Green	Blue	Red
H	0	100	200	300	400
h	20.14	90.00	164.25	237.53	380.14
e	0.8	0.7	1.0	1.2	0.8

e_1 and h_1 are the values of e and h, respectively, for the unique hue having the nearest lower value of h; and e_2 and h_2 are these values for the unique hue having the nearest higher value of h.

Taking H_p to be the part of H after its hundreds digit:

If $H = H_p$, the hue composition is H_p Yellow, $100 - H_p$ Red

If $H = 100 + H_p$, the hue composition is H_p Green, $100 - H_p$ Yellow

If $H = 200 + H_p$, the hue composition is H_p Blue, $100 - H_p$ Green

If $H = 300 + H_p$, the hue composition is H_pRed, $100 - H_p$ Blue

Step 9 Calculate the achromatic response of the sample (A) and reference white (A_w).

$$A = [2R'_a + G'_a + (1/20)B'_a - 2.05]N_{bb}$$

$$A_w = [2R'_{aw} + G'_{aw} + (1/20)B'_{aw} - 2.05]N_{bb}$$

(Use 3.05 instead of 2.05 in both equations if it is requested that $J = 0$ when $Y = 0$.)

Step 10 Calculate lightness (J).

$$J = 100(A/A_w)^{cz}$$

where $z = 1 + F_{LL}n^{1/2}$

Step 11 Calculate brightness (Q).

$$Q = (1.24/c)(J/100)^{0.67}(A_w + 3)^{0.9}$$

Step 12 Calculate saturation (s).

$$s = \frac{5000(a^2 + b^2)^{1/2}e(10/13)N_cN_{cb}}{R'_a + G'_a + (21/20)B'_a}$$

Step 13 Calculate chroma (C).

$$C = 2.44s^{0.69}(J/100)^{0.67n}(1.64 - 0.29^n)$$

Step 14 Calculate colourfulness (M).

$$M = CF_L^{0.15}$$

Reverse CIECAM97s colour appearance model

The reverse CIECAM97s colour appearance model is used to calculate the corresponding tristimulus values (X, Y, Z) from the colour appearance attributes computed from the forward CIECAM97s model, i.e. Q or J, M or C, H or h.

Step R1 Calculate lightness (J) from brightness (Q).

$$J = 100(cQ/1.24)^{1/0.67}/(A_w + 3)^{0.9/0.67}$$

Step R2 Calculate achromatic response (A) from lightness (J).

$$A = A_w(J/100)^{1/cz}$$

Step R3 Use hue composition to determine h_1, h_2, e_1 and e_2 (see Step 8 in the last section).

Step R4 Calculate h.

$$h = \frac{(H - H_1)(h_1/e_1 - h_2/e_2) - 100h_1/e_1}{(H - H_1)(1/e_1 - 1/e_2) - 100/e_1}$$

where H_1 is 0, 100, 200 or 300 according to whether red, yellow, green or blue, respectively, is the hue having the nearest lower value of h.

Step R5 Calculate e.

$$e = e_1(e_2 - e_1)\frac{h - h_1}{h_2 - h_1}$$

where e_1 and h_1 are the values of e and h, respectively, for the unique hue having the nearest lower value of h; and e_2 and h_2 are these values for the unique hue having the nearest higher value of h.

Step R6 Calculate C.

$$C = M/F_L^{0.15}$$

Step R7 Calculate s.

$$s = \frac{C^{1/0.69}}{[2.44(J/100)^{0.67n}(1.64 - 0.29^n)]^{1/0.69}}$$

where $n = Y_b/Y_w$.

Step R8 Calculate a and b.

$$a = \frac{s(A/N_{bb} + 2.05)}{(1 + \tan^2 h)^{1/2}(50000eN_cN_{cb}/13) + s[(11/23) + (108/23)\tan h]}$$

In calculating $(1 + \tan^2 h)^{1/2}$ the result is to be taken as positive if h is equal to or greater than 0, and less than 90; negative if h is equal to or greater than 90 and less than 270; and positive if h is equal to or greater than 270 and less than 360.

$$b = a(\tan h)$$

Step R9 Calculate R_a', G_a', B_a'.

$$R_a' = (20/61)(A/N_{bb} + 2.05) + (41/61)(11/23)a + (288/61)(1/23)b$$

$$G_a' = (20/61)(A/N_{bb} + 2.05) - (81/61)(11/23)a - (261/61)(1/23)b$$

$$B_a' = (20/61)(A/N_{bb} + 2.05) - (20/61)(11/23)a - (20/61)(315/23)b$$

(If 3.05 was used in Step 9, it should also be used in these equations instead of 2.05.)

Step R10 Calculate R', G', B'.

$$F_L R' = 100[(2R_a' - 2)/(41 - R_a')]^{1/0.73}$$

$$F_L G' = 100[(2G_a' - 2)/(41 - G_a')]^{1/0.73}$$

$$F_L B' = 100[(2B_a' - 2)/(41 - B_a')]^{1/0.73}$$

If $R_a' - 1$ is less than 0, then

$$F_L R' = -100[(2 - 2R_a')/(39 + R_a')]^{1/0.73}$$

and similarly for the G' and B' equations.

Step R11 Calculate $R_c Y$, $G_c Y$ and $B_c Y$.

$$\begin{bmatrix} R_c Y \\ G_c Y \\ B_c Y \end{bmatrix} = M_{BFD} M_H^{-1} \begin{bmatrix} R' \\ G' \\ B' \end{bmatrix}$$

where $M_H^{-1} = \begin{bmatrix} 1.91019 & -1.11214 & 0.20195 \\ 0.37095 & 0.62905 & 0.00000 \\ 0.00000 & 0.00000 & 1.00000 \end{bmatrix}$

Step R12 Calculate $(Y/Y_c)R$, $(Y/Y_c)G$ and $(Y/Y_c)B$.

$$(Y/Y_c)R = (Y/Y_c)R_c/[D(R_{wr}/R_w) + 1 - D]$$

$$(Y/Y_c)G = (Y/Y_c)G_c/[D(G_{wr}/G_w) + 1 - D]$$

$$(Y/Y_c)^{1/P}B = [|(Y/Y_c)B_c|]^{1/P}/[D(B_{wr}/B_w^P) + 1 - D]^{1/P}$$

If B_c is negative, $(Y/Y_c)^{1/p}B$ is negative.

$Y_c = 0.43231R_cY + 0.51836G_cY + 0.04929B_cY$

Step R13 Calculate Y'.

$$Y' = 0.43231RY + 0.51836GY + 0.04929BY_c(Y/Y_c)^{1/p}$$

and $(Y'/Y_c)^{1/(p-1)}$.

Step R14 Calculate X''/Y_c, Y''/Y_c and Z''/Y_c.

$$\begin{bmatrix} X''/Y_c \\ Y''/Y_c \\ Z''/Y_c \end{bmatrix} = M_{BFD}{}^{-1} \begin{bmatrix} (Y/Y_c)R \\ (Y/Y_c)G \\ (Y/Y_c)^{1/p}B/(Y'/Y_c)^{1/(p-1)} \end{bmatrix}$$

where $M_{BFD}{}^{-1} = \begin{bmatrix} 0.9870 & -0.1471 & 0.1600 \\ 0.4323 & 0.5184 & 0.0493 \\ -0.0085 & 0.0400 & 0.9685 \end{bmatrix}$

Step R15 Multiply each by Y_c to obtain X'', Y'' and Z'', which equal X, Y and Z, to a very close approximation.

Worked examples for the CIECAM97s colour appearance model

The following examples are given to assist readers to implement the CIECAM97s model. One sample is viewed under the standard illuminant A and four different levels of adapting luminance, L_A.

Input data

(a) Colorimetric data for the test sample, reference whites:

	x	y	Y
Sample under test field	0.3618	0.4483	23.93
Adopted white in test field	0.4476	0.4074	90.00
Reference white in reference field	0.3333	0.3333	100.00

(b) Viewing conditions:

Luminous factor (Y) of the background   18.0
Luminances (L_A, in cd/m^2) of achromatic background 2000, 200, 20 and 2 (four levels)

(c) Model parameters for average surround with sample over 4° viewing field:

F	c	F_{LL}	N_c
1.0	0.69	1.0	1.0

Results

Prediction for the adopted white:				
L_A	2000	200	20	2
Hue angle, h	41.8	57.4	58.8	59.5
Hue quadrature, H	28.3	50.0	52.0	53.0
Lightness, J	100.0	100.0	100.0	100.0
Brightness, Q	70.1	52.7	37.9	26.8
Saturation, s	0.0	0.5	12.6	25.9
Chroma, C	0.1	1.3	12.1	19.8
Colourfulness, M	0.1	1.3	10.8	15.7
Prediction for the test sample:				
L_A	2000	200	20	2
Hue angle, h	190.2	190.0	183.5	175.7
Hue quadrature, H	239.7	239.4	229.9	218.2
Lightness, J	53.0	48.2	45.2	44.2
Brightness, Q	45.8	32.3	22.3	15.5
Saturation, s	120.0	125.9	114.0	96.5
Chroma, C	52.4	53.5	49.5	44.0
Colourfulness, M	58.8	53.5	44.1	34.9

5

Colour notation systems

Peter A. Rhodes

5.1 INTRODUCTION

The problem of ordering colour is a very old one dating back at least 2000 years to the ancient Chinese [1]. In seventeenth century Sweden, Forsius and Brenner separately published the first colour circle and colour chart, but it was not until Newton's time that colour became a subject of serious scientific study. In addition to splitting sunlight with a prism to produce a spectrum, Newton also produced his own colour circle by dividing hue into seven parts: red, orange, yellow, green, blue, indigo and violet. This was later reduced to six divisions by Goethe through his discovery of complementary colours. In the latter half of the nineteenth century, two rival colour theories emerged: trichromatic theory (Helmholtz) and opponent colour theory (Hering). Through the efforts of subsequent researchers we now know that *both* theories apply to human colour vision. So, if colour really is three dimensional, what three scales should be chosen to represent it?

This chapter deals with the use of different colour notation systems and their application to the communication of colour. It begins by highlighting the failings of conventional colour management when applied to cross-media communication. This then leads on to the colour 'language barrier' which is a result of there being no single standard for describing colour. Following this, a representative cross-section of commonly used systems for describing colour are chosen to illustrate the principles, problems and solutions

Colour Engineering, edited by P.J. Green and L.W. MacDonald.
© 2002 John Wiley & Sons Ltd.

associated with their use. The chapter concludes with suggestions as to how these may be interrelated and implemented as computer-based colour atlases.

5.2 COLOUR FIDELITY

Colour fidelity, the subject of several other chapters elsewhere in this book, is both the driving force behind and the limiting factor to colour specification. The need for accurately determining and subsequently reproducing colour across different media or between geographically remote users provides the motivation behind adopting standards for colour specification. Without such standards colour fidelity is very much localised. Conversely, physical specification systems may be constrained by the fidelity of their component colour patches. The reproduction tolerance governs accuracy whilst the colour gamut restricts the available range of colours. Furthermore, many colour specification systems are designed for use only with particular viewing conditions (e.g. illuminant or background) or media (e.g. glossy or matt paint).

One solution to this problem is to adopt the paradigm of a true WYSIWYG (what you see is what you get) colour management system or *WCMS*. In contrast to conventional colour management – which tends to work predominantly with photographic imagery and where the goal is usually to achieve a 'pleasing' colour match – WCMS aims to provide a good visual match across different media. This approach has a number of implications:

- To match colour *visually* (rather than colorimetrically) across multiple media and viewing conditions requires the use of a colour appearance model (see Chapter 4).
- Because *individual* colours rather than entire images are critical, it is not possible to apply gamut mapping techniques (covered in Chapter 13) to deal with out-of-gamut colours. Consequently, a WCMS makes an 'all or nothing' effort, with potentially large errors occurring for tints and shades lying out of range.

There are numerous application domains where WYSIWYG reproduction is essential. One particular example is the automotive industry, which needs to match large areas of colour over different media, including paint, plastic and fabric together with the printed sales brochure. Another demanding area is the textile business where, as well as cross-media matching, there is a strong requirement for accurate global colour communication (e.g. between designer and dye house) involving both standard and in-house colour specifiers and requiring a high degree of colour precision at all stages.

5.3 COLOUR NOTATION SYSTEMS

In order to talk about colour – as with any other form of communication – we first need to adopt a language that both parties are able to comprehend. A scheme for logically ordering and specifying colour stimuli according to three attributes is known as a *colour notation system*. These three attributes constitute the coordinates of the resultant *colour space* and share the same dimensionality as our own tristimulus colour vision. Colour notation systems also encompass *colour order systems* which are typically comprised of material standards such as painted chips arranged in some sort of *colour atlas*. Due to the practical constraints of colorant gamuts, these may depict only a physically realisable subset of a colour notation system. The reader should be aware that it is common for the terms 'colour notation' and 'colour order' to be used interchangeably and for most practical purposes they can be treated as synonymous.

In this section the concepts of colour notation systems will be introduced, categorised and exemplified. To this end, several popular colour order systems are described. The section concludes by addressing the very practical problems associated with interrelating different systems and also in implementing these on computer-controlled displays.

5.3.1 Introduction and definitions

There have been several attempts [2] to define exactly what a *colour order system* actually is. These definitions essentially vary according to the principle upon which the system is based. These bases and examples of systems applying to them are given in Table 5.1. In addition, there are a number of properties which are vitally important to any such system. They include the following:

- to arrange colours in an orderly and continuous fashion;
- to have an accurate physical embodiment (e.g. a colour atlas);
- to incorporate a logical nomenclature;
- to apply perceptually meaningful dimensions; and
- to have scales which are both unique and (ideally) perceptually uniform.

It should be noted that there exist other colour specifiers in everyday use which do not meet the criteria required of colour order systems. Examples of these include RAL, PANTONE and others depicting product ranges such as paint charts. While these are based on subtractive mixing principles, they do not typically employ a continuous ordering nor do they have perceptually based three-dimensional scales. More properly, these should be regarded as *colour naming systems* [2] which are covered later.

There are several advantages [3] to the application of sample-based colour order systems to the dual tasks of colour specification and communication:

Table 5.1 Bases for colour order systems

Category	Example
Additive mixtures of stimuli	Ostwald
Subtractive colorant mixtures	Tintometer
Principles of colour perception or colour appearance:	
• equal spacing of each individual attribute	Munsell
• equal spacing of whole colour space	OSA
• resemblance to elementary colours	NCS

- they can be easily understood;
- they are portable;
- they make it possible to perform side-by-side comparisons;
- the number and arrangement of samples may be adapted to suit a given application; and
- most systems can be related to CIE tristimulus values.

5.3.2 Defining colour notation systems

A further distinction to be used in this chapter is in terms of how a notation system is defined. Three categories [4] will be considered:

- **Device-dependent systems**
  Such systems are directly related to the reproduction system's primaries. In particular, systems based on RGB are covered in the next section. Although they may meet other requirements mentioned in the previous section, they are limited by a lack of perceptual uniformity.
- **Mathematical systems**
  These are formed by the direct mathematical transformation of CIE tristimulus values to produce more uniform colour spaces. Examples of such spaces include CIELAB and CIELUV.
- **Measured aim points**
  In particular for colour order systems existing principally in physical form, colour samples can be measured to establish a database of aim points. As well as numerically describing the system, this approach enables interpolation techniques to be used in order to define very many more colours than would be practically possible using materials.

These categories are particularly useful as a point of reference when it comes to interrelating and implementing the different systems.

5.3.3 Device-dependent specification systems

The two most common imaging devices used for reproducing colour are the computer-controlled CRT display and the colour printer. Unfortunately, their associated device colour spaces (RGB and CMY or CMYK), being hardware-oriented, are not particularly 'user-friendly' due, in the most part, to their lack of intuitive, perceptually based attributes. This has led to the derivation of two colour models, HSV and HLS [5], which are commonly used in computer graphics applications such as CAD and image editing.

The HSV colour model derives the attributes hue, saturation and value from RGB to form the single hexcone colour space illustrated in Figure 5.1. An alternative name for this model is HSB (where B instead stands for brightness). The V scale runs from 0 (black) to 1 (white) along the vertical axis. Hue is measured in degrees with red at $0°$ and complementary colours such as yellow and blue being located diagonally opposite one another on this scale. The device greyscale runs vertically (at $S = 0$) through the centre of the hexcone with saturation increasing in the horizontal plane outwards from this such that the display's primary and secondary colours – red, yellow, green, blue, cyan

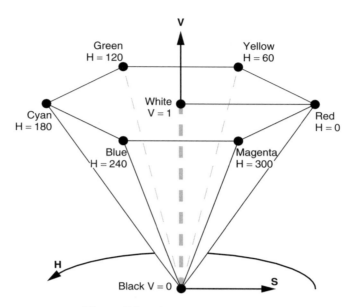

Figure 5.1 The HSV colour model.

and magenta – are located at maximum saturation (i.e. $S = 1$). To compute HSV from normalised RGB, first find the minimum and maximum of the three RGB values and from these determine the range. From these, S is calculated as the ratio of the range to the maximum value, whilst V is simply the maximum RGB value. For non-neutral colours (i.e. for non-zero values of S), the maximum value is also used to determine which $60°$ segment H falls in. The difference between the two remaining R, G or B values is then added to provide a $\pm 30°$ offset.

Although the HLS model is also derived directly from RGB, its three attributes of hue, lightness and saturation together form the *double* hexcone space shown in Figure 5.2. In simple terms, this can be thought of as a deformation of HSV space in order to stretch white to form the upper hexcone at $L = 1$. In this model, the hue scale is identical to that of HSV. The remaining terms can be defined mathematically:

$$L = \frac{\max(R, G, B) + \min(R, G, B)}{2}$$

Due to the double hexcone shape, S depends upon L as follows:

$$S = \begin{cases} \dfrac{\max(R, G, B) - \min(R, G, B)}{2L} & \text{if } L \leqslant 0.5 \\[4mm] \dfrac{\max(R, G, B) - \min(R, G, B)}{2(1 - L)} & \text{otherwise} \end{cases}$$

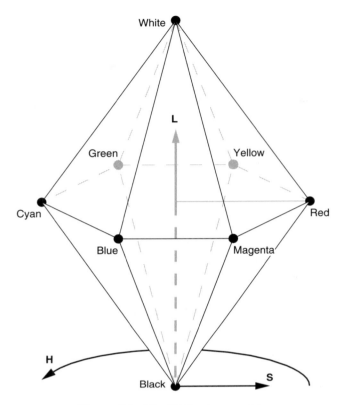

Figure 5.2 The HLS colour model.

The HSV and HLS models are frequently found on computer systems, largely because they are both easy to implement. Another advantage is that they are able to utilise the entire colour gamut of the display. Conversely, their usefulness is limited by their perceptual non-uniformity and lack of portability of their values. A lack of perceptual uniformity makes them more difficult to use on screen and also impacts upon storage because quantisation results in uneven visual results. Portability can, however, be addressed through the use of a colorimetric RGB space such as sRGB, as described in Chapter 17.

5.3.4 CIELAB and CIELUV

The need for perceptually uniform scales was addressed in 1976 by the CIE [6] when it recommended the $CIEL^*a^*b^*$ (CIELAB) and $CIEL^*u^*v^*$ (CIELUV) uniform colour spaces. By virtue of their colorimetric origin, they may also be used for device-independent colour specification. Both CIELAB and CIELUV are purely mathematical and hence neither exists in the form of a physical colour atlas. They have, however, been implemented as computer-based atlases for colour selection. CIELAB and CIELUV are widely used in practice for both colour specification and colour difference calculations. In particular, CIELAB is predominantly used with surface colours whilst CIELUV is normally applied to industries such as television due to its ability to work with additive colour mixtures. Although both CIELAB and CIELUV are adequate for most purposes, more recent research has demonstrated perceptual non-uniformities in CIELAB's hue scale [7]. Furthermore, the white point adaptation present in the CIELUV system – with its subtractive shift rather than multiplicative normalisation – can result in poor visual correspondence and also in predicted corresponding colours lying outside the realisable gamut [2].

5.3.5 TekHVC

The first commercially available system for colour management, TekColor [8], was produced by Tektronix for distribution with their colour printers. A key element of this was a uniform colour space derived from CIELUV called TekHVC [9]. The attributes comprising TekHVC are Hue (H), Value (V) and Chroma (C), although these should not be confused with the device-dependent HLS or HSB models introduced earlier. (TekHVC colour specification *is* device independent.) TekHVC's relationship with CIELUV together with an implementation guide are detailed in the *TekColor System Implementor's Manual* [10]. The basics of this calculation are given below.

$$H = h_{uv} - \theta$$

$$V = L^*$$

$$C = \frac{C_{uv}^* C_f}{13}$$

where the chroma scaling factor, C_f, is 7.50725 and the illumination-dependent hue offset, θ, is designed to rotate h_{uv} such that $0°$ corresponds to an illumination-dependent 'best red' (BR) at $u' = 0.7127$ and $v' = 0.4931$. For a given white point, (u_0', v_0'), defined in terms of CIE 1976 chromaticity coordinates:

$$\theta = \tan^{-1}\left(\frac{v_{br}' - v_0'}{u_{br}' - u_0'}\right)$$

The terms L^*, C_{uv}^* and h_{uv} relate to the corresponding CIELUV coordinates. These equations can be inverted as follows:

$$L^* = V$$

$$C_{uv}^* = \frac{13 \times C}{C_f}$$

$$u^* = C_{uv}^* \cos(H + \theta)$$

$$v^* = C_{uv}^* \sin(H + \theta)$$

The implementation guide also defines both the appearance and operation of any computer-based user interface that uses it for colour selection. An example of such an implementation is shown in Plate 1(d). When used for colour management tasks, the printer gamut boundary would typically be superimposed in outline form over the screen hue leaf, allowing colours to be selected which are reproducible across both device gamuts. TekHVC continues to be used as part of the X Color Management System [11] and, as such, it is a widely used, device-independent, cross-platform colour notation system. One disadvantage of the system (as with CIELAB and CIELUV before) is that a physical embodiment is not available, thereby restricting its application beyond computer-controlled displays.

5.3.6 Munsell

The *Munsell Color System* is the oldest colour order system in terms of continued availability of physical samples. It is also one of the most widely used colour specification systems, having received international acceptance by a number of standards bodies. Furthermore, during its long lifetime it has been thoroughly studied, resulting in the publication of more than 100 research papers. The system was originally devised by the artist and educator Albert Munsell as a teaching aid. In 1905 he published the basic principles [12] from which the first atlas was commercially produced in 1915. The system then underwent revision following extensive work in 1930 by the Optical Society of America to improve its visual spacing. This resulted in the *Munsell Renotation System* [13] in which the aim points were specified colorimetrically using CIE Illuminant C and the CIE 1931 ($2°$) standard observer. The renotation system is now used as the primary standard to which physical specifiers are produced.

The guiding principle of the Munsell system is that of equal visual perception of spacing between adjacent notations in each of its three attributes. These attributes as defined as follows.

- **Munsell Hue (H)**
  The Munsell Hue scale is divided up into five principal hues at 5R (red), 5Y (yellow), 5G (green), 5B (blue) and 5P (purple) as illustrated in Figure 5.3. This is further divided into the five intermediate hues at 5YR, 5GY, 5BG, 5PB and 5RP, giving rise to a total of 10 major hues. For each of these major hues, a total of 10 integral divisions are possible, e.g. 5R, 6R, 7R, 8R, 9R, 10R, 1YR, 2YR, 3YR and 4YR. Chromatic hues are thus written as a combined letter–number pair, whereas achromatic colours simply use the letter 'N' (for neutral).
- **Munsell Value (V)**
  Closely related to CIE lightness, Munsell Value ranges from 0 (black) to 10 (absolute white). Due to practical gamut limitations, the *Munsell Book of Color* atlas depicts a restricted range of 0.5 to 9.5 Value units.
- **Munsell Chroma (C)**
  The Munsell Chroma scale represents the degree of departure of a hue from a neutral grey having the same Value. This ranges from 0 (neutral) and increases with stronger hue content, typically extending up to /10 to /14 according to the particular hue angle and the colour gamut of the medium used.

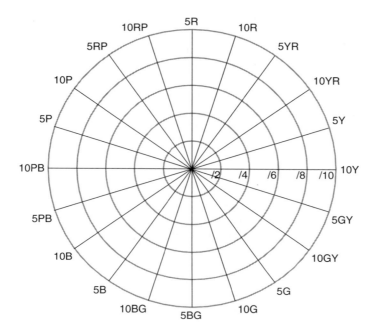

Figure 5.3 The Munsell hue scale.

Note that the equality of visual spacing applies only to individual scales themselves and not between different scales. The scales themselves were designed such that one Value step is visually equivalent to two Chroma steps. Due to the constraints imposed by the cylindrical geometry of the coordinate system, Hue spacing is only equal for constant Chroma. For example, at Chroma five one Value step equates to three steps on the 100-step hue scale.

By combining all three of these attributes in the order 'HV/C', a colour can be precisely specified. Specific examples of Munsell notation include '5GY 8/2' for Hue 5GY, Value 8 and Chroma 2; and 'N 5 /' for a neutral shade with Value 5 (a mid-grey).

As mentioned, Munsell Value is closely related to CIE lightness. In fact, CIE L^* was designed to closely model Value and the approximate relationship between the two is $V = L^*/10$ for Illuminant C, 2° observer. More precisely, the following fifth-order polynomial relates relative luminance, Y, with Value, V:

$$Y = 1.2219V - 0.23111V^2 + 0.23951V^3 - 0.021009V^4 + 0.0008404V^5$$

This can be inverted through iterative methods to obtain the approximation:

$$V = 0.01612Y + 2.5649Y^{1/6} + 1.3455Y^{1/3} + 0.08797Y^{-1} - (2.685 \times 10^{-7})Y^3 - 3.116$$

In general, there is no simple or direct relationship between Munsell and CIE XYZ values. Instead this must be regarded as falling under the 'measured aim point' category introduced above and thus requires interpolation (e.g. trilinear) of the aim point database defined by the renotation system. Reliable extrapolation for very dark colours (/0.2 to /1) is not possible from the renotation data alone and so additional tables [14] must be used. The reverse direction is more complex, though methods do exist. For specific implementation details see Refs 15 and 16.

An example computer-based Munsell atlas is shown in Plate 1(b). A variety of physical Munsell standards, including the atlas itself and its three-dimensional counterpart, the *Munsell Tree*, are available from GretagMacbeth, New Windsor.

5.3.7 The DIN colour system

Work to develop this system began in the 1930s [17] at Deutsches Institut für Normung (the German Standardization Institute) under the direction of Manfred Richter with the intention of superseding the older Ostwald system. The first practical physical embodiment was produced in 1960–62 comprising under 600 matt samples. Since then, it has been colorimetrically specified as DIN 6164 [18] in 1980 and a further expanded glossy edition of the atlas produced in 1978–83. The DIN system's guiding principles are of equality of visual spacing for each of its three variables. Much like the Munsell system, this uniformity does not extend to three dimensions and instead goes only as far as each individual colour attribute described below.

- **Hue number (T)**
  The system defines 24 principal hues starting with yellow at $T = 1$ and continuing in a reverse order via red, blue and green back to yellow. This scale is designed so that perceptually equal hue differences are represented at $S = 6$ and $D = 1$.
- **Saturation degree (S)**
  The saturation degree, ranging from 0 to 15, can be defined as being the perceptual distance from a grey of the same luminance factor. It is calculated as follows:

$$S = \frac{\sqrt{(u' - 0.2105)^2 + (v' - 0.4737)^2}}{r_1}$$

  where (u', v') are the CIE 1976 chromaticity coordinates computed for the colour in question. The origin (0.2105, 0.4737) corresponds to illuminant D65 upon which the standard is based. The term r_1 represents saturation distance and is computed from $r = r_6/6$ where the value for r_6 is obtained – possibly by interpolation – according to the values of T and S from a table published in Ref. 19.
- **Darkness degree (D)**
  Darkness degree describes the relative brightness measured from the optimal colour corresponding to the same chromaticity. On this scale, which by its definition is unique amongst colour order systems (but notionally similar to NCS blackness), $D = 0$ represents ideal white whilst $D = 10$ is ideal black. The concept behind this was based on the earlier ideas of Ostwald who had emphasised that colours of equal saturation are not perceived as being psychologically equivalent if they share the same luminous reflectance. It was therefore suggested that this might instead be achieved if equal

relative luminance factors were applied. The following formula is used in the calculation of D from Y:

$$D = 10 - 6.1723 \log \left(40.7 \frac{Y}{A_0} + 1 \right)$$

where A_0 must be found – again, possibly by interpolation – from the table given in Ref. 19.

When these three attributes are combined, they form a spherical sector colour solid. In DIN notation, a colour is written in the form T:S:D or just T:S to describe its chromaticity alone. For neutral colours ($S = 0$), the symbol 'N' is written instead of hue number.

To perform colour conversions from CIE to DIN coordinates, the following relationship may be used:

$$Y = \frac{A_0}{40.7} (10^k - 1)$$

where $k = (10 - D)/6.1723$. The remainder of the reverse transform is described in Ref. 20.

5.3.8 Natural colour system

Following the concept of opponent hues that was promoted by the German psychologist Ewald Herring in the mid-1870s, further work by Tryggve Johansson, Sven Hesselgren and (later) Anders Hård in Sweden led to the development of the *Natural Colour System* or NCS. This was published in 1972 as a Swedish standard for colour notation [21] followed in 1979 by an atlas [22] containing 1412 colour samples. In 1983, the tristimulus values defining the NCS aim points were published [23]. More recently, the atlas itself went through further revision (mainly to improve the accuracy of its samples), resulting in the production of the second edition.

The system's guiding principle is that of resemblance to the six elementary colours: red (R), yellow (Y), green (G), blue (B), white (W) and black (S). One might ask why do we need *six* variables when our own colour vision is trichromatic? This is because the colours red–green, blue–yellow and black–white form opponent hue pairs which cannot be perceived simultaneously. NCS therefore uses three variables to describe a colour: hue (ϕ), blackness $(s)^*$ and chromaticness (c). NCS hue is determined by a shade's percentage resemblance to the nearest two chromatic elementary colours as illustrated in Figure 5.4. Using this scale pure red, for example, has a redness (denoted by a lower-case letter r) of 100%. An orange colour might exhibit a redness of $r = 50\%$ and a yellowness of $y = 50\%$, giving rise to the notation Y50R.

White (W) and black (S) form the system's two achromatic elementary colours. The blackness scale (s) is defined to be a colour's resemblance to the perfect black (S). Thus, a mid-grey might have a blackness of $s = 50$. Finally, chromaticness (c) is the resemblance

---

* Note that s ('swarthy') is used for black rather than B which represents the elementary blue.

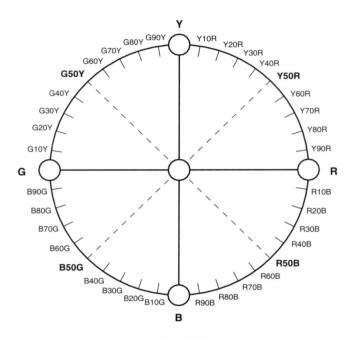

Figure 5.4   The NCS hue scale.

to a colour of the same hue having the maximum chromatic content. This can also be expressed mathematically as the sum of redness, yellowness, greenness and blueness:

$$c = r + y + g + b$$

One property of NCS coordinates is that their sum always equals 100, i.e.

$$s + w + r + y + g + b = s + w + c = 100$$

Consequently, it is sufficient to quote two out of the three numbers. To illustrate NCS notation, an example is given in Figure 5.5.

For the specific case of neutral colours, the following equation links blackness and relative luminance:

$$Y = \frac{56v}{1.56 - v}$$

where $v = (100 - s)/100$. In comparison with the Munsell system, NCS does not possess equal visual spacing but instead relies on colour appearance concepts. Although its hue scale includes five principal hues, Munsell does not incorporate the notion of resemblance to elementary colours. In addition, whilst lightness (Value) does not directly appear in NCS, it is possible to derive an analogous scale. In common with the Munsell system, there is no simple analytical relationship between NCS coordinates and the CIE system.

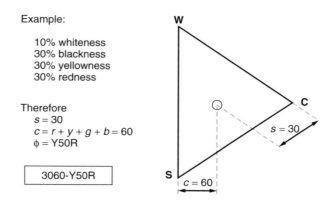

Example:

10% whiteness
30% blackness
30% yellowness
30% redness

Therefore
$s = 30$
$c = r + y + g + b = 60$
$\phi = Y50R$

3060-Y50R

Figure 5.5 An example NCS notation calculation.

However, as with Munsell, it is possible to interpolate the aim points specified in Ref. 23 using computer programs both to display the atlas and also to determine intermediate colours [24]. A page from a computer implementation of the atlas is shown in Plate 1(c). The physical atlas is still in production and a wide range of NCS products are available from the Scandinavian Colour Institute based in Stockholm.

5.3.9 The OSA-UCS colour system

The Optical Society of America's Committee on Uniform Colour Scales developed their colour system from 1947 to 1974 [25]. The guiding principle of OSA-UCS is that of uniform visual spacing (like Munsell). Furthermore by adopting a regular rhombohedral lattice spacing of its samples, their system was able to realise (almost) global as well as local uniform colour spacing (unlike Munsell). Consequently, within its colour space the distance between *any* two points should represent the magnitude of the perceived colour difference between those colours so represented. The particular choice of geometry was necessary to achieve the closest possible uniform packing in three dimensions; however, the system made no attempt to incorporate correlates of hue, chroma or saturation. Instead, opponent scales were chosen. Within this system, a colour notation is written in the form $L:j:g$ where L represents lightness of a colour, j describes its yellowness (positive) or blueness (negative), and g describes its greenness (positive) or redness (negative). These range approximately as follows: L from -7 to $+5$, j from -6 to $+11$ and g from -10 to $+6$. The lightness scale is designed such that $L = 0$ when the lightness is the same as that of the reference grey (a medium grey having 30% reflectance factor).

An atlas, completed in 1976, was produced by the OSA in the form of 558 paint chips presented on a mid-grey surround to be viewed under D65. OSA-UCS is specified colorimetrically in terms of aim points [25] but can be related to the CIE system. Given tristimulus values (x, y, Y), the luminous reflectance, Y_0, of a grey colour perceived to have an equal lightness to the target is computed thus:

$$Y_0 = Y(4.4934x^2 + 4.3034y^2 - 4.276xy - 1.3744x - 2.5643y + 1.8103)$$

Next, an intermediate lightness coordinate, Λ, is found:

$$\Lambda = 5.9[Y_0^{1/3} - \tfrac{2}{3} + 0.042(Y_0 - 30)^{1/3}]$$

The target colours tristimulus values are used to derive R, G, B:

$$R = 0.7990X + 0.4194Y - 0.1648Z$$

$$G = -0.4493X + 1.3265Y + 0.0927Z$$

$$B = -0.1149X + 0.3394Y + 0.7170Z$$

Next, a factor C (which applies 'crispening' to chromaticity differences according to the proximity of lightness to the 30% reflectance grey surround) is calculated as follows:

$$C = \Lambda/[5.9(Y_0^{1/3} - \tfrac{2}{3})]$$

Finally, L, j and g are found:

$$L = (\Lambda - 14.4)/2^{1/2}$$

$$j = C(1.7R^{1/3} + 8G^{1/3} - 9.7B^{1/3})$$

$$g = C(-13.7R^{1/3} + 17.7G^{1/3} - 4B^{1/3})$$

From these can be derived OSA-UCS hue $= \tan^{-1}(g/j)$ and chroma $= \sqrt{j^2 + g^2}$.

The inverse transformation from L, j, g to X, Y, Z is somewhat more complicated. For details, see Ref. 26.

5.3.10 Colour naming systems

Instead of using complex scientific or mathematical concepts to denote colour, it is also possible to use a nomenclature based on linguistic description. The motivation behind such an approach is that colour naming systems using familiar words should be readily understood by all without further training. This advantage has to be balanced by the limitation that naming systems have definite limits to their precision. To this end, the *Universal Color Language* (UCL), proposed in 1965 by Kenneth Kelly [27], defines colour naming in six levels according to desired tolerance. All of these naming levels involve the division of the solid formed by the Munsell colour space. The simplest levels, 1–3, describe colour using names, whilst levels 4–6 apply numeric descriptions. Level 1, for example, divides the solid up into black, grey and white together with 10 sectors according to the Munsell Hue. In some cases, where there exists an everyday term, each sector is further divided in Value: for example pink and red, or orange and brown. At the opposite end of the scale, level 6 achieves maximum precision by using either CIE

(x, y, Y) or interpolated Munsell coordinates, although this can hardly be referred to as a colour naming system.

The work of the Inter-Society Color Council and the National Bureau of Standards in developing their ISCC-NBS Dictionary of Color Names [28–30] formed the basis of level 3 of UCL. This method divides up the Munsell solid into cylindrical blocks, illustrated in Figure 5.6, which are bounded by upper and lower limits in Munsell Hue, Value and Chroma. Named regions can span multiple merged blocks, thus leading to some very irregular shapes. To simplify the situation, centroids were created representing the merged volumes of all elementary blocks having the same name. Using the published data containing 267 names together with centroid values for these blocks, an arbitrary shade can thus be named according to whichever is the closest centroid. Closeness can be determined either visually or by calculation using a colour difference formula. Where a colour lies evenly between two or more of the 267 central colours, it then becomes necessary to consider the elementary bounding limits. These centroid values have also been reproduced by NBS as physical charts which facilitate the rapid visual matching of samples to names. Some examples of ISCC-NBS names include deep greenish blue, light olive brown and vivid reddish purple.

In addition to ISCC-NBS, there exist very many more colour naming systems, often based on more arbitrary principles. As mentioned earlier, these include commercial product ranges (such as paint or textile samplers) and professional colour specifiers such as those produced by Focaltone, PANTONE, RAL, Toyo and Trumatch. In general, these are very convenient and relatively inexpensive. However, as pointed out by Fairchild [2], they are not perceptually ordered and their viewing conditions are not specified. Furthermore, few are defined colorimetrically, and of those that are, fewer still are reproduced accurately. With these provisos, it is possible to relate such systems to CIE XYZ either by buying licensed sample libraries or, where not available, by direct sample measurement. In the same manner as described for ISCC-NBS centroids, given an arbitrary colour it is simply a matter of searching through a database of samples to locate the colour having the

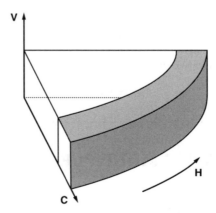

Figure 5.6 A cylindrical sectioned block in Munsell space.

smallest colour difference from the target. Unlike the colour notation systems described in the preceding sections, it is generally not possible to interpolate between samples.

5.3.11 Other colour notation systems

In this chapter it has been possible to mention only a limited number of colour notation systems. Some of the others are now briefly introduced. For further reading on the subject, see Refs 31 and 32.

The German chemist Wilhelm Ostwald developed his *Ostwald System* [33] using spinning disc colorimetry. This system's guiding principle was of equality of visual spacing (like Munsell or DIN) and complementarity of hues (like NCS). Three variables were used to express the notation: lightness (a log scale providing approximately equal steps using Weber and Fechner laws), saturation (relative to certain 'full' or chromatic colours) and hue (comprised of the four main hues: red, yellow, green and blue). The system was first published in 1915 and the first atlas produced in 1942 as the *Colour Harmony Manual*. This had several unique features including double-sided removable colour chips (one side was matt, the other full gloss). The third edition used hexagonal chips, permitting closer packing of samples arranged in an equilateral triangle depicting constant hue. Although colorimetric data are provided in Ref. 33 and elsewhere, no simple analytical relationship between the Ostwald and CIE systems has yet been found.

The *Colorcurve System* [34] is a much more recent development that represents a hybrid between a colour communication system and a colour mixture system. CIELAB rectangular (a^*, b^*) coordinates are used as the basis for an additive colour mixing system. All the aim points were formulated with real pigments so that the system could be reproduced with known reflectance curves. The master atlas contains around 1200 samples at 18 lightness levels. Since samples are defined by their reflectance, different materials can be used, illumination is not critical and metamerism can be avoided. Computer software is available with the system that computes the reflectance values of any point within the Colorcurve space given CIELAB or tristimulus values.

The *Coloroid System* [35, 36] was created by Antal Nemcsics and his co-workers at the Technical University of Budapest, Hungary, for use in environmental colour design. Its guiding principle is that of equality of aesthetic spacing. This, it is believed, refers to equality of spacing which leads to global visual uniformity as derived from extensive visual scaling experiments. Coloroid variables are hue (A), saturation (T) and lightness (V). A simple relationship between CIE Y and V exists:

$$V = 10Y^{1/2}$$

and further details of the relationship with the CIE system are given in Ref. 35.

5.4 INTERRELATING MULTIPLE NOTATION SYSTEMS

As already mentioned, the principal goal of colour notation systems is to facilitate the specification and communication of colour information. We have seen in the preceding text that there are many different ways to describe colour and these are incompatible with each other. This then begs the questions: *why are there so many systems* and *why not just use one?* It is argued that existing notation systems will not be supplanted by a single universal system for a number of reasons, which include the following:

- Different systems have already been adopted as either national or industry standards.
- Many users are highly experienced with one particular system and retraining them would be expensive and time consuming.
- Historical data in one or other system are difficult or in some cases, such as colour difference, impossible to translate.

Since the adoption of any universal notation system is unlikely at present, the solution to the immediate problem of colour communication is to interrelate existing systems. To this end, tables of data have been published listing equivalent notations in several different specification systems, for example NCS and Munsell [37] and computer software developed to automate the inter-conversion [38] (although source code has not been published). Rather than implementing N^2 bidirectional mappings between N notation systems, such approaches usually simplify the problem by using the CIE XYZ system as a central hub [39], as illustrated in Figure 5.7.

This approach suffers from two limitations. The first, which is unavoidable, is that for many colour order systems necessarily defined by aim points there is no simple or direct means to derive an inverse mapping. While interpolation techniques can be used to compute XYZ values for a given notation, going backwards is not so straightforward. Secondly, the measurement data defining different systems are often acquired under dissimilar conditions. Discrepancies arise due to the use of dissimilar instruments and illumination. Instruments can disagree due to such factors as differences in integrating sphere size, specular component inclusion and measurement geometry. Taking a specific example, the Munsell system was designed to be viewed under Illuminant C whilst NCS adopts

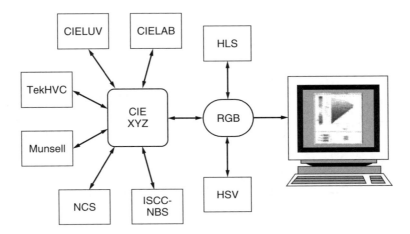

Figure 5.7   Interrelating colour notations via CIE XYZ.

Illuminant D65. Illumination differences cause an appearance mismatch whereas measuring atlas chips using undefined light sources can give rise to metamerism as they are not guaranteed to be colour constant. The solution proposed by Smith *et al.* [38] is to obtain aim point data for all possible light sources. This, however, both is impracticable and does not solve the potentially serious problem of metamerism.

Work by this author [26] also used the concept of mapping to a central hub of CIE XYZ values. Using this approach, differences in illumination conditions and media are compensated for through the application of a colour appearance model (introduced in Chapter 4). To overcome the difficulties due to inter-instrument agreement and metamerism, the reflectance of primary standards can be measured with a single instrument. These data are then used together with each system's specified illuminant to compute an initial set of XYZ values which are passed through the colour appearance transform illustrated in Figure 5.8. Using the measured conditions and illumination as input, this transform then 'perceptually normalises' the colour data to a single viewing condition, yielding another set of CIE tristimulus values, XYZ'. This output data may then be used either:

- to compute the particular screen RGB values necessary to display the colour (via the display's characterisation model), or
- by reversing this transform using an alternative set of viewing parameters, to find the corresponding notation in a different colour specification system, thereby interrelating the two systems.

5.4.1 Implementing computer-based systems

As physical atlases are convenient, portable, easy to understand and relatively cheap, why bother to use a computer system to do the same task? The following list gives some possible motivating factors.

- **Cost** – a set of atlases for one of the advanced colour order systems is already significantly more expensive than a basic computer system. As the cost of computer hardware

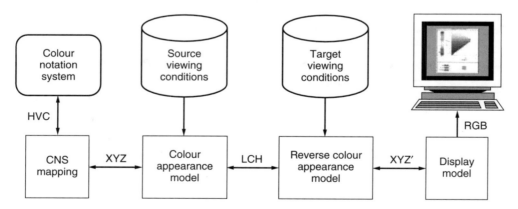

Figure 5.8 Using a colour appearance model to interrelate colours.

is constantly falling whilst the cost of producing tight-tolerance colour samples is not, computer-based systems are becoming a cost-effective alternative. Furthermore, computer systems have become ubiquitous throughout both office and home, and therefore deploying colour communication systems is largely a matter of software distribution.

- **Range** – the number of coloured chips in an atlas is limited by both manufacturing costs and practical constraints. (An atlas containing 16 million colours would be rather heavy.) By using interpolation, computer equivalents have no such limitation. In addition, an electronic atlas is able to represent (although not accurately display) shades which fall outside its colour gamut.

- **Interrelation** – as mentioned in the previous section, computer software can be used to instantly convert colours from one notation system to another. To do this with physical samples could easily take an experienced worker several minutes of searching for each desired shade.

- **Accuracy** – physical coloured chips have a limited lifespan: they fade with time or become scratched and dirty. A computer monitor can provide an equally accurate representation and through proper monitor characterisation and periodic recalibration does not suffer from these limitations.

- **Portability** – physical atlases are inherently portable (if somewhat heavy in certain cases); however, LCD and alternative low-power display technology is advancing such that the use of portable computers is becoming a viable alternative. An additional consideration is that of illumination. To avoid metamerism, physical chips usually need to be viewed under a specific light source (which might not be portable). Screen colours are self-luminous and so can avoid the problem.

- **Communication** – colour atlases facilitate the rapid exchange of colour information by notation alone. Without such systems, physically transporting colour samples can be very time consuming and expensive. Likewise, perhaps the biggest advantage of computer-based systems is the ability to take advantage of global communication networks to instantly send colour electronically around the world.

There are numerous software applications implementing individual colour notation systems, such as *Color Cleaver* [40] (which reproduces OSA-UCS) and *Adobe Photoshop* (which includes PANTONE). As mentioned in the last section, there are very few that implement multiple system interrelation. To help illustrate some of the implementation issues, specific reference is made to the *ColourTalk* system [41, 42] developed by this author and currently applied within the textile industry. This software incorporates both on-screen visualisation of existing colour notation systems and also the transparent inter-conversion between different notations.

When developing computer-based colour specification software, the implementor would do well to pay regard to existing colour order systems. In particular, this includes mimicking the same layout of coloured patches and background. Not only does this make such software familiar to existing users but it is also vital in preserving the same colour appearance. By ensuring that sample shape and size correspond to surface colours, differences in viewing geometry can be avoided that would otherwise lead to an appearance mismatch between the two media. Patches that are positioned too close together can exhibit a heightened perceived edge contrast – a phenomenon known as the 'Mach-band' effect.

The appearance of a coloured patch is also influenced by its background. Systems such as CIELAB and CIELUV do not have physical embodiments; however, it is stipulated that they should be viewed against a white to mid-grey background. Where available, the background employed by physical specifiers should be preserved on screen to avoid simultaneous colour contrast.

Within ColourTalk's colour picker window (Plates 1(a)–(c)), the display is divided up into three areas as shown. The top left area permits coarse hue and chroma selection by depicting a plane of constant lightness with the current hue and chroma indicated by a movable cursor. The right-hand side shows a plane of constant hue. Within this area, coarse lightness and chroma selection can be made. The currently selected values are indicated by the highlighted patch. Finally, in the lower left portion of the display, fine colour selection can be carried out within the notation system's micro space. This area is comprised of two orthogonal planes about the target colour whose colour differences can be adjusted using the scroll bars in order to achieve a zoom control.

ColourTalk adopts the 'cut and paste' paradigm found in the majority of word-processing programs to interrelate between different colour notation systems. Thus all the user needs to do is to copy a specific colour from one colour picker window and paste it into another to have its notation immediately and transparently converted. The same approach can be used to locate the closest matching shade from arbitrary collections of colours; however, in this case the match is most likely to be inexact. As a result, it is necessary to present both numerical and visual feedback on the difference between target and matched colours.

One final topic for consideration is colour gamut. All reproduction processes have limits to their range of colours and this will inevitably mean that some colours present within the gamut of colour order systems fall outside the screen gamut. As a result of the WYSIWYG approach, it is vital that the user is given some visual indication of out-of-gamut colours. Displaying out-of-gamut colours in an on-screen atlas alongside those in gamut will only mislead the user. For this reason, all colour specifiers within ColourTalk present colours within the monitor gamut. (This is no different from physical atlases that limit their sample ranges to the gamut of available pigments.) However, it is sometimes of critical importance to be able to work with out-of-gamut colours; just because a particular display cannot reproduce them does not mean that they cannot have an accurate numeric representation inside the computer. ColourTalk therefore takes the approach of clearly signalling to the user that a chosen colour is out of gamut. This is done by highlighting the colour with a black border. The colour itself is then presented as a hopefully close in-gamut colour by clipping to the monitor's gamut boundary. Normally, this would be bad practice because the border would cause undesirable colour contrast effects. However, in this case, since the displayed colour is wrong anyway, this is permissible.

5.5 SUMMARY

This chapter has outlined the issues surrounding colour specification, covering both physical colour atlases as well as computer-oriented colour spaces. Various types of colour notation systems have been identified and exemplified. In order to address the practical limitations of physical atlases and the difficulty in communicating colour, the chapter has also discussed implementing electronic notation systems and methods for interrelating incompatible systems. Unfortunately, there are many notation systems in common use, but it is hoped that this chapter has provided a good foundation in the principles of colour specification. Further information on colour notation can be found in Refs 31, 32, 43, 44 and 45.

REFERENCES

1. Tonnquist, G. (1993) 25 Years of Colour with the AIC – and 25,000 Without. *Color Research and Application*, **18**(5), 135–146, John Wiley and Sons.
2. Fairchild, M. D. (1998) Color-order systems. In *Color Appearance Models*, pp. 111–130. Reading, MA: Addison Wesley Longman.
3. Hunt, R. W. G. (1998) *Measuring Colour*, 3rd edn. Fountain Press.
4. Rhodes, P. A. (1998) Colour management for the textile industry. In *The Colour Image Processing Handbook*, S. J. Sangwine and R. E. N. Horne (ed.), pp. 307–331. Chapman and Hall, London.
5. Foley, J. D., van Dam, A., Feiner, S. K. and Hughes, J. F. (1990) Achromatic and colored light. In *Computer Graphics*, 2nd edn., pp. 590–595. Addison-Wesley.
6. CIE (1986) *Colorimetry*, 2nd edn. CIE Publication No. 15.2. Vienna, Austria: CIE.
7. Hung, P. C. and Berns, R. S. (1995) Determination of constant hue Loci for a CRT Gamut and Their Predictions using Color Appearance Spaces. *Color Research and Application*, **20**, 285–295, John Wiley and Sons.
8. Tektronix (1990) *TekColor Color Management System: System Implementor's Manual*. Tektronix, Inc.
9. Taylor, J. M., Murch, G. N. and McManus, P. A. (1989) TekHVC: A Uniform Perceptual Color System for Display Users. *Proceedings of the SID*, **30**(1), 15–21.
10. Tektronix, The TekHVC Color Space. In *TekColor System Implementor's Manual*, pp. 2.1–2.22. Tektronix, Inc., Beaverton, Oregon.
11. Tabayoyon, A. and Taylor, J. M. (1991) A technical introduction to the X Color Management System. *The X Resource*, **0**.
12. Munsell, A. H. (1905–16) *A Color Notation*. Boston, MA: Ellis.
13. Newhall, S. M., Nickerson, D. and Judd, D. B. (1943) Final Report of the OSA Subcommittee on the Spacing of Munsell Colors. *Journal of the Optical Society of America*, **33**, 385–418.
14. Judd, D. B. and Wyszechi, G. W. (1956) Extension of the Munsell Renotation System to Very Dark Colors. *Journal of the Optical Society of America*, **46**, 281–285.
15. Rheinboldt, W. C. and Menard, J. P. (1960) Mechanized Conversion of Colorimetric Data to Munsell Renotations. *Journal of the Optical Society of America*, **50**(8), 802–807.
16. Simon, F. T. and Frost, J. A. (1987) A New Method for the Conversion of CIE Colorimetric Data to Munsell Notation. *Color Research and Application*, **12**(5), 256–260, Wiley-Interscience.
17. Richter, M. and Witt, K. (1986) The Story of the DIN Color System. *Color Research and Application*, **11**(2), 138–145, Wiley-Interscience.
18. Deutsche Normen, DIN (1980) *DIN 6164 Part 2: DIN Colour Chart. Specification of Colour Samples*. Berlin: Beuth Verlag.
19. Deutsche Normen, DIN (1980) *DIN 6164 Part 1: DIN Colour Chart. System Based on the 2° Standard Colorimetric Observer*. Berlin: Beuth Verlag.
20. Witt, K. (1979) Beziehung zwischen der Kennzeichnung von Farbarten im Farbsystem DIN 6164 und der im Normvalenzsystem [Relationship between identification of chromaticities in the colour system specified in DIN 6164 and that in the standard colorimetric system]. *Farbe und Lack*, **85**, 459–463 (In German).

21. Swedish Standards Institution, SIS (1972–73) *Swedish Standard for Colour Notation, SS 01 91 00.*
22. Swedish Standards Institution, SIS (1979) *Swedish Standard Colour Atlas, SS 01 91 02.*
23. Swedish Standards Institution, SIS (1982) *Swedish Standard SS 01 91 03 CIE Tristimulus Values and Chromaticity Coordinates for Colour Samples in SS 01 91 02 (SIS).*
24. Derefeldt, G., Hedin, C. -E. and Sahlin, C. (1990) NCS colour space for VDU colours. *Displays*, **11**, 8–29, Butterworth Scientific Limited.
25. MacAdam, D. L. (1974) Uniform Color Scales. *Journal of the Optical Society of America*, **64**(12), 1691–1702.
26. Rhodes, P. A. (1995) *Computer Mediated Colour Fidelity and Communication*. PhD Thesis, Loughborough University of Technology.
27. Kelly, K. L. (1965) A Universal Color Language. *Color Eng.*, **3**(2), 16–21.
28. Kelly, K. L. and Judd, D. B. (1955) *The ISCC-NBS Method of Designating Colors and a Dictionary of Color Names*. NBS Circular 553, National Bureau of Standards, Washington, DC.
29. Kelly, K. L. (1958) Central notations for the revised ISCC-NBS color-name blocks. *Research of the National Bureau of Standards*, **61**(5), 427–431. National Bureau of Standards, Washington, DC.
30. Kelly, K. L. and Judd, D. B. (1976) *Color: Universal Language and Dictionary of Names*. Special Publication 440, National Bureau of Standards, Washington, DC.
31. Billmeyer, F. W. Jr. (1987) Survey of Color Order Systems. *Color Research and Application*, **12**(4), 173–186, John Wiley and Sons.
32. Gouras, P. (ed.) (1991) Colour appearance systems. In *Vision and Visual Dysfunction*, **6**, 218–259, New York: MacMillan Press.
33. Granville, W. C. (1994) The Color Harmony Manual, a Color Atlas Based on the Ostwald Color System. *Color Research and Application*, **19**(2), 77–98, John Wiley and Sons.
34. Stanziola, R. (1992) The Colorcurve System. *Color Research and Application*, **17**(4), 263–272, John Wiley and Sons.
35. Nemcsics, A. (1980) The Coloroid Color System. *Color Research and Application*, **5**, 113–120, Wiley-Interscience.
36. Nemcsics, A. (1987) Color Space of the Coloroid Color System. *Color Research and Application*, **12**(3), 135–146, Wiley-Interscience.
37. Billmeyer, F. W. Jr. and Bencuya, A. K. (1987) Interrelation of the Natural Color System and the Munsell Color Order System. *Color Research and Application*, **12**, 243–255, John Wiley and Sons.
38. Smith, N. S., Whitfield, T. W. A. and Wiltshire, T. J. (1990) A Colour Notation Conversion Program. *Color Research and Application*, **15**(6), 338–343, Wiley-Interscience.
39. Rhodes, P. A., Luo, M. R. and Scrivener, S. A. R. (1990) Colour model integration and visualisation. In *Proceedings of IFIP INTERACT '90: Human-Computer Interaction*, D. Diaper, *et al.*, (ed.), pp. 725–728. Cambridge: Elsevier Science Publishers B.V. (North-Holland).
40. Color Cleaver, *Slicing and Dicing the World of Color*. Sperryville, VA: Bronson Color Company.
41. Rhodes, P. A., Scrivener, S. A. R. and Luo, M. R. (1992) ColourTalk – a system for colour communication. *Displays*, **13**(2), 89–96, Butterworth Scientific Limited.
42. Rhodes, P. A. and Luo, M. R. (1996) A System for WYSIWYG Colour Communication. *Displays*, **16**(4), 213–221, Elsevier Science B.V.
43. Robertson, A. R. (1984) Colour Order Systems: An Introductory Review. *Color Research and Application*, **9**(4), 234–240, John Wiley and Sons.
44. Tonnquist, G. (1986) Philosophy of Perceptive Color Order Systems. *Color Research and Application*, **11**(1), 51–55, John Wiley and Sons.
45. Billmeyer, F. W. Jr. (1985) *AIC Annotated Bibliography on Color Order Systems*. Mimeoform Services, Beltsville.

Overview of characterization methods

Phil Green

6.1 INTRODUCTION

Many colour engineering tasks involve transporting colours between colour imaging devices through the medium of CIE colorimetry, and for this it is necessary to create a model of the relationship between the coordinates of the device and the corresponding CIE colorimetry. Device coordinates are normally scalars that either record the output of an image capture device, or represent colorant amounts that will be used to drive a colour output device. For input devices, the CIE colorimetry to be associated with these device scalars will normally be the tristimulus values or CIELAB coordinates of the colours on the original media, while for output devices it will be the colorimetry of the colours reproduced on the device. Most models are developed by first measuring a sample of the colours on the media which is to be characterized, and then defining a general relationship between the two colour spaces which can be used to transform any colour from one space to the other [9].

No one type of device model gives optimum results with all types of device, and a wide range of different models have been developed. There are three basic approaches to generating device models:

Colour Engineering, edited by P.J. Green and L.W. MacDonald.
© 2002 John Wiley & Sons Ltd.

1. Physical models (or colour-mixing models) which include terms for various physical properties of the device, such as absorbance, scattering and reflectance of colorant and substrates.
2. Numerical models in which a series of coefficients is defined, usually by regression from a set of known samples, with no prior assumptions about physical behaviour of the device or associated media.
3. Look-up tables which define the conversion between a device space and a CIE colour space at a series of coordinates within the colour space, and interpolate the values for intermediate coordinates. The entries in a look-up table can be determined either by direct measurement or through a physical or numerical model.

In practice colour transformations frequently involve an element of two or more of these approaches. For example, a numerical method may be used to define the relationship between colorimetric and printer values, with the precise dot areas to image found by a physical model, while the resulting function is used to calculate the entries in a colour look-up table. The resulting look-up table will then be used to perform actual colour image transformations, as this will normally be more computationally efficient than computing the transform on a pixel-by-pixel basis. Alternatively a look-up table may be generated directly from measurement data, in which case a larger number of measurements will be required.

The chapters in this section describe methods for modelling a range of different classes of device and media. Many of the methods used are fundamental to all types of model, and the purpose of this chapter is to serve as an introduction to such methods, particularly those of the numerical and look-up table type.

Physical models for different media and devices are then discussed in the following chapters in this section. In describing numerical models in this chapter the focus is mainly on linear regression methods, although it should be noted that a variety of other methods of generating device models have also been described, including spectral methods, radial basis functions and neural networks [3, 12, 15, 29, 39].

Objectives of device characterization can be summarized as:

- *Accuracy.* The model should predict colours with minimal errors over the entire colour space. The average colour difference between predicted and measured colours should ideally be no greater than the noise present within the system.
- *Visual acceptability.* There should be no artefacts caused by discontinuities in the model, and the distribution of errors should not lead to shifts in colour attributes that are considered visually significant by observers.
- *Computational simplicity.* This may be important where a device model is to run on desktop computers or must be frequently recalculated to compensate for changes such as device drift or media changes.
- *The minimum number of measurements.* This is desirable to simplify the modelling process, particularly where a numerical model has to be redefined for different reproduction conditions such as a change of substrate.
- *Analytic invertibility.* Device models may need to compute both forward transforms (e.g. into colorant amounts) and also reverse transforms (e.g. from colorant amounts

into another domain such as CIE colorimetry). If a model is not analytically invertible, alternatives are to use numeric methods to invert the transform, or to determine the model for both directions. In both cases the forward and reverse models are likely to give different errors, so that a colour that is transformed from one colour space to another and then back again will not have identical coordinates to the original.

6.2 NUMERICAL MODELS

In creating a numerical model the device space is first sampled colorimetrically so that the relationship between device scalars and tristimulus values is known for a given set of colours. This set is chosen so that the whole space of device coordinates is represented, and the known values that result are then used to derive a numerical model of the relationship between device space and CIE colour space.

For input devices, this process uses colour patches that have been measured so that their tristimulus values are known. For display and hard copy devices, sample colour patches of a range of device values are output and the resulting colours measured by spectrophotometer or other colour measurement instrument.

It is often possible to define a relationship between the CIE measurement for a colour and its corresponding device coordinate through a 3×3 matrix:

$$\mathbf{C} = \begin{bmatrix} a_{1,1} & a_{1,2} & a_{1,3} \\ a_{2,1} & a_{2,2} & a_{2,3} \\ a_{3,1} & a_{3,2} & a_{3,3} \end{bmatrix} \mathbf{D} \qquad (6.1)$$

where $\mathbf{C}$ is the vector of CIE colour space values (e.g. XYZ); $\mathbf{D}$ is the vector of values in a device colour space and $\mathbf{A}$ is a 3×3 matrix of coefficients. For example:

$$\begin{bmatrix} X \\ Y \\ Z \end{bmatrix} = \mathbf{A} \begin{bmatrix} R \\ G \\ B \end{bmatrix} \qquad (6.2)$$

where $\mathbf{A}$ is the coefficient matrix.

If the relationship between the device and the CIE space is constant (for example, if the spectral sensitivities of a scanner are linearly related to 1931 Standard Observer matching functions), the values of the coefficients in $\mathbf{A}$ are constant for all combinations of device coordinates, and can be found by using three known samples and solving three simultaneous equations. Where, as is more often the case, this relationship is not constant, the precise values of $\mathbf{A}$ will be different for each pair of device coordinates and corresponding CIE values. The single $\hat{\mathbf{A}}$ that will give the best fit to all the real matrices $\mathbf{A}$ must then be found, usually by some form of regression.

The steps in creating a numerical model can be described as:

1. Select the regression method.
2. Select the domains in which to represent the source data and the device coordinates.

3. Select the black printer model if required.
4. Select the training set of device coordinates which will form the test target.
5. Reproduce the test target and obtain measurement data.
6. Form the regression equation with the vectors of measurement data and device coordinates.
7. Compute the model coefficients.
8. Evaluate the model and add corrections if required.

6.2.1 Regression methods used in characterization

First-order model

In performing the regression, the elements of $\begin{bmatrix} X \\ Y \\ Z \end{bmatrix}$ and $\begin{bmatrix} R \\ G \\ B \end{bmatrix}$ in equation (6.2) above become vectors of device coordinates and corresponding CIE values, rather than scalars.

A least-squares method is frequently used to determine the best-fit approximation of the elements of $\mathbf{A}$. A system of linear equations can be written as follows [28]:

$$\mathbf{X}_n = \mathbf{D}_{n \times 3}\mathbf{A}_{3 \times 1} + \mathbf{E}_n \tag{6.3}$$

where $\mathbf{X}_n$ is the vector of n measured values of X, $\mathbf{D}_{n \times 3}$ is the matrix of n device values, $\mathbf{A}_{3 \times 1}$ are the three coefficients corresponding to row 1 of $\mathbf{A}$, and $\mathbf{E}_n$ is the vector of residual errors. By matrix algebra, the least-squares solution that leads to the vector $\hat{\mathbf{A}}$ that minimises the errors in $\mathbf{E}_n$ is given by:

$$(\mathbf{X} - \mathbf{D}\hat{\mathbf{A}}_\mathbf{X})^\mathrm{T}(\mathbf{X} - \mathbf{D}\hat{\mathbf{A}}_\mathbf{X}) \tag{6.4}$$

where $^\mathrm{T}$ denotes the matrix transpose.

$\hat{\mathbf{A}}$ is then found by:

$$\hat{\mathbf{A}}_\mathbf{X} = (\mathbf{D}^\mathrm{T}\mathbf{D})^{-1}\mathbf{D}^\mathrm{T}\mathbf{X} \tag{6.5}$$

The vectors defining the remaining values of $\mathbf{A}$ for $\mathbf{Y}$ and $\mathbf{Z}$ are found in the same way:

$$\hat{\mathbf{A}}_\mathbf{Y} = (\mathbf{D}^\mathrm{T}\mathbf{D})^{-1}\mathbf{D}^\mathrm{T}\mathbf{Y}$$

$$\hat{\mathbf{A}}_\mathbf{Z} = (\mathbf{D}^\mathrm{T}\mathbf{D})^{-1}\mathbf{D}^\mathrm{T}\mathbf{Z} \tag{6.6}$$

Then:

$$\mathbf{A} = \begin{bmatrix} \hat{\mathbf{A}}_\mathbf{X} \\ \hat{\mathbf{A}}_\mathbf{Y} \\ \hat{\mathbf{A}}_\mathbf{Z} \end{bmatrix} \tag{6.7}$$

The most costly part of equation (6.5) is the matrix inversion. If there is matrix algebra support in the environment in which the regression is performed, then equation (6.5) can be rewritten to make use of the matrix left division operator:

$$\hat{\mathbf{A}}_\mathbf{X} = (\mathbf{D}^\mathrm{T}\mathbf{D})\backslash\mathbf{D}^\mathrm{T}\mathbf{X} \tag{6.8}$$

This directly computes the solution using Gaussian elimination, avoiding the need to find the explicit inverse. If the left division operator is unavailable, then an alternative method is to solve using LU factorization.

Higher-order models

When modelling many devices the first-order model does not yield a sufficiently accurate transformation, and a higher-order polynomial regression is required [6]. Thus in a second-order model equation (6.2) becomes:

$$
\begin{bmatrix} X \\ Y \\ Z \end{bmatrix} = \begin{bmatrix} a_{1,1} & a_{1,2} & a_{1,3} & a_{1,4} & a_{1,5} & a_{1,6} \\ a_{2,1} & a_{2,2} & a_{2,3} & a_{2,4} & a_{2,5} & a_{2,6} \\ a_{3,1} & a_{3,2} & a_{3,3} & a_{3,4} & a_{3,5} & a_{3,6} \end{bmatrix} \begin{bmatrix} R \\ G \\ B \\ R^2 \\ G^2 \\ B^2 \end{bmatrix} \tag{6.9}
$$

The coefficients of the 3×6 matrix A are found in the same way as in equation (6.5).

In principle, the polynomial order can be up to $n - 1$ (where n is the number of discrete samples), although since each additional order adds a possible bend to the function (such bends being known as local maxima or minima, or relative extrema) it is desirable to keep the polynomial order as low as possible. In general, if the relationship between a set of dependent and independent variables is monotonic, a second or third order polynomial should be adequate [28]. As the number of coefficients increases, the sizes of the vector **D** and the matrix **A** are adjusted accordingly. equation (6.3) has the general form $\mathbf{D}_{n \times p}$ and $\mathbf{A}_p$ where p is the number of terms (i.e. the number of colours) in the vector **D**.

Further refinements include adding cross-products to the matrix **D** and varying the coefficients according to the region of colour space [6, 18, 19, 20]. Using colorimetric density as the destination domain and colorant c, m, y as the source, this can be written as:

$$
\begin{bmatrix} D_r \\ D_g \\ D_b \end{bmatrix} = \mathbf{B} \begin{bmatrix} c \\ m \\ y \\ c^2 \\ m^2 \\ y^2 \\ cm \\ cy \\ my \end{bmatrix} \tag{6.10}
$$

where **B** is a 9×3 matrix of coefficients to be found by the regression, again using the same form of regression equation as equation (6.5).

In a third-order model with cross-products the vector of device values would be:

$$
[c \ m \ y \ c^2 \ m^2 \ y^2 \ cm \ cy \ my \ c^2m \ c^2y \ m^2y \ cm^2 \ cy^2 \ my^2 \ c^3 \ m^3 \ y^3 \ cmy] \tag{6.11}
$$

and the coefficient matrix is of size 3×19.

In some situations the tristimulus values are known and the corresponding device scalars are sought. This can be achieved by inverting $\mathbf{A}$ in equation (6.2):

$$\begin{bmatrix} R \\ G \\ B \end{bmatrix} = \mathbf{A}^{-1} \begin{bmatrix} X \\ Y \\ Z \end{bmatrix} \tag{6.12}$$

So far we have considered methods applicable to colour systems with three components. If the number of components in the source and destination colour space do not match (for example, in the case of a CMYK printer), there may be many possible solutions to equation (6.5) above and further stages are required in building a numerical model. This is discussed in more detail below.

Choosing the polynomial order

The optimum polynomial order for characterizing a given device will vary according to the characteristics of the device output and the measurement accuracy. For a device with a linear response, a low polynomial order will suffice. Where there are complex interactions between channels and a highly non-linear response, a higher-order polynomial will be required. However, aside from the risk of generating local extrema with a higher-order polynomial, there is also an increased likelihood of simply fitting the noise present in the measurement data.

Constant offset

To allow for the possibility that a constant device offset will produce a better fit in the regression, a 1 can be added to the vector of device scalars in equations 6.9–6.11, and a corresponding term added to the coefficient matrix.

Spline methods

The principle of using a polynomial function to fit a set of measurement data can be extended by determining coefficients for multiple segments of the polynomial. This is known as a piecewise polynomial or spline. For a univariate function, a two-dimensional array of coefficients is generated instead of a single vector.

For colour data a multivariate regression is required, and this can be achieved by creating spline functions that are tensor products of univariate splines.

Spline methods have the advantage that they can pass through each data point if required (thus reducing the size of maximum errors relative to other methods), the degree of smoothing can be controlled, and the estimation of boundary values can be significantly improved through setting the boundary conditions of the spline.

In evaluations of spline methods in printer characterization, their performance was similar to [5] or slightly better than [13] third-order polynomials.

6.2.2 Domain

If the domains of the source and destination data are chosen to be as correlated as possible, the errors of a numerical model will tend to be minimized. For example, in printer characterization the relative log tristimulus values, or colorimetric densities [24], are often better correlated than reflectance with CMY (where the latter is defined either as fractional amounts of CMY or as densities with subsequent look-up tables to convert to dot areas). Colorimetric density is defined as follows:

$$
\begin{bmatrix} D_r \\ D_g \\ D_b \end{bmatrix} = \log \begin{bmatrix} \dfrac{X_0}{X} \\ \dfrac{Y_0}{Y} \\ \dfrac{Z_0}{Z} \end{bmatrix}
\tag{6.13}
$$

where $[D_r,\ D_g,\ D_b]$ are the colorimetric densities, $\{X_0\ Y_0\ Z_0]$ are the tristimulus values of the media white point and $[X,\ Y,\ Z]$ the tristimulus values of the print.

Then the regression seeks to find the coefficient matrix $\mathbf{A}$ in:

$$
\begin{bmatrix} D_r \\ D_g \\ D_b \end{bmatrix} = \mathbf{A} \begin{bmatrix} c \\ m \\ y \end{bmatrix}
\tag{6.14}
$$

where $[c,\ m,\ y]$ are fractional amounts of cyan, magenta and yellow.

In practice, CIELAB and XYZ often give poor performance when used as the regression domain for printer models as there is insufficient correlation with CMYK ink amounts. This is shown in Table 6.1, where a third-order polynomial regression with cross-products was performed on a test set of 323 CMY colours (no black) printed on newsprint.

Ink amounts correlate well with density (log reflectance), and similarly with log tristimulus values. By comparison, the correlation with linear XYZ is extremely poor. The correlation with CIELAB is better, probably as a result of the similarity between a log function and the cube root function use to derive L^*, a^* and b^* from tristimulus values; however, it is still poor in comparison with colorimetric density, since CIELAB has a different origin for L^*, a^* and b^* and has a mix of negative and positive values for a^* and b^*.

6.2.3 Training sets

The selection of the set of device coordinates to use either as the entries in a look-up table or as a training set in a regression affects both the magnitude and distribution of errors. If the training set consists of a large number of device coordinates distributed throughout

Table 6.1 Performance of a third-order polynomial regression using different domains

Domain	Mean	Median	Max.	95th percentile
Colorimetric density	0.73	0.68	3.06	1.33
CIELAB	15.11	8.38	100.02	59.99
XYZ	27.08	11.66	267.66	115.24

Table 6.2 Performance of device models using three different training sets: inclusion of additional colours from the gamut boundary has reduced the maximum error of the model

	Set A	Set B	Set C
Mean ΔE_{ab}	2.03	1.69	1.72
Max. ΔE_{ab} at boundary	14.77	12.20	4.55

colorant space, it can tend to reduce errors; however, it is desirable to minimize the number of colours in order to avoid the need for large numbers of measurements. Ensuring that the spacing of the colours selected is reasonably perceptually uniform reduces the need for large numbers of colours [23]. The ISO 12640 [17] test target was designed with this consideration in mind [20]. The ISO 12640 and 12641 [16] targets are illustrated in Plate 2(a) and (b) and described in more detail in Chapter 18.

In a typical test target the colours of many patches can be accurately predicted from their neighbouring patches, and it has been argued that the number of patches required to characterize a colour device can be further reduced if the spacing of the patches is carefully chosen [32].

In a polynomial regression, the largest errors tend to be at the gamut boundary and often in colorant primaries. This is because through the regression the values for any point in colour space are estimated from the known points that lie either side of it; for points on the boundary there are no known points beyond and hence the estimate for such colours has a higher degree of error. The effect of increasing the number of patches at the colorant boundary was explored in Ref. 10. A second-order polynomial regression was used to characterize the process, using three different training sets: (A) a basic set of 107 colour patches; (B) the 107 basic colours together with an additional 215 patches from the ISO 12640 characterization target; and (C) the 107 colours basic colours together with 215 colours from a gamut boundary target. The boundary target is illustrated in Plate 2(c).

The results show that both the larger data sets improved the performance of the model, but use of the colorant boundary data resulted in a smaller maximum error at the gamut boundary. Similar results were obtained in a comparison of characterization methods on an inkjet printer.

The measurement data which forms the training set should be representative of the device performance with the selected media and colorant. The greater the variation in the performance of the process, the more important it is to base the measurement data on a number of samples. If the targets measured are not fully representative of the process, there is a danger that the resulting model will perform poorly and restrict the gamut available [10, 41].

6.3 LOOK-UP TABLES WITH INTERPOLATION

A look-up table is a set of key-value pairs of coordinates in different domains, such as device scalars and their corresponding values in a CIE colour space. Look-up tables are often employed where the two domains are not trivially related, as it will usually be faster to locate the values and perform any interpolation required than to evaluate a complex function.

In the simplest look-up table and interpolation method the input colour space is considered as a cube with the three primaries and their secondary and tertiary combinations located at the vertices. The corresponding output values of the vertices are determined by measurement. The output value corresponding to any combination of the primaries can then be calculated using the distance from the vertices as weights.

Linearly interpolating between a source and destination space using only these eight coordinates can give rise to errors that are unacceptably large. Such errors can be reduced by partitioning the input cube into sub-cubes and determining output values for the vertices of each of the sub-cubes. This method can be extended to a comprehensive sampling of the device space by measuring a large number of samples and using these directly as the entries in the look-up table. This approach is sometimes referred to as 'full characterization' [23].

6.3.1 Interpolation

The purpose of an interpolation algorithm is to estimate the output value of a coordinate whose input value is known, given two or more coordinates for which both input and output values are known. The relative distances of the point to be found from the known points are used as weights in determining the value of the unknown point.

Linear interpolation to determine the y coordinate of a point x,y from two points x_1,y_1 and x_2,y_2 where x is known is given by:

$$y = y_1 + (y_2 - y_1)\frac{x - x_1}{x_2 - x_1} \qquad (6.15)$$

The interpolation error at a given coordinate is the distance between the coordinate determined by the interpolation and the true value of the function being interpolated at that coordinate. Naturally the magnitude of the error will increase as this function becomes more non-linear. Non-linear interpolation algorithms, such as cubic spline interpolation, can give better results than linear interpolation, although at higher computational cost.

A two-dimensional input space requires a bilinear interpolation in which two additional known points are used.

Input colour spaces are normally three-dimensional, and require the extension of the linear interpolation to a trilinear interpolation from eight vertices of a three-dimensional cube (Figure 6.1). A trilinear interpolation [36, 37] essentially iterates the bilinear interpolation to determine two points on opposite surfaces of a cube, and then once more to determine the location of the unknown coordinate on the line connecting these two points, for example:

$$X = (1 - c)(1 - m)(1 - y)X_1 + c(1 - m)(1 - y)X_2 + (1 - c)m(1 - y)X_3$$

$$+ cm(1 - y)X_4 + (1 - c)(1 - m)yX_5 + c(1 - m)yX_6 + (1 - c)myX_7 + cmyX_8$$

$$(6.16)$$

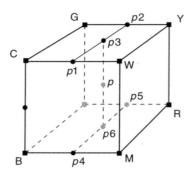

Figure 6.1 The value of point p is found by successively interpolating to find the values of $p1$, $p2$, $p3$, $p4$, $p5$ and $p6$[26].

where X_{1-8} are the X tristimulus values of the cube vertices and $[c\ m\ y]$ the device coordinates of the colour to be interpolated. The trilinear interpolation requires eight vertices and is computationally intensive.

Three-dimensional linear interpolation can also be carried out on polar coordinates as shown by the following equations which calculate the chroma and lightness of a new coordinate whose hue angle h_p is known, from the coordinates of two known colours $L_1C_1h_1$ and $L_2C_2h_2$:

$$C_p = C_1 + (C_2 - C_1)\frac{\Delta h_p}{\Delta h} \tag{6.17}$$

$$L_p = L_1 + (L_2 - L_1)\frac{\Delta h_p}{\Delta h} \tag{6.18}$$

where Δh_p is the hue angle difference between h_p and h_1, and Δh is the hue angle difference between h_1 and h_2.

A number of other geometric interpolation algorithms have been proposed which segment the input space into alternative polyhedra with fewer vertices. Foremost are the prism [25, 35] and tetrahedron [31, 27, 8]. Prism, pyramid and tetrahedron algorithms require that the input space cube is segmented into two, three and six pieces respectively. The tetrahedron, being the polyhedron with the smallest number of vertices, requires the minimum of computation when carrying out the interpolation. However, the use of any algorithm in which the input space is partitioned makes it necessary to search the input space to determine which polyhedron the input coordinate is located within, and thus introduces an additional step in the algorithm. Figure 6.2 shows a cube partitioned into tetrahedra.

Regular tetrahedra can be considered the ideal from the point of view of minimizing the distance between a coordinate to be interpolated and a LUT node, but are not space-filling. The difference between a given irregular tetrahedron and a regular tetrahedron can be described by a 'normalized shape ratio' [27] (the ratio of the radius of the inscribed sphere to the radius of the circumscribing sphere).

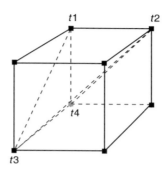

Figure 6.2 Partitioning a colour cube into tetrahedra.

Kasson [27] tested a range of methods of determining tetrahedral partitioning of the sub-cube, and found that disphenoid (equal-sized face and solid angle) and Kanamori (sub-cube pairs with mirror-image tetrahedra) methods both had similar accuracy to trilinear interpolation (although with less computational effort).

In a barycentric interpolation [8], a function is associated with each of the vertices of a tetrahedral surface. This function defines the distance of the coordinate from the vertex. For each interpolated point, four barycentric coordinates are needed.

The division of the gamut solid into uniform sub-cubes (whether based on evenly spaced device coordinates, or on a uniform division of XYZ or CIELAB space) prior to partitioning into tetrahedra has disadvantages: for XYZ or CIELAB the gamut solid is not rectilinear, and thus a considerable amount of grid points are wasted; and for device coordinates, a uniform spacing will not sample the device gamut optimally. In addition, the vertices in a cubic gamut solid will form an irregular polyhedron in the destination colour space unless the transformation is linear, thus making it difficult to invert the transformation.

It is possible to omit the cubic division and partition the space directly into tetrahedra [4], which will require more computation but has the advantage of being efficient at run-time, especially for higher-dimensional transformations. This approach makes the location of the tetrahedron vertices more flexible, and can thus increase the frequency of vertices in those regions of colour space in which it is appropriate.

An alternative approach, called sequential linear interpolation (SLI) [2], has been proposed in which the grid points can be optimally allocated according to the local characteristics of the transformation function. This technique is further developed in Ref. 1, which offers a method for minimizing the maximum interpolation error through an iterative minimax grid point allocation. Spline functions with SLI can be used to smooth the characterization data and reduce the error arising from measurement noise [7].

Rolleston [38] describes a different, non-geometric method in which there is no partitioning of the colour space, but instead Sheperd's Interpolation is used to define the mapping from tristimulus space to device coordinates in terms of an array of error vectors. This method has the advantage that sample points can be unevenly spaced to optimally sample the printer gamut, but is not appropriate if the error vectors are large.

In some situations, such as the implementation of a gamut mapping algorithm that is not applied uniformly throughout colour space, it is desirable to vary the way in which LUT nodes are located according to the region of colour space. This can be achieved with tetrahedral interpolation where the tetrahedra are not constrained to sub-cubes, as well as SLI and Sheperd's Interpolation, but not with a sub-cube partitioning.

6.3.2 LUT implementation

The risk of interpolation error or of introducing discontinuities into the output data can be minimized by the way that the LUT implementation is designed.

The performance of a LUT can be improved by transforming the input data so that it is more perceptually uniform prior to input into the LUT. This can be achieved by applying a function such as a log or cube root transformation, or alternatively a one-dimensional transfer function can be derived empirically. If the data has already undergone such a transformation (such as gamma-correction of scanner output) such a step is not required.

In the characterization process there will typically be from 120 to 512 colours measured, and for some applications it is unrealistic to require more than this. A larger LUT will give rise to smaller interpolation errors at the transformation stage, but at the cost of increasing memory requirements. If the input data are adequately linearized prior to entry into the LUT, fewer lattice points are needed [20, 32].

Lack of precision of the LUT entries can lead to quantization effects. Even if the input data are limited to 8 bits per channel, it is preferable to use 16-bit precision in the table if quality requirements are high. The conversion to 16 bits can be carried out in the linearization stage described above.

Although computing colour transformations is likely to be faster with a LUT than with a function that must be evaluated for each pixel, a LUT cannot readily be modified at run-time. Where parameters such as GCR amount need to be adjusted, the LUT values must be recalculated unless the workflow allows the transformation to be divided into a series of discrete stages.

A look-up table embodying a device model is the basis for colour transformations with ICC colour profiles. Before colours can be transformed into device coordinates with a device model, they should already have been transformed into the destination gamut and viewing condition – thus gamut mapping and appearance model transforms must be applied to the source colour values before the conversion to device space. In the present ICC architecture, the device model, gamut mapping and appearance model are folded together in a single look-up table, although proposals have been made to enable separate transforms for these stages in the colour imaging chain [22, 30].

6.4 EVALUATING MODEL ACCURACY

It is often possible to generate a model with mean errors of the order of $1 \Delta E_{ab}$, at least as a goal. However, the mean error is not a sufficient measure of performance by itself, as the errors are highly unlikely to be normally distributed, and in fact a χ^2 function often models the distribution of errors in colour samples better than the normal distribution [7], as illustrated in the example in Figure 2.10 in Chapter 2. Other summary measures should be used where possible, such as the maximum and median or RMS error, the error at the 95th percentile [34], and the coefficient of skewness [40]. Error histograms can also be useful in illustrating the distribution of errors within a sample.

Errors can also be analysed in terms of the separate ΔL^*, ΔC_{ab}^* and ΔH^* differences, and this analysis can often be used to detect any systematic tendencies in the error distribution and thus develop corrections to improve the model and further reduce any residual error.

Where a test target has been reproduced a number of times and a summary statistic is required for the set of reproductions, the χ^2 function can be utilized. If it is assumed that $(\Delta E_{ab}^*/\overline{\sigma})^2$ follows the χ^2 distribution, where $\overline{\sigma}$ is the mean standard deviation of a series of samples, then $\overline{\sigma}$ can be taken as a measure of the error distribution. The error standard deviation σ is found for each reproduction and the mean calculated to give $\overline{\sigma}$.

When reporting errors it is often useful to clarify whether absolute errors are given, e.g. $|\Delta E_{ab}^*|$, or to include a sign in order to indicate the direction of error.

Visual tests of reproductions made with a device model should also be carried out, as these will reveal discontinuities in the model that may not appear in the colorimetric analysis. A model that appears to have the lowest error may in fact produce less visually acceptable results depending on how the errors are distributed – for example, a shift in hue is likely to be less acceptable than a shift in chroma that has the same numerical ΔE_{ab}^* difference.

The error metric selected should normally be reasonably perceptually uniform. Hence consideration should be given to expressing the errors in an advanced colour difference formula such as CIEDE2000, with $l{:}c$ weightings optimized for the application. However, a comparison of CIELAB and CIE$_{94}$ [11] showed that for modelling acceptability thresholds in the graphic arts CIE$_{94}$ did not perform significantly better than CIELAB. In some situations, other error metrics may be more appropriate [14, 15].

It should also be noted that a highly accurate model may merely be fitting the noise present in the measurement data, and a degree of smoothing in the characterization is often desirable.

The test set used to evaluate model accuracy should where possible be different from the training set used to derive it, as otherwise the errors reported give a misleading impression of the accuracy of the model.

Within the ISO 12640 test target, the extended elements (S9–10) can be used as a test set for a model derived using the basic elements (S7–8). Alternatively the ISO 12641 target can be used to provide the test set as in columns 1–12 it includes a sampling of

Table 6.3 Performance of a regression model on two test sets: the first test set is limited to the colours in the training set and gives a misleading impression of the model accuracy

Test set	Mean	Median	Max.	95th percentile
ISO 12640 (training set)	1.08	0.96	3.62	2.22
ISO 12640 (all patches)	2.83	2.63	9.60	5.43

the colorant space with an approximately uniform perceptual spacing [33], whose colours will have different device coordinates from those in the ISO 12640 set. The difference between model performance when evaluated with the training set and with a different data set as test data is illustrated in the performance of a two-stage regression model [10] summarized in Table 6.3. As can be expected, the errors appear to be smaller when the test set is the same as the training set.

REFERENCES

1. Agar, A. and Allebach, J. P. (1996) A minimax method for sequential linear interpolation on non-linear colour transformations. *Proc. IS&T/SID Colour Imaging Conf.*, **4**, 1–5.
2. Allebach, J. P., Chang, J. Z. and Bouman, C. A. (1993) Efficient implementation of non-linear colour transformations. *Proc. IS&T/SID Colour Imaging Conf.*, **1**, 143–148.
3. Artusi, A. and Wilkie, A. (2001) Color printer characterization using radial basis function networks. *Proc. Conf. Col. Imaging: Device-Independent Color, Color Hardcopy and Graphic Arts*, **6**, 70–80.
4. Bell, I. E. and Cowan, W. (1993) Characterizing printer gamuts using tetrahedral interpolation. *Proc. IS&T/SID Colour Imaging Conf.*, **1**, 108–113.
5. Bell, I. E. and Cowan, W. (1994) Device characterization using spline smoothing and sequential linear interpolation. *Proc. IS&T/SID Colour Imaging Conf.*, **2**, 29–33.
6. Clapper, F. R. (1961) An empirical determination of halftone colour reproduction requirements. *Proc. TAGA Conf.*, 31–41.
7. Dolezalek, F. K. (1994) Appraisal of production run fluctuations from color measurements in the image. *Proc. TAGA Conf.*
8. Genetten, K. D. (1993) RGB to YCMK conversion using Barycentric interpolation. *Proc. SPIE Conf.*, **1909**, 116–126.
9. Green, P. J. (1999) *Understanding Digital Color*, 2nd edn. GATF Press, Sewickley, PA.
10. Green, P. J. (2001) A test target for defining media gamut boundaries. *Proc. SPIE Conf. Col., Imaging: Device-Independent Color, Color Hardcopy and Graphic Arts*, **6**, 105–113.
11. Green, P. J. and Johnson, A. J. (2000) Issues of measurement and assessment in hard copy colour reproduction. *Proc. SPIE Conf. Col., Imaging: Device-Independent Color, Color Hardcopy and Graphic Arts*, **5**.
12. Hardeberg, J. Y., Schmitt, F. J. and Brettel, H. (2000) Multispectral image capture using a tunable filter. *Proc. SPIE Conf. Col., Imaging: Device-Independent Color, Color Hardcopy and Graphic Arts*, **5**, 77–88.
13. Herzog, P. and Roger, T. (1998) Comparing different methods of printer characterization. *Proc. IS&T 51st PICS Conf.*, 23–29.
14. Holm, J. and Susstrünk, S. (1994) An EIQ – subjective image quality correlation study. *Proc. IS&T's 47th Conf.* 634–640.
15. Holm, J., Tastl, I. and Hordley, S. (2000) Evaluation of DSC (Digital Still Camera) Scene Analysis Error Metric: Part I. *Proc. 8th IS&T/SID Color Imaging Conf.*, 279–287.
16. ISO 12641. *Graphic technology – Prepress digital data exchange – Colour targets for input scanner calibration.*
17. ISO 12642. *Graphic technology – Prepress digital data exchange – Input data for characterization of 4-colour process printing targets.*
18. Johnson, A. J. (1977) *A Study of the Preferred Tone Reproduction Characteristics for Colour Reproduction.* Pira report PR/143.
19. Johnson, A. J. (1982) *Defining Optimum Photomechanical Colour Reproduction.* Pira report PR/170.
20. Johnson, A. J. (1995) *Colour Management in Graphic Arts and Publishing*, Pira International, Leatherhead.

21. Johnson, A. J. (1996) Methods for characterizing colour printers. *Displays*, **16**, 193–202.
22. Johnson, A. J. (2000) An effective colour management architecture for Graphic Arts. *Proc. TAGA Conf.*
23. Johnson, A. J., Luo, M. R., Lo, M.-C. and Rhodes, P. A. (1997) *Aspects of colour management. Part 1: Characterisation of three-colour imaging devices*. Unpublished paper.
24. Johnson, A. J., Tritton, K. T., Eamer, M. and Sunderland, B. H. W. (1982) *Systematic Lithographic Colour Reproduction*. Pira report PR/171.
25. Kanamori, K. and Kotera, H. (1992) Color correction technique for hard copies by 4-neighbours interpolation. *J. IS&T*, **36**, 73–80.
26. Kang, H. R. (1996) Color Technology for Electronic Imaging Devices. *SPIE* Press, Bellingham, WA.
27. Kasson, J., Plouffe, W. and Nin, S. (1993) A tetrahedral interpolation technique for colour space conversion. *Proc. SPIE Conf. Device-independent colour imaging*, **1909**, 127–138.
28. Kleinbaum, D. G., Kupper, L. L., Muller, K. E. and Nizam, A. (1998) *Applied Regression Analysis and Other Multivariable Methods*, 3rd edn. Duxbury Press.
29. Koenig, F. and Herzog, P. G. (2000) Spectral scanner characterization using linear programming. *Proc. SPIE Conf. Col. Imaging: Device-Independent Color, Color Hardcopy and Graphic Arts*, **5**, 36–46.
30. Kohler, T. (2000) The next generation of colour management system. *Proc. 8th IS&T/SID Color Imaging Conf.*, 61–64.
31. Korman, N. I. and Yule, J. A. C. (1971) Digital computation of dot areas in a colour scanner. *Proc. 11th IARIGAI conf.*
32. Mahy, M. (2000) Analysis of colour targets for output characterization. *Proc. 8th IS&T/SID Color Imaging Conf.*, 348–355.
33. Maier, T. O. and Rinehart, C. E. (1990) Design criteria for an input scanner test object. *J. Phot. Sci.*, **38**, 169–172.
34. Morovic, J. (1998) *To develop a universal gamut mapping algorithm*. Ph.D. Thesis, University of Derby.
35. Motomura, H., Fumoto, T., Yamada, O., Kanaori, K. and Kotera, H. (1994) CIELAB to CMYK color conversion by prism and slant prism interpolation method. *Proc. 2nd IS&T/SID Color Imaging Conf.*, 156–158.
36. Nin, S., Kasson, J. and Plouffe, W. (1992) Printing CIELAB images on a CMYK printer using tri-linear interpolation. *Proc. SPIE Conf. Colour hardcopy and the graphic arts*, **1670**, 316–324.
37. Pugsley, P. (1974) Image reproduction methods and apparatus. British Patent 1369702.
38. Rolleston, R. (1994) Using Sheperd's Interpolation to build color transformation tables. *Proc. IS&T/SID Colour Imaging Conf.*, **2**, 74–77.
39. Tominaga, S. (1993) A neural network approach to color reproduction in color printers. *Proc. IS&T/SID Colour Imaging Conf.*, **1**, 173–177.
40. Viggiano, J. A. S. (1999) *Statistical distribution of CIELAB color difference*. Unpublished paper, http://www.rit.edu/~jasvppr/Statistics.pdf.
41. Werfel, M. (2001) QUIZ-Quality initiative newspaper printing. *Proc. 28th IARIGAI Conf., Advances in Colour Reproduction*.

Methods for characterizing displays

Roy Berns and Naoya Katoh

7.1 INTRODUCTION

Viewing a display image is much like seeking the time of day. Ask one person for the time and your need has been met. Ask two people and you are not sure which person, if either, has the correct time. Create an image on one display and take the data file to a second imaging system. When viewed on the second system, it may match your memory of the first system. (We are aided in this task by chromatic adaptation, the psychological tendency to expect color constancy, and poor color memory.) However, if the two systems are placed adjacent to one another, one's confidence quickly vanishes because they are likely have different tone and color reproduction characteristics. The purpose of this chapter is to give background on why this occurs and methods to calibrate and characterize displays to avoid this problem. References 1–8 are suggested for further study.

7.2 GOG MODEL FOR CRT DISPLAYS

An uncharacterized image is defined by three image planes of digital information, d_r, d_g, and d_b, where the subscripts refer to the red, green, and blue channels. When an

---

Colour Engineering, edited by P.J. Green and L.W. MacDonald.
© 2002 John Wiley & Sons Ltd.

image is characterized, it is possible to define the image in terms of spectral radiance or tristimulus values. Berns *et al.* [1] have derived the relationship between digital and spectral data based on historical literature and hardware common to digitally controlled CRT displays. This relationship is given in equation (7.1) for the red channel. Similar expressions can be written for the green and blue channels.

$$
L_{\lambda,\mathrm{r}} =
\begin{cases}
k_{\lambda,\mathrm{r}} \left[a_\mathrm{r} \left[(v_{\max} - v_{\min}) \left(\dfrac{\mathrm{LUT}_\mathrm{r}(d_\mathrm{r})}{2^N - 1} \right) + v_{\min} \right] + b_\mathrm{r} - v_{\mathrm{c,r}} \right]^{\gamma_\mathrm{r}}; \\[6pt]
\qquad\qquad v_{\mathrm{c,r}} \leqslant a_\mathrm{r} \left[(v_{\max} - v_{\min}) \left(\dfrac{\mathrm{LUT}_\mathrm{r}(d_\mathrm{r})}{2^N - 1} \right) + v_{\min} \right] + b_\mathrm{r} \\[6pt]
0; \qquad v_{\mathrm{c,r}} > a_\mathrm{r} \left[(v_{\max} - v_{\min}) \left(\dfrac{\mathrm{LUT}_\mathrm{r}(d_\mathrm{r})}{2^N - 1} \right) + v_{\min} \right] + b_\mathrm{r}
\end{cases}
\tag{7.1}
$$

LUT represents the video look-up table, N is the number of bits in the digital-to-analog converter (DAC), $v_{\min}$ and $v_{\max}$ are voltages dependent on the computer video signal generator, a_r and b_r are the CRT video amplifier gain and offset, $v_{\mathrm{c,r}}$ is the cut-off voltage defining zero beam current, γ_r is an exponent accounting for the nonlinearity between amplified video voltages and beam currents, and $k_{\lambda,r}$ is a spectral constant accounting for the particular CRT phosphors and faceplate combination. Equation (7.1) is generic in that it considers signal processing common to all computer-controlled CRT displays. The common components include the DAC, video LUTs, video signal generator, and video amplifier. Because the spectral radiance depends on properties of both the graphics display controller and CRT, characterization of a display's spectral or colorimetric properties as a function of digital input must not be made independent of the graphics display controller. The terminology 'display system' will be used, accordingly.

By normalizing radiometric measurements relative to the maximum radiant output, equation (7.1) reduces to:

$$
L_{\lambda,\mathrm{r}} =
\begin{cases}
L_{\lambda,\mathrm{r,max}} \left[k_{\mathrm{g,r}} \left(\dfrac{\mathrm{LUT}_\mathrm{r}(d_\mathrm{r})}{2^N - 1} \right) + k_{\mathrm{o,r}} \right]^{\gamma_\mathrm{r}}; & \left[k_{\mathrm{g,r}} \left(\dfrac{\mathrm{LUT}_\mathrm{r}(d_\mathrm{r})}{2^N - 1} \right) + k_{\mathrm{o,r}} \right] \geqslant 0 \\[10pt]
0; & \left[k_{\mathrm{g,r}} \left(\dfrac{\mathrm{LUT}_\mathrm{r}(d_\mathrm{r})}{2^N - 1} \right) + k_{\mathrm{o,r}} \right] < 0
\end{cases}
\tag{7.2}
$$

Constants $k_{\mathrm{g,r}}$ and $k_{\mathrm{o,r}}$ are referred to as the system gain and system offset, respectively. This three-parameter gain, offset, and gamma model will be referred to as the 'GOG' model. It is the method recommended by the CIE [14]. $L_{\lambda,\mathrm{r,max}}$ defines the maximum spectral radiance of the red channel for a given CRT setup.

It is useful to define a radiometric scalar, R, according to equation (7.3):

$$
L_{\lambda,\mathrm{r}} = R L_{\lambda,\mathrm{r,max}}
\tag{7.3}
$$

A properly set-up display will, in theory, exhibit additivity between its three channels. Thus using the scalars and considering the three channels simultaneously results in equations (7.4) and (7.5), where n represents successive wavelength intervals.

$$
R = \begin{cases}
\left[k_{\mathrm{g,r}} \left(\dfrac{\mathrm{LUT_r}(d_\mathrm{r})}{2^N - 1} \right) + k_{\mathrm{o,r}} \right]^{\gamma_r}; & \left[k_{\mathrm{g,r}} \left(\dfrac{\mathrm{LUT_r}(d_\mathrm{r})}{2^N - 1} \right) + k_{\mathrm{o,r}} \right] \geqslant 0 \\[4mm]
0; & \left[k_{\mathrm{g,r}} \left(\dfrac{\mathrm{LUT_r}(d_\mathrm{r})}{2^N - 1} \right) + k_{\mathrm{o,r}} \right] < 0
\end{cases}
$$

$$
G = \begin{cases}
\left[k_{\mathrm{g,g}} \left(\dfrac{\mathrm{LUT_g}(d_\mathrm{g})}{2^N - 1} \right) + k_{\mathrm{o,g}} \right]^{\gamma_g}; & \left[k_{\mathrm{g,g}} \left(\dfrac{\mathrm{LUT_g}(d_\mathrm{g})}{2^N - 1} \right) + k_{\mathrm{o,g}} \right] \geqslant 0 \\[4mm]
0; & \left[k_{\mathrm{g,g}} \left(\dfrac{\mathrm{LUT_g}(d_\mathrm{g})}{2^N - 1} \right) + k_{\mathrm{o,g}} \right] < 0
\end{cases} \tag{7.4}
$$

$$
B = \begin{cases}
\left[k_{\mathrm{,g,b}} \left(\dfrac{\mathrm{LUT_b}(d_\mathrm{b})}{2^N - 1} \right) + k_{\mathrm{o,b}} \right]^{\gamma_b}; & \left[k_{\mathrm{g,b}} \left(\dfrac{\mathrm{LUT_b}(d_\mathrm{b})}{2^N - 1} \right) + k_{\mathrm{o,b}} \right] \geqslant 0 \\[4mm]
0; & \left[k_{\mathrm{g,b}} \left(\dfrac{\mathrm{LUT_b}(d_\mathrm{b})}{2^N - 1} \right) + k_{\mathrm{o,b}} \right] < 0
\end{cases}
$$

$$
\begin{bmatrix} L_{\lambda=1,\mathrm{pixel}} \\ \cdot \\ \cdot \\ \cdot \\ L_{\lambda=n,\mathrm{pixel}} \end{bmatrix}
=
\begin{bmatrix}
L_{\lambda=1,\mathrm{r,max}} & L_{\lambda=1,\mathrm{g,max}} & L_{\lambda=1,\mathrm{b,max}} \\
\cdot & \cdot & \cdot \\
\cdot & \cdot & \cdot \\
\cdot & \cdot & \cdot \\
L_{\lambda=n,\mathrm{r,max}} & L_{\lambda=n,\mathrm{g,max}} & L_{\lambda=n,\mathrm{b,max}}
\end{bmatrix}
\begin{bmatrix} R \\ G \\ B \end{bmatrix} \tag{7.5}
$$

Because of the additive nature of the CRT display, equation (7.5) can be replaced with a colorimetric definition shown in equation (7.6):

$$
\begin{bmatrix} X_{\mathrm{pixel}} \\ Y_{\mathrm{pixel}} \\ Z_{\mathrm{pixel}} \end{bmatrix}
=
\begin{bmatrix}
X_{\mathrm{r,max}} & X_{\mathrm{g,max}} & X_{\mathrm{b,max}} \\
Y_{\mathrm{r,max}} & Y_{\mathrm{g,max}} & Y_{\mathrm{b,max}} \\
Z_{\mathrm{r,max}} & Z_{\mathrm{g,max}} & Z_{\mathrm{b,max}}
\end{bmatrix}
\begin{bmatrix} R \\ G \\ B \end{bmatrix} \tag{7.6}
$$

Equation (7.4) relates digital and radiometric data. This relationship has been referred to as a tone-reproduction curve (TRC), such as in ICC color management [20] or optoelectronic transfer function (OETF). OETF will be used in this chapter. Thus the OETF is defined by the 'GOG' model.

In practice, the system gain, system offset, and gamma values are estimated statistically. Accordingly, the terms can be generalized:

$$
\psi = (a\xi + b)^{\gamma} \tag{7.7}
$$

It is useful to think about the relationship between digital and spectral or colorimetric data as a two-stage process. There is a nonlinear relationship between digital counts and the radiometric scalars (equations (7.4)) followed by a linear transformation between the device's scalars and spectral radiance output (equation (7.5)) or CIE tristimulus values (equation (7.6)) [21].

7.3 THE EXTENDED GOGO MODEL

The GOG model described in Section 7.2 is based on the assumption that both the proportionality and additivity rules hold, particularly that individual channel ramps have identical chromaticities. In some cases these assumptions do not hold. Quite often the model does not perform well where some amount of luminance can be detected at the black level, commonly caused by incorrect setting of the brightness and contrast controls on the monitor [15]. That is, the system gain is less than unity and the system offset is greater than zero [equation (7.4)]. Here, the black level is defined as the luminance level with the digital signal counts of all channels set to zero, and is expressed as in equation (7.8), which is the sum of the minimum excitation from red, green and blue phosphors.

$$\begin{bmatrix} X \\ Y \\ Z \end{bmatrix}_{k,min} = \begin{bmatrix} X \\ Y \\ Z \end{bmatrix}_{r,phosphor,min} + \begin{bmatrix} X \\ Y \\ Z \end{bmatrix}_{g,phosphor,min} + \begin{bmatrix} X \\ Y \\ Z \end{bmatrix}_{b,phosphor,min} \qquad (7.8)$$

Even when the input data has a non-zero signal value for the red component and zero signal values for green and blue, some luminance from the green and blue channels may also be produced. We call these components unwanted emissions. The measurement value for the single channel can be expressed as:

$$\begin{bmatrix} X \\ Y \\ Z \end{bmatrix}_{r,meas} = \begin{bmatrix} X \\ Y \\ Z \end{bmatrix}_{r,phosphor} + \left(\begin{bmatrix} X \\ Y \\ Z \end{bmatrix}_{g,phosphor,min} + \begin{bmatrix} X \\ Y \\ Z \end{bmatrix}_{b,phosphor,min} \right)$$

$$\begin{bmatrix} X \\ Y \\ Z \end{bmatrix}_{g,meas} = \begin{bmatrix} X \\ Y \\ Z \end{bmatrix}_{g,phosphor} + \left(\begin{bmatrix} X \\ Y \\ Z \end{bmatrix}_{b,phosphor,min} + \begin{bmatrix} X \\ Y \\ Z \end{bmatrix}_{r,phosphor,min} \right) \qquad (7.9)$$

$$\begin{bmatrix} X \\ Y \\ Z \end{bmatrix}_{b,meas} = \begin{bmatrix} X \\ Y \\ Z \end{bmatrix}_{b,phosphor} + \left(\begin{bmatrix} X \\ Y \\ Z \end{bmatrix}_{r,phosphor,min} + \begin{bmatrix} X \\ Y \\ Z \end{bmatrix}_{g,phosphor,min} \right)$$

In spectral notation, it can be expressed as:

$$L(\lambda)_{r,meas} = L(\lambda)_{r,phosphor} + (L(\lambda)_{g,phosphor,min} + L(\lambda)_{b,phosphor,min})$$

$$L(\lambda)_{g,meas} = L(\lambda)_{g,phosphor} + (L(\lambda)_{r,phosphor,min} + L(\lambda)_{b,phosphor,min}) \qquad (7.10)$$

$$L(\lambda)_{b,meas} = L(\lambda)_{b,phosphor} + (L(\lambda)_{r,phosphor,min} + L(\lambda)_{g,phosphor,min})$$

When we measure the luminance of the single channel, the measurement device (or the eye) detects the luminance from all channels, and there is no way to separate the unwanted emissions from the true measurement value, expressed as integrated tristimulus values. As seen from equation (7.10), both the proportionality and additivity rules fail when the measurement data are used for characterization. The theoretical GOG model, equations (7.4)–(7.6), is no longer able to describe the CRT monitor characteristics in such situations. Woolfe, *et al.* performed principal component analysis on the measurement spectral data to separate these unwanted emission components [17]. Their method requires accurate spectral measurements to perform the statistical analysis.

Where unwanted emissions from other channels are present, the measured tristimulus values $(X_{r,meas}, Y_{r,meas}, Z_{r,meas})$ of the red channel can be expressed as below:

$$X_{r,meas} = X_{r,phosphor} + X_{g,phosphor,\min} + X_{b,phosphor,\min}$$

$$= X_{r,phosphor,\max} \left(a_r \cdot \frac{\text{LUT}(d_r)}{2^n - 1} + b_r \right)^{\gamma_r} + X_{g,phosphor,\min} + X_{b,phosphor,\min}$$

$$\tag{7.11a}$$

Similarly, for Y and Z:

$$Y_{r,meas} = Y_{r,phosphor,\max} \left(a_r \cdot \frac{\text{LUT}(d_r)}{2^n - 1} + b_r \right)^{\gamma_r} + Y_{g,phosphor,\min} + Y_{b,phosphor,\min}$$

$$Z_{r,meas} = Z_{r,phosphor,\max} \left(a_r \cdot \frac{\text{LUT}(d_r)}{2^n - 1} + b_r \right)^{\gamma_r} + Z_{g,phosphor,\min} + Z_{b,phosphor,\min}$$

$$\tag{7.11b}$$

where:

$$\begin{cases} X_{r,phosphor,\min} = X_{r,phosphor,\max} \cdot b_r^{\gamma_r} \\ Y_{r,phosphor,\min} = Y_{r,phosphor,\max} \cdot b_r^{\gamma_r} \\ Z_{r,phosphor,\min} = Z_{r,phosphor,\max} \cdot b_r^{\gamma_r} \end{cases} \text{when } b_r > 0, \text{ and} \quad \begin{cases} X_{r,phosphor,\min} = 0 \\ Y_{r,phosphor,\min} = 0 \\ Z_{r,phosphor,\min} = 0 \end{cases} \text{when } b_r \leqslant 0$$

$$\begin{cases} X_{g,phosphor,\min} = X_{g,phosphor,\max} \cdot b_g^{\gamma_g} \\ Y_{g,phosphor,\min} = Y_{g,phosphor,\max} \cdot b_g^{\gamma_g} \\ Z_{g,phosphor,\min} = Z_{g,phosphor,\max} \cdot b_g^{\gamma_g} \end{cases} \text{when } b_g > 0, \text{ and} \quad \begin{cases} X_{g,phosphor,\min} = 0 \\ Y_{g,phosphor,\min} = 0 \\ Z_{g,phosphor,\min} = 0 \end{cases} \text{when } b_g \leqslant 0$$

$$\begin{cases} X_{b,phosphor,\min} = X_{b,phosphor,\max} \cdot b_b^{\gamma_b} \\ Y_{b,phosphor,\min} = Y_{b,phosphor,\max} \cdot b_b^{\gamma_b} \\ Z_{b,phosphor,\min} = Z_{b,phosphor,\max} \cdot b_b^{\gamma_b} \end{cases} \text{when } b_b > 0, \text{ and} \quad \begin{cases} X_{b,phosphor,\min} = 0 \\ Y_{b,phosphor,\min} = 0 \\ Z_{b,phosphor,\min} = 0 \end{cases} \text{when } b_b \leqslant 0$$

By normalizing by the maximum measurement value, we get:

$$R'(X) = \frac{X_{r,meas}}{X_{r,meas,\max}} = \frac{X_{r,phosphor} + X_{g,phosphor,\min} + X_{b,phosphor,\min}}{X_{r,phosphor,\max} + X_{g,phosphor,\min} + X_{b,phosphor,\min}}$$

$$= \frac{X_{r,phosphor,\max} \left(a_r \cdot \dfrac{\text{LUT}(d_r)}{2^n - 1} + b_r \right)^{\gamma_r} + X_{g,phosphor,\max} \cdot b_g^{\gamma_g} + X_{b,phosphor,\max} \cdot b_b^{\gamma_b}}{X_{r,phosphor,\max} + X_{g,phosphor,\max} \cdot b_g^{\gamma_g} + X_{b,phosphor,\max} \cdot b_b^{\gamma_b}}$$

$$\tag{7.12a}$$

Similarly, for Y and Z:

$$R'(Y) = \frac{Y_{r,phosphor,\max}\left(a_r \cdot \dfrac{\mathrm{LUT}(d_r)}{2^n - 1} + b_r\right)^{\gamma_r} + Y_{g,phosphor,\max} \cdot b_g^{\gamma_g} + Y_{b,phosphor,\max} \cdot b_b^{\gamma_b}}{Y_{r,phosphor,\max} + Y_{g,phosphor,\max} \cdot b_g^{\gamma_g} + Y_{b,phosphor,\max} \cdot b_b^{\gamma_b}}$$

$$R'(Z) = \frac{Z_{r,phosphor,\max}\left(a_r \cdot \dfrac{\mathrm{LUT}(d_r)}{2^n - 1} + b_r\right)^{\gamma_r} + Z_{g,phosphor,\max} \cdot b_g^{\gamma_g} + Z_{b,phosphor,\max} \cdot b_b^{\gamma_b}}{Z_{r,phosphor,\max} + Z_{g,phosphor,\max} \cdot b_g^{\gamma_g} + Z_{b,phosphor,\max} \cdot b_b^{\gamma_b}}$$

$$(7.12b)$$

From equations (7.12a) and (7.12b), it may be seen that these OETFs have the second offset term outside the parentheses, thus requiring a more elaborate model known as the GOGO (gain-offset-gamma-offset) model [16]:

$$\psi = (a \cdot \xi + b)^{\gamma} + c \tag{7.13}$$

As the X value for the red channel is larger than Y or Z value, the Y value for the green channel is larger than X or Z value, and the Z value for the blue channel is larger than X or the Y value, we have:

$$X_{r,phosphor,\max} \gg X_{g,phosphor,\max}, \ X_{b,phosphor,\max}$$
$$Y_{g,phosphor,\max} \gg Y_{r,phosphor,\max}, \ Y_{b,phosphor,\max}$$
$$Z_{b,phosphor,\max} \gg Z_{r,phosphor,\max}, \ Z_{g,phosphor,\max}$$

and if we assume that black is close to an achromatic color, we have:

$$b_r^{\gamma_r} \cong b_g^{\gamma_g} \cong b_b^{\gamma_b}$$

Therefore, it is not guaranteed that second offset terms in these three TRCs will be the same. It is usually the case that second offset term in equation (7.12b) is larger than that of equation (7.12a), in which case we no longer have $R'(X) = R'(Y) = R'(Z)$, since the spectral radiance at any level of phosphor excitation cannot be related to the maximum spectral radiance by a scalar. This indicates that we cannot simply use the additivity rule as in equation (7.4) with this form of TRC. However, if we examine equations (7.11a) and (7.11b) again, we could reorganize the equation as:

$$X_{r,meas} = X_{r,phosphor,\max}\left(a_r \cdot \frac{\mathrm{LUT}(d_r)}{2^n - 1} + b_r\right)^{\gamma_r} + X_{g,phosphor,\min} + X_{b,phosphor,\min}$$

$$= X_{r,phosphor,\max}\left(a_r \cdot \frac{\mathrm{LUT}(d_r)}{2^n - 1} + b_r\right)^{\gamma_r}$$

$$+ (X_{r,phosphor,\min} + X_{g,phosphor,\min} + X_{b,phosphor,\min}) - X_{r,phosphor,\min}$$

$$= \left\{ X_{r,phosphor,\max}\left(a_r \cdot \frac{\mathrm{LUT}(d_r)}{2^n - 1} + b_r\right)^{\gamma_r} - X_{r,phosphor,\min} \right\} + X_{k,\min}$$

$$(7.14a)$$

Similarly, for Y and Z:

$$Y_{r,meas} = \left\{ Y_{r,phosphor,max} \left(a_r \cdot \frac{\text{LUT}(d_r)}{2^n - 1} + b_r \right)^{\gamma_r} - Y_{r,phosphor,min} \right\} + Y_{k,min}$$

$$Z_{r,meas} = \left\{ Z_{r,phosphor,max} \left(a_r \cdot \frac{\text{LUT}(d_r)}{2^n - 1} + b_r \right)^{\gamma_r} - Z_{r,phosphor,min} \right\} + Z_{k,min}$$

(7.14b)

Therefore, we get:

$$X_{r,meas} - X_{k,min} = X_{r,phosphor,max} \left(a_r \cdot \frac{\text{LUT}(d_r)}{2^n - 1} + b_r \right)^{\gamma_r} - X_{r,phosphor,min}$$

$$= X_{r,phosphor} - X_{r,phosphor,min}$$

$$Y_{r,meas} - Y_{k,min} = Y_{r,phosphor,max} \left(a_r \cdot \frac{\text{LUT}(d_r)}{2^n - 1} + b_r \right)^{\gamma_r} - Y_{r,phosphor,min}$$

$$= Y_{r,phosphor} - Y_{r,phosphor,min}$$

(7.15)

$$Z_{r,meas} - Z_{k,min} = Z_{r,phosphor,max} \left(a_r \cdot \frac{\text{LUT}(d_r)}{2^n - 1} + b_r \right)^{\gamma_r} - Z_{r,phosphor,min}$$

$$= Z_{r,phosphor} - Z_{r,phosphor,min}$$

At the maximum value, we have:

$$X_{r,meas,max} - X_{k,min} = X_{r,phosphor,max} - X_{r,phosphor,min}$$

$$Y_{r,meas,max} - Y_{k,min} = Y_{r,phosphor,max} - Y_{r,phosphor,min}$$

(7.16)

$$Z_{r,meas,max} - Z_{k,min} = Z_{r,phosphor,max} - Z_{r,phosphor,min}$$

If we now divide both sides of equations (7.14a) and (7.14b) by the maximum value, we obtain:

$$R''(X) = \frac{X_{r,meas} - X_{k,min}}{X_{r,meas,max} - X_{k,min}}$$

$$= \frac{X_{r,phosphor,max}}{X_{r,phosphor,max} - X_{r,phosphor,min}} \left(a_r \cdot \frac{\text{LUT}(d_r)}{2^n - 1} + b_r \right)^{\gamma_r}$$

(7.17a)

$$- \frac{X_{r,phosphor,min}}{X_{r,phosphor,max} - X_{r,phosphor,min}}$$

$$= \left(a_r' \cdot \frac{\text{LUT}(d_r)}{2^n - 1} + b_r' \right)^{\gamma_r} - c_r$$

where:

$$a'_r = a_r \left(\frac{X_{r,phosphor,\max}}{X_{r,phosphor,\max} - X_{r,phosphor,\min}} \right)^{1/\gamma_r} = \frac{a_r}{(1 - b_r^{\gamma_r})^{1/\gamma_r}}$$

$$b'_r = b_r \left(\frac{X_{r,phosphor,\max}}{X_{r,phosphor,\max} - X_{r,phosphor,\min}} \right)^{1/\gamma_r} = \frac{b_r}{(1 - b_r^{\gamma_r})^{1/\gamma_r}}$$

$$c_r = \frac{X_{r,phosphor,\min}}{X_{r,phosphor,\max} - X_{r,phosphor,\min}} = \frac{b_r^{\gamma_r}}{1 - b_r^{\gamma_r}} = (b'_r)^{\gamma_r}$$

Similarly, for Y and Z:

$$R''(Y) = \left(a'_r \cdot \frac{\mathrm{LUT}(d_r)}{2^n - 1} + b'_r \right)^{\gamma_r} - c_r$$

$$R''(Z) = \left(a'_r \cdot \frac{\mathrm{LUT}(d_r)}{2^n - 1} + b'_r \right)^{\gamma_r} - c_r$$

(7.17b)

Therefore, we now have $R'' = R''(X) = R''(Y) = R''(Z) = \left(a'_r \cdot \frac{\mathrm{LUT}(d_r)}{2^n - 1} + b'_r \right)^{\gamma_r} - c_r$. TRCs for the green and blue channels can be obtained similarly as below:

$$R'' = \left(a'_r \cdot \frac{\mathrm{LUT}(d_r)}{2^n - 1} + b'_r \right)^{\gamma_r} - c_r$$

$$G'' = \left(a'_g \cdot \frac{d_g}{2^n - 1} + b'_g \right)^{\gamma_g} - c_g$$

(7.18)

$$B'' = \left(a'_b \cdot \frac{d_b}{2^n - 1} + b'_b \right)^{\gamma_b} - c_b$$

We can now use the additivity rules with R'', G'', B'' values as shown below:

$$\begin{bmatrix} X \\ Y \\ Z \end{bmatrix}_{meas} = \begin{bmatrix} X_{r,phosphor,\max} - X_{r,phosphor,\min} & X_{g,phosphor,\max} - X_{g,phosphor,\min} & X_{b,phosphor,\max} - X_{b,phosphor,\min} \\ Y_{r,phosphor,\max} - Y_{r,phosphor,\min} & Y_{g,phosphor,\max} - Y_{g,phosphor,\min} & Y_{b,phosphor,\max} - Y_{b,phosphor,\min} \\ Z_{r,phosphor,\max} - Z_{r,phosphor,\min} & Z_{g,phosphor,\max} - Z_{g,phosphor,\min} & Z_{b,phosphor,\max} - Z_{b,phosphor,\min} \end{bmatrix} \begin{bmatrix} R'' \\ G'' \\ B'' \end{bmatrix} + \begin{bmatrix} X_{k,\min} \\ Y_{k,\min} \\ Z_{k,\min} \end{bmatrix}$$

$$= \begin{bmatrix} X_{r,phosphor,\max} - X_{r,phosphor,\min} & X_{g,phosphor,\max} - X_{g,phosphor,\min} & X_{b,phosphor,\max} - X_{b,phosphor,\min} & X_{k,\min} \\ Y_{r,phosphor,\max} - Y_{r,phosphor,\min} & Y_{g,phosphor,\max} - Y_{g,phosphor,\min} & Y_{b,phosphor,\max} - Y_{b,phosphor,\min} & Y_{k,\min} \\ Z_{r,phosphor,\max} - Z_{r,phosphor,\min} & Z_{g,phosphor,\max} - Z_{g,phosphor,\min} & Z_{b,phosphor,\max} - Z_{b,phosphor,\min} & Z_{k,\min} \end{bmatrix} \begin{bmatrix} R'' \\ G'' \\ B'' \\ 1 \end{bmatrix}$$

(7.19)

From equation (7.16), the above can be expressed only in terms of the measurement data, instead of theoretical values, as follows:

$$
\begin{bmatrix} X \\ Y \\ Z \end{bmatrix}_{meas} = \begin{bmatrix} X_{r,max,meas} - X_{k,min} & X_{g,max,meas} - X_{k,min} & X_{b,max,meas} - X_{k,min} & X_{k,min} \\ Y_{r,max,meas} - Y_{k,min} & Y_{g,max,meas} - Y_{k,min} & Y_{b,max,meas} - Y_{k,min} & Y_{k,min} \\ Z_{r,max,meas} - Z_{k,min} & Z_{g,max,meas} - Z_{k,min} & Z_{b,max,meas} - Z_{k,min} & Z_{k,min} \end{bmatrix} \begin{bmatrix} R'' \\ G'' \\ B'' \\ 1 \end{bmatrix} \quad (7.20)
$$

This indicates that use of the GOGO model as given by equation (7.18) and the 3×4 matrix in equation (7.20) is necessary when the black level is detectable.

Reducing the monitor offset ('brightness' control) such that there isn't any measured black level reduces equations (7.18)–(7.20) to the theoretical CRT model, expressed by equations (7.4)–(7.6).

7.4 CALIBRATION, CHARACTERISATION, AND GAMMA

Calibration refers to achieving a predefined state for a display system. For example, one could specify a system gain of unity, system offset of zero, system gamma of 1.25, peak white with chromaticities equivalent to CIE Illuminant D_{65}, and peak white luminance of 75 cd/m^2. One then characterizes the system's properties (i.e. the model parameters of equations (7.4) and (7.5) or (7.6)), followed by digital or analog adjustments based on the differences between the characterization and the reference setup. Characterization is always required in order to define an image's color in a 'device-independent' fashion. The method of calibration depends on how the display system will be used. It may be more efficient for the system manufacturer to develop a color rendering dictionary (PostScript Level II) or color profile (ICC version 3.0) than the user. The profile is defined for a given setup and the user's task is then to validate or calibrate the system to the predefined setup.

As shown in equation (7.1), γ defines the inherent nonlinearity of all vacuum tubes, including CRTs, and traces back to derivations of Child and Langmuir around 1910 [1].* Unfortunately, γ is used today to represent the entire OETF expressed in equations (7.4) and (7.18). Suppose a monitor has its gun amplifier's offset adjusted (most commonly the 'contrast' control) with a slight negative bias to insure 0 DAC values yield 0 spectral radiance. This is shown in Figure 7.1. Below the monitor's cut-off voltage, the spectral radiance is 0. If one estimates the system gamma ignoring the system gain and offset terms for data above cut-off using least squares, significant errors result for dark colors as shown

---

* Recent measurements suggest that the Langmuir–Child law only applies to current density (where the cross-sectional area of the electron beam is held constant), not total current as assumed in equation (7.1). To compensate for this difference a more complicated model is required [9]. However, this departure from the Langmuir–Child law occurs at drive voltages close to the cut-off voltage, where typical inter-reflections and radiometric measurement uncertainty mask this error. Furthermore, any measurable errors are well below visual threshold, particularly when displaying pictorial images.

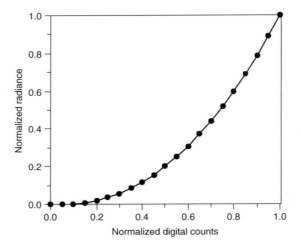

Figure 7.1   Data for a theoretical monitor setup with a slight negative system offset.

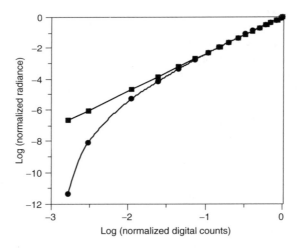

Figure 7.2   Estimated values (line with filled squares) and actual values (filled circles) for monitor data shown in Figure 7.1 when system gain and offset terms are ignored in the estimation of Γ.

in log-log axes Figure 7.2. In fact, this curvature has led to the incorrect conclusions that two γ values are required to describe this nonlinearity, or polynomial expansions of γ or that γ can be varied by changing the gun amplifier's settings. It is critical that one is aware of how the term 'gamma' is interpreted. Recognize that one can change the video look-up table values, 'contrast,' and 'brightness' in order to vary the display system's OETF despite the invariance of the cathode ray tube's γ.

7.4.1 Determining the colorimetry of each channel

The first step in characterizing a display system is to determine the matrix elements of equation (7.6), i.e. the tristimulus matrix. One method is to use published data of each channel's chromaticities and measurements of each channel's luminance as shown in equation (7.8). This reduces the cost of the instrumentation dramatically by requiring only a photometer. However, chromaticities usually are published for a given class of CRT; measurements of individual monitors can have very large difference from the published values. Hence this method is not recommended.

$$\begin{bmatrix} X_{r,max} & X_{g,max} & X_{b,max} \\ Y_{r,max} & Y_{g,max} & Y_{b,max} \\ Z_{r,max} & Z_{g,max} & Z_{b,max} \end{bmatrix} = \begin{bmatrix} (x_r/y_r) & (x_g/y_g) & (x_b/y_b) \\ 1 & 1 & 1 \\ (z_r/y_r) & (z_g/y_g) & (z_b/y_b) \end{bmatrix} \begin{bmatrix} L_{r,max} & 0 & 0 \\ 0 & L_{g,max} & 0 \\ 0 & 0 & L_{b,max} \end{bmatrix}$$

(7.21)

A second method is to use a colorimeter or spectroradiometer and measure each channel's peak output. The spectroradiometer should have good photometric linearity and wavelength accuracy within ± 0.5 nm for pictorial imagery [2]. Users of these devices should own radiance or irradiance standards and line sources to calibrate their instruments periodically or send them back to the manufacturer for periodic recalibration (see Chapter 2). Hopefully, the instrument's software is taking suitable account of the instrument's wavelength, sampling increment, range, and band-pass when calculating tristimulus values. Filter colorimeters should have spectral responsivities that are linear transformations of CIE color matching functions (all colorimeters attempt to match the 1931 2° observer). As a rule of thumb, the greater the number of filters, the closer the fit. It is possible to use a three-filter device with poor fit to estimate tristimulus value for displays with identical channel spectral properties by assuming that the inaccurate device is measuring the radiometric scalars. Multiplying the scalars by the tristimulus matrix obtained using an accurate device will result in reasonable tristimulus estimates [10].

Ideally, the colorimetric measurements should be verified with chromatic sources similar in spectral shape to CRT displays and not used to calibrate the measuring instrument. Single phosphor displays driven at high refresh rates are excellent verification sources.

The use of the linear matrix (equation (7.6)) has the underlying assumptions of channel independence (output from one channel does not affect another channel), spatial independence (output from one spatial location does not affect another spatial location), and constancy of each channel's chromaticities. Most monitors adhere to these assumptions to varying degrees. The following tests are recommended before trying to use this tristimulus matrix.

Before testing

Degauss the monitor, and roughly calibrate the display system. In particular, set the gains and offsets of the three channels' gun amplifiers to achieve approximately the required correlated color temperature, peak luminance, and black level. (The load placed on the

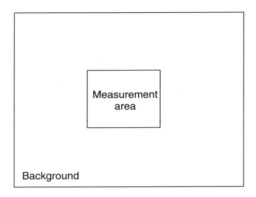

Figure 7.3   Image for evaluating channel and spatial independence.

power supplies and the beam's current and spatial properties have a large influence on a system's adherence to these assumptions.)

Channel independence test

Display an image as shown in Figure 7.3. The background is set to $d_r = d_g = d_b$ with a luminance of 20% of the peak white. The central square is arbitrary in size, though it is important that its area is larger than the measuring instrument's aperture in order to minimize inter-reflections by the background.* Display and measure the tristimulus values of each channel's maximum output and the system's peak white. If the system exhibits channel independence, the sum of the three channels should equal the measured peak white. Typically, a lack of channel independence is due to overdriving the gun amplifiers. This is remedied by decreasing each channel's gain (often labeled as the 'brightness' control).

Spatial independence test

The central stimulus is set to the system's peak white. The background is first set to $d_r = d_g = d_b = 0$ and the central stimulus is measured. The background is next set also to the peak white and the central stimulus re-measured. Differences between the two measurements indicate a lack of spatial independence. A lack of spatial independence is usually caused by power supply limitations, the extent of which is image dependent.

Discussion

All display systems deviate from the theoretically perfect system. Other assumptions and tests are described elsewhere [2]. One must decide whether the deviations are significant.

---

* A spectroradiometer or colorimeter can be used to measure potential inter-reflections by setting the central square to $d_r = d_g = d_b =$ min and making measurements in succession of the central square with the background set to $d_r = d_g = d_b = 0$ and $d_r = d_g = d_b = 20\%$ of peak white. 'Min' represents the minimum digital value that an instrument can measure with high precision. The measurement difference is an estimate of inter-reflections. This can be accounted for in the channel independence test.

These errors should be below $1 \Delta E^*_{ab}$ when using the monitor's peak white as an estimate of X_n, Y_n, and Z_n in order to calculate CIELAB coordinates. If they are much larger, one has the options of reducing the peak white's luminance, changing the image display (e.g. smaller pictorial images surrounded by gray), replacing the monitor, developing a correction algorithm where the matrix elements vary as a function of digital values (color) or image content (spatial), or replacing the matrix with a three-dimensional look-up table. The first three solutions are recommended. Clearly, each channel's colorimetric values should be determined using an image as similar as possible to the intended application. The image depicted in Figure 7.3 could be used to provide a global characterization for pictorial imagery based on the assumption that natural images, on average, spatially integrate to medium gray.

The above discussion has considered spatial uniformity. In general, the spectral radiance reduces from the centre to edges of most displays due to the increased path length of the electron beam. Monitor electronics and faceplate characteristics compensate for this with varying degrees of success. This non-uniformity mainly affects luminance and can be compensated by altering the tristimulus matrix (or luminance matrix shown in equation (7.21)) as a function of spatial location. The OETF parameters tend to not vary significantly as a function of spatial location [1, 2]. Care should be exercised when correcting spatial non-uniformities to ensure that one is not trading off low spatial frequency artifacts for higher spatial frequency artifacts that are more readily observed. As illustrated by the band-pass nature of the human luminance sensitivity function [11], an image slowly varying in luminance will appear uniform. If a correction introduces abrupt changes, which can occur with eight-bit-per-channel systems, objectionable banding may result. It is more practical to select a monitor with reasonable uniformity than to correct a poor monitor through software.

7.4.2 Determining the OETF of each channel

Determining the nonlinear relationship between the digital and radiometric values for each channel is the next requirement in order to characterize the overall colorimetry of the display system. As shown in equation (7.1), this relationship is complex; but fortunately, by making relative measurements, the relationship reduces to either the GOG or GOGO model. In theory, a minimum of two (GOG) or three (GOGO) measurements are required to calculate the three model parameters for a given channel because the gain and offset values must sum to unity (a result of the normalization). In practice, it seems prudent to collect more data and to estimate the parameters statistically using least squares; this reduces adverse effects that might result if any of the data points were noisy or in error. When estimating model parameters, the values of each video LUT should be known.

Another method to determine the OETF is via direct measurement. For an eight-bit DAC, 256 measurements are required. The dangers with this technique are twofold. First, images take a certain time to stabilize; it is critical to ensure the system is temporally stable before making each measurement. For some display systems, this could increase the total measurement time dramatically. Fortunately, because the relationship is monotonic, one can reduce the number of required measurements by subsampling and then

using linear or nonlinear interpolation to estimate missing values. Seventeen measurements tend to be sufficient for this task [3–5, 18]. The second danger results from many
measurement devices having insufficient dynamic range and numerical precision to measure the low-luminance colors. Both the direct measurement and interpolation methods
are susceptible to this because measurements at 0 digital value are required. (This limitation in measurement equipment is minimized using a model; measurement errors from
an instrument with poor precision can be avoided.)

Having decided upon the method of estimating the OETF, one must decide which
colors to display. The obvious choice is to ramp each channel separately where colors
range from black to the full primary colour, i.e. red, green, or blue. Only a single channel
measurement device is required as shown in equation (7.22) for the red channel, where r_λ
is the detector's relative spectral responsivity. The channel's scalar is directly measured
at maximum output.

$$R = \int_\lambda L_{\lambda,\text{red}} r_\lambda \, d\lambda \tag{7.22}$$

The radiometric instrument should have good spectral response in the region of the
channel's maximum output. Most monitors use phosphors peaking around 450, 530, and
630 nm. Often photometers are used, though care must be taken to ensure reasonable
signal-to-noise ratios for the blue and red channels. If tristimulus colorimeters or spec-
troradiometers are used, the scalars can be calculated from tristimulus values as shown
in equations (7.23). Using all three tristimulus values rather than only Y (luminance)
improves signal-to-noise properties, since X and Z are closer than Y to the red and blue
phosphor peak wavelengths.

$$R = X/X_{\max}$$

$$G = Y/Y_{\max}$$

$$B = Z/Z_{\max} \tag{7.23}$$

A related scheme is to ramp each channel as above, but have the other two channels
set to a fixed level of output. For example, by having the green and blue channels fixed
at their peak output, ramping the red channel results in colors ranging from cyan to
white. This practice may alleviate dynamic range limitations though not signal-to-noise
limitations. The scalars for the ramped channel are calculated by subtracting the minimum
value from each measurement.

When one is using either a three-channel device, such as a television color analyzer or
tristimulus colorimeter, or a spectroradiometer, the three channels can be ramped simultaneously where the colors range from black to white. The scalars are estimated using
the inverse tristimulus matrix as shown in equation (7.24). This reduces the number of
measurements by two-thirds. However, it becomes more difficult to analyze whether each

channel has constant chromaticities.

$$
\begin{bmatrix} \hat{R} \\ \hat{G} \\ \hat{B} \end{bmatrix} = \begin{bmatrix} X_{r,phosphor,\max} & X_{g,phosphor,\max} & X_{b,phosphor,\max} \\ \quad - X_{r,phosphor,\min} & \quad - X_{g,phosphor,\min} & \quad - X_{b,phosphor,\min} \\ Y_{r,phosphor,\max} & Y_{g,phosphor,\max} & Y_{b,phosphor,\max} \\ \quad - Y_{r,phosphor,\min} & \quad - Y_{g,phosphor,\min} & \quad - Y_{b,phosphor,\min} \\ Z_{r,phosphor,\max} & Z_{g,phosphor,\max} & Z_{b,phosphor,\max} \\ \quad - Z_{r,phosphor,\min} & \quad - Z_{g,phosphor,\min} & \quad - Z_{b,phosphor,\min} \end{bmatrix}^{-1}
$$

$$
\times \left[\begin{bmatrix} X \\ Y \\ Z \end{bmatrix}_{\text{measured}} - \begin{bmatrix} X_{k,\min} \\ Y_{k,\min} \\ Z_{k,\min} \end{bmatrix} \right] \tag{7.24}
$$

7.5 NUMERICAL EXAMPLE OF A COLORIMETRIC CHARACTERIZATION

A Power Macintosh 7100/66 with a $16''$ Sony Trinitron monitor was used in conjunction with Adobe Photoshop 3.0. To recreate typical usage, the room was darkened and a pictorial image supplied by Photoshop was displayed; the image was viewed with Photoshop's gray background. The monitor's 'brightness' and 'contrast' controls were adjusted until the image's tone reproduction was pleasing. The adjustments were made by an observer with extensive photographic experience. Following several degaussings with the monitor's built-in degaussing coil, an image was created in Photoshop as shown in Figure 7.3. A Minolta CA-100 Color Analyzer was used in its colorimeter mode (Y, x, y) and its measurement head placed in contact with the faceplate. (By assuming the radiant output is diffuse, the CA-100 reports its measurements in units of cd/m^2.) Its 'universal' synchronization mode was used (100 ms sampling rate). The monitor was temporally stable, easily achieved by leaving it turned on. The colorimeter was warmed up for several hours before performing a zero calibration (in order to subtract its dark current from any reported readings).

Measurements first were made of each channel's peak output and the peak white. In order to evaluate channel independence, the measured and estimated peak white were compared with one another as shown in Table 7.1. The color difference between the measured and estimated values was $0.75 \Delta E_{ab}^*$; these results are very acceptable. These data are used to build the tristimulus matrix as shown in equation (7.25) where inter-reflection flare was assumed to be negligible.

$$
\begin{bmatrix} X_{\text{pixel}} \\ Y_{\text{pixel}} \\ Z_{\text{pixel}} \end{bmatrix} = \begin{bmatrix} 22.35 & 16.68 & 11.24 \\ 11.90 & 37.40 & 4.56 \\ 1.17 & 8.15 & 62.82 \end{bmatrix} \begin{bmatrix} R \\ G \\ B \end{bmatrix} \tag{7.25}
$$

Table 7.1 Measured and estimated peak white luminance
and chromaticities

Channel	Y (cd/m^2)	x	y
Red	11.9	0.631	0.336
Green	37.4	0.268	0.601
Blue	4.56	0.143	0.058
Measured peak white	53.8	0.286	0.307
Calculated peak white	53.86	0.285	0.306
Difference	−0.06	0.001	0.001

The second test was the spatial independence test where the central square was set to the peak white and the background was changed between black and white. The chromaticities remained unchanged while the luminance increased by 1.2 cd/m^2 corresponding to $0.87\Delta E^*_{ab}$. The background was also set to each channel's peak output with similar performance. These results indicate reasonable spatial independence.

Because the monitor exhibited reasonable independence, each channel's OETF was characterized by measuring grays to minimize the number of measurements. Seventeen neutrals ranging from black to white were measured. Each color was temporally stable, at most, after 15 seconds within the colorimeter's precision. Luminance varied up to ± 0.004 cd/m^2; chromaticities varied ± 0.001. This variability likely was a result of a lack of synchronization with the monitor's refresh rate. The two darkest colors were below the instrument's dynamic range as denoted by a flashing instrument display. The chromaticities were converted to tristimulus values and, using the inverse matrix of equation (7.25), each channel's scalars were estimated. Because this imaging system had a small amount of channel dependence, the scalars for the peak white did not equal unity (1.003, 0.999, and 0.990 for the red, green, and blue channels, respectively). Therefore, the scalars estimated using the tristimulus matrix were normalized by each channel's maximum value.

The colorimetric measurements and estimated normalized scalars are shown in Table 7.2 and plotted in Figure 7.4. Because the three channels did not have the identical nonlinear relationships, the chromaticities of the neutral scale were not constant. From examining the data and corresponding plot, it appears that the system was set up with a positive offset (because the measured values are all above 0) and that the video look-up tables may have values that are not smooth. On Macintosh systems, the video LUTs have entries that partially correct for the inherent CRT γ. For all Macintosh computers, the aim system 'gamma' is 1.8 when characterized without the gain and offset parameters. (It is unclear why Apple selected 1.8 as its default 'gamma'; it has no perceptual basis.) For example, if it is assumed that all CRTs have $\gamma = 2.2$ and perfect setup (eliminating the need for gain and offset terms), the video LUT can be directly calculated:

$$\text{integer}[255(d/255)^{1/(2.2/1.8)}]$$

Table 7.2 Measured luminance and chromaticities and estimated scalars for 17-step neutral scale

Digital counts	Y (cd/m^2)	x	y	R	G	B
0	0.1	0.272	0.299	0.002	0.002	0.002
16	0.13	0.251	0.245	0.002	0.002	0.004
32	0.51	0.257	0.242	0.010	0.009	0.016
48	1.58	0.269	0.267	0.030	0.028	0.040
64	3.3	0.275	0.281	0.063	0.059	0.075
80	5.49	0.277	0.288	0.103	0.100	0.118
96	8.15	0.279	0.293	0.151	0.149	0.169
112	11.4	0.281	0.296	0.213	0.209	0.230
128	15	0.282	0.299	0.278	0.277	0.296
144	18.8	0.283	0.301	0.347	0.348	0.366
160	23.1	0.283	0.302	0.426	0.428	0.446
176	27.8	0.284	0.303	0.515	0.515	0.532
192	32.6	0.284	0.304	0.601	0.606	0.620
208	37.8	0.285	0.305	0.701	0.702	0.712
224	43	0.285	0.306	0.795	0.800	0.807
240	48.3	0.285	0.306	0.890	0.900	0.903
255	53.8	0.286	0.307	1.000	1.000	1.000

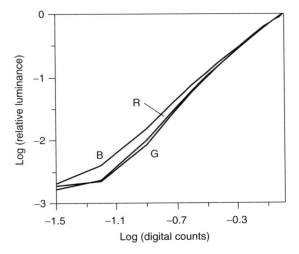

Figure 7.4 Logarithm of each channel's scalar vs. logarithm of digital counts for data shown in Table 7.2.

However, since display systems rarely have a perfect setup, errors result for small digital values. Thus, in practice, a 'gamma' of 1.8 is achieved by measuring the digital count luminance relationship of each channel for a representative monitor (13″ Trinitron, 16″ Trinitron, etc.) with the video LUTs set to values that do not alter the input digital values. The difference between the actual measurements normalized by their maximum output and $(d/255)^{1.8}$ is used to build the LUT entries. Consequently, if the particular display representing that class of monitors had a significant offset, the video LUT would have a kink at the digital values corresponding to the video amplifier offset. Although this LUT will achieve the aim tone reproduction characteristics for that monitor, it will only properly 'gamma correct' monitors that have the identical setup, a rare occurrence. The kink at 16 digital counts shown in Figure 7.4 demonstrates this problem.

As defined in equation (7.18), the four system parameters of a, b, c and γ for each channel were first estimated using only five measurements (32, 64, 112, 176, and 255 digital counts) as recommended by Berns et al. [1]. For all three channels, negative offsets, b, were estimated (-0.10, -0.12, and -0.08 for the red, green, and blue channels, respectively); offsets, c, were negligible. This was unexpected and contradicted the measurements shown in Table 7.2 where the values never reached zero. Upon closer inspection, it was noticed that there was an offset to all the measurements, suggesting optical flare or an incorrectly zeroed instrument. The measurements were repeated using the Minolta Color Analyzer and also using an LMT C1200 Colorimeter (with sufficient dynamic range to measure 0 digital counts) with similar results, thus eliminating the suspicion that the instrument was not properly zeroed. The optical flare was a result of using the image shown in Figure 7.3. Through inter-reflections, the background was contributing optical flare of about $0.10\,\mathrm{cd/m^2}$ to the central field.

Given that for any pixel, neighboring pixels will contribute flare, this flare is probably realistic and should be included in the characterization based on the assumption that, on average, the inter-reflections are equal to a middle gray. As a consequence, equation (7.19) was used rather than equation (7.6). The minimum tristimulus values (at 0 digital counts) were subtracted from each measured value and the tristimulus matrix recalculated as shown in equation (7.26).

$$\begin{bmatrix} X_{\text{pixel}} \\ Y_{\text{pixel}} \\ Z_{\text{pixel}} \end{bmatrix} = \begin{bmatrix} 0.09 \\ 0.01 \\ 0.14 \end{bmatrix}_{\substack{\text{inter-}\\ \text{reflection}\\ \text{flare}}} + \begin{bmatrix} 22.26 & 16.59 & 11.15 \\ 11.80 & 37.30 & 4.46 \\ 1.03 & 8.01 & 62.67 \end{bmatrix} \begin{bmatrix} R \\ G \\ B \end{bmatrix} \qquad (7.26)$$

This new matrix (inverted) was used to estimate each channel's scalar. Following normalization, the scalars ranged between zero and unity. The four model parameters were estimated using the five measurements, all of the measurements where the Minolta has sufficient dynamic range (32 digital counts and above), and just two measurements at 48 and 176 digital counts. For these estimations of OETF, the parameters a and b were defined using one degree of freedom; in other words, a was estimated and b was calculated by $1 - a$. A simplex algorithm available with the SYSTAT [12] statistics software package was used to perform the nonlinear optimizations. Starting values of 0.8 for a,

Table 7.3 GOG Model parameters and model fit for neutral data listed in Table 7.2

	k_g	γ	n	Average ΔE_{ab}^*	Maximum ΔE_{ab}^*	Average $\|\%\Delta Y\|$	Maximum $\|\%\Delta Y\|$	Average $\sqrt{\Delta u'^2 + \Delta v'^2}$	Maximum $\sqrt{\Delta u'^2 + \Delta v'^2}$
R	*	1.86	17	2.32	9.95	29.19	221.76	0.013	0.073
G	*	1.86							
B	*	1.77							
R	1.1	1.61	15	0.95	5.04	5.84	51.73	0.012	0.092
G	1.13	1.56							
B	1.1	1.55							
R	1.1	1.61	5	0.8	3.65	4.54	36.12	0.01	0.092
G	1.11	1.58							
B	1.08	1.57							
R	1.1	1.60	2	0.78	2.68	4.52	29.87	0.008	0.092
G	1.11	1.58							
B	1.08	1.55							

0.1 for c, and 2.0 for γ always yielded convergence. A γ as also estimated using all 17 measurements and not including the system gain and offset parameters.

Model parameters a and γ along with residual errors for all of the 17 data points are given in Table 7.3. In all cases, parameter c was approximately 0.002. The average and maximum ΔE_{ab}^* were calculated as perceptual metrics where the measured peak white was used to define X_n, Y_n, and Z_n. The luminance errors were expressed as the average and maximum $|\%\Delta Y|$. The average and maximum chromaticity errors were calculated by $\sqrt{\Delta u'^2 + \Delta v'^2}$. Ignoring model parameters a, b and c resulted in the worst estimates. Including these model parameters improved the performance by a factor of two. Reducing the number of data points to estimate each channel's OETF model parameters had a minimal effect, although there was a trend where the performance improved as the number of data points reduced. All of the estimates had their largest errors at 32 digital counts.

There are several 'lessons' resulting from the above characterization. The first concerns the importance of considering inter-reflections from neighboring image areas. The above simplistic approach, where inter-reflections are assumed to equal that resulting from a middle-gray background, seems like a practical and useful approach. The alternative would be a calculation that is image dependent, a very computationally intensive exercise. The second lesson concerns the utility of having a physical model that describes the relationship between input digital counts and output spectral radiance. The disconnect between the colorimetric measurements and the negative offset parameter b due to inter-reflection errors would not have been found if one directly built three LUTs. Also, the negative offsets indicate that the first 24 or so digital counts will all map to the black level, i.e. clipping. This is easily determined by using the GOGO model parameters and calculating when the scalars turn positive. Better tone reproduction will result by adjusting the monitor's 'contrast' to achieve b values close to -0.01. The third lesson has to do

Table 7.4 Model performance of independent data set

	n	Average ΔE_{ab}^*	Maximum ΔE_{ab}^*	Average $\|\%\Delta Y\|$	Maximum $\|\%\Delta Y\|$	Average $\sqrt{\Delta u'^2 + \Delta v'^2}$	Maximum $\sqrt{\Delta'^2 + \Delta v'^2}$
Simple Γ	17	2.17	3.76	5.86	14.73	0.004	0.010
GOG	15	0.54	1.06	0.76	2.55	0.001	0.003
GOG	5	0.72	1.23	1.41	3.38	0.001	0.003
GOG	2	1.05	1.66	2.44	6.00	0.002	0.005
3@LUT	17	0.42	1.06	0.46	1.47	0.001	0.003

with measuring neutrals rather than individual channels. The excellent modeling results given in Table 7.4 and the savings in time are obvious benefits. The fourth lesson has to do with video LUTs containing unknown values. The GOG model assumes that the video LUTs either do not alter the input values or have entries in them that are calculated using parameters identical to a given model (such as would result from just an exponential correction). If the LUTs have different functions, the GOG model will not give accurate results, especially for neutral colors.

7.5.1 Characterization summary

The following is a list of practices to characterize computer-controlled CRT displays.

1. Purchase the best colorimetric or spectral measurement system one can afford.
2. Purchase a monitor with its peak white chromaticities close to calibration specifications. If there isn't a specification, select a monitor with chromaticities near D_{65}. (Our adaptation is most complete here. Thus the monitor's peak output will appear 'white' rather than bluish or yellowish; the gray scale similarly will appear neutral throughout its range.)
3. When using Macintosh systems, use software such as Knoll Gamma to rewrite the video LUTs to smooth values. Changing any of the setting from default ('gamma adjust' $= 0$, 'black point' $= 0$, etc.) will easily accomplish this. It is also likely that many of the ICC visual profiling tools also result in smooth values.
4. Approximately set up the display system visually. First turn the CRT's controls to their minimum. Turn off ambient lighting and display a 'black' image ($d_r = d_g = d_b = 0$). Adjust the nested gun amplifier offset control ('brightness') to just above the setting when you can begin to perceive light. Display a text image and adjust the nested gun amplifier gain ('contrast') to maximize luminance without a loss of image sharpness. In most cases this is its maximum setting.
5. Display an image as shown in Figure 7.3 with the background set to 0 digital counts. Measure the tristimulus values for the peak output of each channel (red, green, and blue), the peak white, the darkest neutral that can be measured with good precision and accuracy, and an intermediate neutral.

6. Calculate the tristimulus matrix and its inverse. Calculate whether the measured white is estimated by the addition of the three channels. Use the inverse matrix to estimate each channel's scalars from the neutral measurements. Using a mathematics or statistics package, calculate the GOGO model parameters where $b = 1 - a$ for each channel. This is a direct calculation solving simultaneous equations with three unknowns and three data points. (This is suggested rather than using nonlinear optimization to avoid potential convergence problems.)

7. If the display does not exhibit good channel independence (additivity), reduce the gain control. Adjust the offset control in the proper direction so that the b model offsets are close to zero but still negative (between about -0.01 and -0.03). Repeat steps 4–6 in order to achieve a proper setup. (The idea is that neutral colors are used to set up the display properly.)

8. Change the background to an approximate middle gray (20% of the monitor peak white). One can use the GOGO parameters to determine the appropriate digital values. Re-measure the darkest neutral and evaluate whether there is any inter-reflection flare present.

9. Measure red, green, and blue ramps. The number of measurements depends on whether one wants to build three LUTs from direct measurements or from enumerating the GOGO model once the parameters are estimated. In the former case, 17 even increments in conjunction with piecewise linear or cubic interpolation result in an acceptable description of the display system's three OETFs. When estimating the GOGO parameters, collect between five and 17 measurements.

10. Plot the chromaticities of each ramp. If inter-reflections are negligible and if the black level, that is the 'brightness' control, has been properly set, the chromaticities should be constant. If there is a systematic deviation as the digital values decrease, black-level flare needs to be subtracted from the measurements. (Many instruments have insufficient precision to measure low-luminance stimuli. In this case, the flare can be statistically estimated. Essentially, the flare is estimated minimizing the variance in all three channel's chromaticities [19]. The flare-free data are used to build LUTs or estimate model parameters.

7.6 CONCLUSIONS

Colorimetrically characterizing and calibrating computer-controlled displays requires a fair amount of work initially in order to gain experience and learn what are reasonable trade-offs between instrumentation, characterization methodologies, monitor properties, and performance. Once the procedures are optimized for a given display system, the time required for characterization and calibration is modest. If one is serious about achieving high-accuracy color reproduction, the effort is worthwhile.

REFERENCES

1. Berns, R. S., Motta, R. J. and Gorzynski, M. E. (1993) CRT colorimetry, part I: theory and practice. *Col. Res. Appl.*, **18**, 299–314.

2. Berns, R. S., Motta, R. J. and Gorzynski, M. E. (1993) CRT colorimetry, part II: metrology. *Col. Res. Appl.*, **18**, 315–325.
3. Post, D. L. and Calhoun, C. S. (1989) An evaluation of methods for producing desired colors on CRT monitors. *Color Res. Appl.*, **14**, 172–186.
4. Lucassen, M. P. and Walraven, J. (1990) Evaluation of a simple method for color monitor recalibration. *Color Res. Appl.*, **15**, 321–326.
5. Post, D. L. (1992) Colorimetric, measurement, calibration, and characterization of self-luminous displays. In *Color in Electronic Displays*, H. Widdel and D. L. Post (eds), New York: Plenum Press.
6. Travis, D. (1991) *Effective Color Displays Theory and Practice*. Academic Press.
7. Sproson, W. N. (1983) *Colour Science in Television and Display Systems*. Adam Hilger.
8. Commission Internationale de l'Eclairage (CIE) (1994) *Colorimetry of self-luminous displays – a bibliography*. Publication no. CIE 87, Vienna, Austria: CIE.
9. Olson, T. (1994) *Behind gamma's disguise*. Society of Motion Picture and Television Engineers, Preprint No. 136-22.
10. American Society for Testing and Materials (1992) *Practice for obtaining colorimetric data from a visual display unit using tristimulus colorimeters*, ASTM Designation, E1455–92.
11. Kelly, D. H. (1994) *Visual Science and Engineering: Models and Applications*, Chapter 3. Marcel Dekker, New York.
12. Wilkinson, L. (1992) *SYSTAT: The System for Statistics*. Evanston, IL: SYSTAT, Inc.
13. Stokes, M., Fairchild, M. D. and Berns, R. S. (1992) Colorimetric quantified visual tolerances for pictorial images. In *Comparison of Color Images Presented in Different Media*, Vol. 2, M. Pearson (ed.), Proceedings 1992, pp. 757–777. Technical Assoc. Graphic Arts and Inter-Soc. Color Council.
14. CIE Publication 122 (1996) *The Relationship between Digital and Colorimetric Data for Computer-Controlled CRT Displays*
15. Katoh, N., Deguchi, T. and Berns, R. S. (2001) An Accurate Characterization of CRT Monitor: (I) Verifications of Past Studies and Clarifications of Gamma, *Optical Review*, **8**(5), 305–314.
16. Katoh, N., Deguchi, T. and Berns, R. S. (2001) An Accurate Characterization of CRT Monitor: (II) Proposal for an Extended CIE Method and its Verification, *Optical Review*, **8**(5), 397–408.
17. Woolfe, G. L., Berns, R. S., and Alessi, P. J. (1997) An Improved Method for CRT Characterization based on Spectral Data, *Proc. CIE Expert Symposium '97*, pp. 38–45.
18. IEC 61966-3 "Color measurement and management in multimedia systems and equipment, Part 3: equipment using CRT displays," (2000).
19. Fernandez, S. and Taplin, L. (2001) Personal communication.
20. ICC (2001) File format for colour profiles ICC.1:2001–04.
21. Berns, R. S. (1997) 'A generic approach to color modeling', *Color Res. Appl.*, **22**, 318–325.

Methods for characterizing colour scanners and digital cameras

Tony Johnson

8.1 INTRODUCTION

Whilst colour scanners are now used to capture images for output on a variety of media, their development has essentially taken place within a Graphic Arts environment. Thus many of the techniques developed for obtaining high quality colour reproductions have evolved from that background. In recent years these procedures have been somewhat generalized so that they can be used to define data in a 'colour space' more appropriate to multi-media applications, but the basic techniques used to define the colour transformation remain essentially unaltered, particularly where high quality is paramount. For that reason this introduction commences from a Graphic Arts perspective, before going on to discuss how this has been developed for multi-media applications.

Before commencing this introductory discussion it is worth clarifying some issues of terminology. Three terms will be used frequently in this chapter and it is important that the reader understands my definition of them in order to be clear on the intent. The terms are calibration, characterization and colour transformation.

1. *Calibration*. This defines the setting up of a scanner (or any other process) such that it gives repeatable data day in and day out. Calibration of a scanner or digital camera is

Colour Engineering, edited by P.J. Green and L.W. MacDonald.
© 2002 John Wiley & Sons Ltd.

an essential first step before any imaging process is carried out. In a photo-multiplier scanner it usually consists of setting the amplifier offset and slope such that white and black areas of known density produce a specific voltage; with a CCD scanner or camera it may also mean compensating for any non-uniformity in sensitivity of the individual elements of the array.

2. *Characterization.* This defines the relationship between the device 'colour space' and the CIE system of colour measurement. Thus, for a scanner or camera it normally defines the relationship between the voltages quantized as data recorded on disc and the CIE measurements of the colours scanned. For a hard copy output system it defines the relationship between the data on disc and the resultant CIE measurements of the colours achieved when these values are used to obtain an image on this output system. The characterization may be defined as a mathematical model based on a set of equations or a definition of a discrete number of points which constitute a look-up table.

3. *Colour transformation.* This is a general term for a mathematical relationship which enables data from one 'colour space' to be transformed into another. In the CIE system it may be used for the definition of a uniform colour space, for example, or in the context of this chapter it may be used for transforming data from one device 'colour space' to another. Thus it may be used for defining the transformation which converts the scanner voltages into the data needed to drive the output device in order to achieve a high quality colour reproduction. It is clear that a characterization provides a special-case colour transformation.

During much of this chapter the emphasis will be on scanners. The procedure for characterizing cameras, *when used as scanners*, is identical, but when the camera is used for imaging natural scenes the situation may get rather more complex because of metamerism. However, to be successful in such situations the camera really has to have a spectral sensitivity which is linearly related to that of the CIE standard colorimetric observer. If this is the case, and providing that any content-dependent image processing can be avoided, the test images described for scanner characterization will prove equally effective for characterizing cameras, and the procedure can be used as described in this paper. If the sensitivity is not that close, no specific test image can be recommended; the user will have to define his own to include samples typical of those being imaged and then use the procedures discussed in this paper for characterization. However, with such a camera colour image quality is likely to be poor for many colorants encountered. Evaluation of the colorimetric quality of cameras can be carried out, as described below, using the test images described in this chapter. It is recommended that those from a number of manufacturers be used to try and provide metameric origination.

8.1.1 Colour management of device-dependent and device-independent systems

Obtaining high quality colour reproductions is a fundamental requirement of any Graphic Arts pre-press system. It is not something which most scanner and system operators give much thought to; it comes from using their experience, in conjunction with a number of default colour transformations, for the printing processes they are preparing the reproductions for. However, to make this procedure work it is important for consistency to be maintained in the process. To this end calibration techniques have evolved to enable each stage of the process to be monitored and controlled.

A common Graphic Arts procedure is to directly compute, at the scanner, the amount of ink required to produce a specific colour when printed. The colour scanners and systems designed for this industry have enabled the user to manipulate the ink amounts, often differently in different colour regions, in order to achieve the desired colour transformation. During the past 50 years many attempts have been made to develop procedures for characterizing a system (which define the colour directly) but for a variety of reasons these have not proved successful. However, the trend towards multi-media publishing, the computer power now readily available and a better understanding of how to overcome the limitations of the CIE colour measurement system for Graphic Arts applications has caused re-assessment of this position.

With the development of colour management systems we are now seeing increasing use of colorimetric definitions for the data, rather than a device-specific definition. When data is encoded in this form it is often known as device-independent data; when it defines the ink amounts it is known as device-dependent data.

This terminology is actually rather misleading. With our current state of knowledge (particularly with regard to appearance matching and gamut compression) truly device-independent data cannot be achieved when the highest quality levels are required. However, it is to be expected that this will change in the future and the quality obtained from truly device-independent systems will improve as our level of knowledge increases. Furthermore, as we shall see later, such data is now commonly encoded in a device-dependent form, together with a tag which provides the device independence by specifying how it should be converted into a colorimetric domain. Nevertheless, the term device-independent data is a convenient shorthand for differentiating between colour encoding schemes in which a colour transformation has to be carried out prior to making separations or plates for printing and those which can be used directly.

There are clearly advantages and disadvantages to both approaches. Device-dependent data is generally an efficient way of describing an image for a specific process since no transformation is required at output. Furthermore, where more than three colorants are used and no unique solution exists for achieving a specific colour, such encoding will ensure that user requirements are maintained with minimal encoding. As an example, a border which should be printed in black ink will be when encoded in device-dependent form, but will be produced in cyan, magenta, yellow and black for device-independent systems unless some special encoding technique is employed. The most important advantage, however, is that quality is likely to be optimized for that specific process in terms of both colour accuracy and quantization artefacts.

Device-independent data is seen as being more appropriate to multi-media environments providing that the data has sufficient precision for the colour space chosen. If the colour is defined in an unambiguous way, it follows that for each device on which the image is rendered the same colour should be achieved, providing that each device is properly calibrated, a satisfactory colour transformation is provided and the colour gamuts permit. However, it should be noted that the same can be achieved just as easily by transmitting device-dependent data together with a procedure (or data to define a procedure) which specifies how it may be converted for another device. This is then no different from the device-independent mode of working and in many ways offers the best of both worlds.

Colour management is the process of encoding data in a recognizable way and then transforming the data from one data encoding scheme to that which is appropriate for the device on which the colour is being rendered. It should be noted that many colour management systems do not encode the data in a colorimetric form. They obtain their device independence from a tag that accompanies the image which specifies the conversion to a colorimetric definition from the device-dependent data in which the image is encoded (be it monitor drive voltages, scanner RGB values or even ink amounts).

Attempts are being made to 'standardize' these tags so that they can be interpreted by any colour management system and used to define the colour transformation necessary to produce pleasing reproductions. These tags are becoming known in this context as device profiles. The standardization work needed to achieve openness comes from an industry consortium known as the International Color Consortium (ICC) and this initial effort is now gaining wide acceptance.

8.1.2 Principles of scanner characterization

We need to recognize that there are at least two distinct ways in which a scanner may be operated. The characterization procedure for each is different. The two methods are defined as follows:

1. *Colour digitizer.* In this mode the scanner has a simple objective: to capture the colour information of the original image being scanned for subsequent processing elsewhere. This is the device-independent mode of working and the output data must, therefore, bear some logical relationship to the original.
2. *'Gamut compressed' colour digitizer.* In this mode the scanner is effectively operated in a device-dependent manner. The application package associated with the scanner defines and applies the colour transformation which converts the RGB data directly into that required by the output device. It may directly define the colorant amount required (e.g. CMYK printing) or the exposure levels required (e.g. RGB transparency recorders) or the gamma-corrected drive voltages required (e.g. CRT displays). However, a special-case characterization may be defined in which the required gamut compression is applied to the original data but the data may still be transmitted as XYZ, $L^*a^*b^*$ or $L^*u^*v^*$ colour data rather than, say, colorant amount specification. This is the device-dependent mode of working and the output data in this mode must, therefore, bear some logical relationship to the reproduction.

Clearly such 'gamut-compressed' or device-dependent data can, in general, be transformed back into the original data. However, the transformation is unlikely to be simple and may produce artefacts depending upon the precision of the data and the accuracy of the transformation. But this is no different from the problems of transforming device-independent data to a specific output device.

The most efficient method for characterizing a scanner or digital camera is to image a set of colours of known tristimulus values. Using techniques to be described later, the data obtained from this imaging process can be compared to the tristimulus values of the test image and a colour transformation defined. Whether the characterization so defined is of the type 1 or 2 above depends on the values chosen for the reference tristimulus values. In the next section we will discuss a special image defined for this purpose.

8.2 USE OF THE ANSI/ISO SCANNER CHARACTERIZATION TARGET

Characterization of input devices can be conveniently achieved by use of the ISO 12641 targets. The target is described in more detail in Chapter 18.

To minimize metameric effects the response of the colour scanner should yield a linear relationship to the XYZ values of the input target. That is, RGB values obtained from the scanner should be linearly related to XYZ, which implies that the response of the scanner itself, V_r, V_g and V_b, should be linearly related to XYZ for each of the colour patches.

$$\begin{bmatrix} V_r \\ V_g \\ V_b \end{bmatrix} = \begin{bmatrix} a_{11} & a_{12} & a_{13} \\ b_{21} & b_{22} & b_{23} \\ c_{31} & c_{32} & c_{33} \end{bmatrix} \begin{bmatrix} X \\ Y \\ Z \end{bmatrix} \tag{8.1}$$

However, this is very seldom the case, even in high-end scanners, often because of the need to achieve high signal-to-noise ratios for specific output devices. This is the reason why the test image needs to be reproduced on each of the film materials. Few scanners are designed as colorimeters, since filters are normally chosen to minimize cross-talk and are therefore relatively narrowband when compared to colour matching functions. Since the dyes used in the various materials are generally metameric to each other, it is necessary, for any scanners with particularly narrowband filters, to have separate transformations for each material. Thus, the manufacturer of the colour scanner may provide some processing of the signal to achieve a specific relationship of the scanner output values to the target colours for each material. This may also include the application processing used to convert to the device-dependent output such as CMYK.

If the processing has *not* included any 'correction' for device dependency or gamut compression, or to a uniform colour space such as $L^*a^*b^*$, it should have modified V_r, V_g and V_b to the values to be stored on disk (S_r, S_g and S_b) such that:

$$\begin{bmatrix} S_r \\ S_g \\ S_b \end{bmatrix} = \begin{bmatrix} a_{14} & a_{15} & a_{16} \\ b_{24} & b_{25} & b_{26} \\ c_{34} & c_{35} & c_{36} \end{bmatrix} \begin{bmatrix} X \\ Y \\ Z \end{bmatrix} \tag{8.2}$$

However, the processing necessary to achieve the above characterization matrix, which is correct for every patch in the target, is likely to be highly non-linear and so the matrix, if attained, may be no more than an approximation.

It is a derivation of this matrix that may be defined in the ICC device profile discussed above as the procedure for processing the scanner data into its device-independent form.

The required profile (i.e. that which *must* be provided) is the relationship between the scanner data, S_r, S_g, and S_b, and the tristimulus values scanned by it as defined by the inverse of the simple linear matrix described above. Alternatively, because of the non-linearity mentioned earlier, a more complex characterization relationship may be defined as a look-up table which is effectively the relationship between the 'raw' scanner data, V_r, V_g and V_b, and the tristimulus values scanned by it.

As stated earlier, with many current scanners, the initial processing also includes the application processing for gamut compression and a transformation to CMYK or non-colorimetric device-dependent RGB. This then makes it impossible to approximate the characterization of the scanner by such a procedure as the simple matrix transformation specified above. For device-dependent output the relationship between CMYK or RGB (or tristimulus values of the output system) and *XYZ* of the input target is far more complex. Given specific rules for gamut compression, defined by a function f, then for every patch of the colour target we obtain the relationship:

$$(X, Y, Z)_g = f(X, Y, Z) \tag{8.3}$$

where $(X, Y, Z)_g$ represents the tristimulus values after gamut compression. For *CMYK* or non-colorimetric *RGB* data, the device characterization defines another function h, such that:

$$(X, Y, Z)_g = h(C, M, Y, K) \text{ or } h(R, G, B) \tag{8.4}$$

Thus the scanner output S_r, S_g, S_b, or S_c, S_m, S_y, S_k, is related to the original colour patches by the function:

$$S_r, S_g, S_b = (h - 1)^* f(X, Y, Z) \tag{8.5}$$

or

$$S_c, S_m, S_y, S_k(h - 1)^* f(X, Y, Z) \tag{8.6}$$

Via the work of the International Standards community, the data to provide the function h is being defined for various 'standard' printing processes and by device manufacturers; on the other hand, f is not agreed in any way and seems unlikely to be in the short term. This represents one of the areas where device manufacturers or colour management vendors can add value to their products.

8.3 USE OF THE TARGET IN SCANNER CHARACTERIZATION

The target has a number of applications. In particular it can be used for defining the scanner characterization (or more general colour transformation) as well as part of a scanner evaluation. In general the characterization either will be provided by the manufacturer of the scanner or will occur within the colour management system. The detail of any particular characterization procedure cannot be specified in this chapter; it depends upon the particular application. However, the sort of procedures which may be used for this will be discussed later in this chapter. In general terms, following calibration of the scanner, the image will be scanned and output to disk using default systems for processing to S_r, S_g, S_b (or S_c, S_m, S_y, S_k). This is followed by an analysis of the scanned data obtained from the target which is then compared to the tristimulus values of the target, thereby defining the characterization. It should be noted that whilst the encoded data has generally been scaled into 0–255 (8 bits), often with unknown 0% and 100% (white and black points), the target essentially contains this 0% and 100% information (patches 1 and 22 on the grey scale). The former is likely to be used for normalizing the data; the latter for scaling it.

The colour management applications package(s) provided would then use this characterization to obtain the 'correct' colour output. Note that in general this output includes gamut compression, and since this function is not generally agreed, any assessment must be subjective by the user. However, currently this is primarily determined at the time of output and is discussed in more detail in Chapter 13.

Some of the evaluation procedures for which the target may be used are worthy of a brief discussion. They are generally worth undertaking prior to a characterization procedure to ensure that the scanner is capable of giving acceptable quality.

Below we review analysis of the data obtained after calculating the colour transformation between the scanner data and the tristimulus values. Derivation of simple transformations was covered above, and more complex methods are described later under 'Characterization procedures'. The same sort of analysis can also be used to evaluate colour management systems themselves if the system enables the user to interrogate the intermediate tristimulus data.

In 'Evaluation of "gamut compressed" colour digitizer' we briefly discuss techniques for evaluation of scanners providing device-dependent data. Later we briefly mention two other tests which the calibration target can be used for, namely evaluation of signal-to-noise ratio and scanner repeatability.

8.3.1 Evaluation of colour digitizers

All scanners can be evaluated for their colorimetric capabilities if the data is obtainable. Such evaluation is worthwhile when trying to determine whether a poor quality of colour reproduction is caused by the scanner or by the colour management system. In general, if a scanner is a reasonable colour digitizer, it makes the colour management easier and more likely to be successful. Problems arise with such an evaluation for any scanner or colour management system which produces device-dependent data and for which the 'raw' *RGB* data is not easily accessible. Clearly, if these vendors wish to sell their product as a simple colour digitizer, they need to provide processing capabilities for this task. Where available this can be evaluated. However, what they are likely to provide in many cases is processing into gamut compressed data which is covered below.

For evaluating scanners which produce RGB data directly, for no specific output device, the testing would be based on the evaluation of linearity as defined above. To be a good colour digitizer, when no information exists with regard to the output device, the resultant RGB data should be a simple linear transformation of the *XYZ* values of the original. (In the event that the scanner processing nominally converts the data to CIELAB or CIELUV, or some similar uniform colour space, this would need to be converted to *XYZ* prior to evaluation.)

For those scanners which include an applications package for calibration, or for the derivation of tristimulus values, this must be used to calibrate the system first. For those which do not, and are therefore expected to work with any colour management package, the scanner data is used directly in the evaluation. Any instructions given by the scanner vendor for calibration should be carried out immediately prior to any testing.

After obtaining a table of colour values, the next step is to test these colour values for a linear relationship to the XYZ values of the standard target. In general, this should yield very high r^2 values (>0.95 in view of the number of data points involved) if a linear relationship is to be assumed.

In many cases, the regression analysis will show that the first-order linear relationship is not satisfactory and a higher-order relationship or even non-linear processing will be required to correctly map the scanner data to the standard. (Note that for such scanners a separate characterization may well be required for each type of original.) A test for second- or third-order relationships would then be appropriate as discussed below under 'Characterization procedures'.

For the case where a linear relationship is established, the characterization of the scanner is straightforward. This provides the values of the coefficients in the characterization matrix and calculation of the inverse provides the transformation to be applied to the image data. If a good correlation is found to a higher-order function, this defines a similar, albeit larger and non-invertible, matrix for the material on which the target is made. If no such correlation is found, it is necessary to consider a non-linear transformation for characterization. The techniques described in the next section are then appropriate.

If reasonable correlation is found with either a first- or a second-order function, one can begin to evaluate the performance of the scanner (and the utilized applications software) by plotting the lightness (L^*), hue angle (h_{ab}) and chroma (C^*) values of the target together with the results obtained after characterization. Since we have three-dimensional data, it is normal to plot this in two dimensions at a constant level of the third. The target is arranged with 12 hue angles and three levels of lightness. Undoubtedly the most useful plot is to show a^* vs. b^* at each of the three lightness levels. However, an alternative is to calculate h_{ab}, C_{ab}^* and L^* and show L^* and C_{ab}^* for each hue angle. (Alternatively u', v' and h_{uv}, C_{uv}^* and L^* can be used.)

Clearly, this gives a useful plot for 'visual' analysis, but it really does not quantify the errors, for which the colour differences between the predicted and actual values should be computed and analyzed statistically.

8.3.2 Evaluation of 'gamut compressed' colour digitizer

As described earlier, we cannot necessarily expect a linear transformation between the data used to drive an output device and the targets' tristimulus values. We therefore need to define a test to be applied to such data which gives an indication of its value. One useful procedure is to look at the 'monotonicity' and the 'separability' of the data. Can all steps in the target be logically differentiated? No ideal separation can be defined since it will be application dependent. Transparency recorders, for example, require log data for much of the range (but not at the toe and shoulder), CRTs require a power function (gamma correction) and printing inks require a function which is not defined in a simple expression but is approximately logarithmic.

Nevertheless it is clear that if the applications package is to sensibly manipulate the data, all steps must be separable and monotonic in certain axes. The grey scale is an obvious example; unless a bizarre GCR function has been employed all channels will be monotonic and every step should be differentiated. Such information provides useful insight as to whether or not any applications package can possibly produce sensible output data from the input data and, indeed, whether the original tristimulus values can be recreated. Clearly, it is still an issue whether any applications package can produce the correct output in such a situation, but this allows the user to separate the issues.

Where the applications package is an integral part of the scanner processing, this test is somewhat irrelevant. With some high-end scanners it is possible to set up in null mode (i.e. with no processing at all except, perhaps, logging) but not all. In this context only one test seems reasonable: to output the resultant target on the device for which it was set up and measure the colours. Clearly, for this to work the data must be separable, but that is not key; it is the output results which are significant.

As discussed earlier, the inevitable gamut compression means that a colour match cannot be achieved; however, plotting the resultant hue angle, chroma and lightness values (in CIELAB or CIELUV) and comparing against the original provides some useful information. We can expect that hue errors should be minimal and chroma and lightness reproduced in a logical way, as discussed in Chapter 13 on gamut mapping. However, such information would normally be used to improve the scanner calibration so any data obtained may just tell us that the scanner has not been optimized. Conversely, if the scanner has been optimized, such information could be used to define optimum gamut compression.

Clearly the results of such testing cannot provide conclusive data about 'scanner quality'; the applications package, scanner characteristics and subjective evaluation of gamut compression make too many variables. However, plotting the data in the way proposed will provide useful and interesting information which may well lead to pertinent conclusions and reveal issues for vendors and users alike.

8.3.3 Signal-to-noise ratio

Another feature of the target is its ability to provide information on the colour scanner's signal-to-noise ratio. Since the scanner will make many measurements on each patch of the target (in many cases over 100 measurements are made per patch, even when staying away from the edges of the patch), the opportunity exists to average these measurements and determine the signal-to-noise performance of the scanner. Since such an analysis includes the noise associated with the uniformity of the patches of the target, it can only be a comparative test between scanners unless it is possible to make two scans with an identical start-of-scan position. By subtracting one scan from the other the effect of the original can be minimized.

8.3.4 Scanner repeatability

By repeating some of the tests over a number of days it is possible to obtain information about scanner repeatability and whether differences from scan to scan are predictable from some simple measure (e.g. the white point difference).

8.4 CHARACTERIZATION PROCEDURES

Five possible procedures for characterizing scanners and cameras (and thereby defining device profiles for use by colour management systems) are summarized below. The methods are equally applicable to the definition of any general colour transformation, such as those between device spaces, but for the purposes of this paper discussion will be restricted to characterization only. The first three of the procedures can easily make use of the ISO characterization target described earlier; the fourth really requires a custom-made target. The procedures can be summarized as:

1. Least-squares fitting (using linear mathematics).
2. Non-linear transformations, based on such procedures as those achieved on many device-dependent scanners.
3. Modelling of dye combinations.
4. Look-up table with interpolation.
5. Defining sensor spectral response.

General principles of the characterization procedures in 1–4 above are described in Chapter 6, and here we discuss their application to scanners and cameras. Spectral methods include the use of basis functions to derive a transform from the sensor spectral response, and such methods are more commonly used to characterize digital cameras. Further detail of procedures for spectral characterization are given in Chapter 9.

8.4.1 Analytical solution to first-order equations

If there is reason to believe that the scanner produces good quality colorimetric data, the transformation matrix is a simple first-order matrix as described above and it may be computed from a limited number of points of the target only. The points chosen do not matter very much. The white point of the sample (step 1 in the grey scale), together with two coloured patches of a different hue, provide enough points for a solution.

To define the matrix, it is necessary to define the scanner values for each of these steps. This leads to nine equations, three each in X, Y and Z. These can be solved as simultaneous equations to compute the coefficient for each row. For example:

$$X_1 = a_{11}R_1 + a_{12}G_1 + a_{13}B_1$$

$$X_2 = a_{11}R_2 + a_{12}G_2 + a_{13}B_2 \qquad (8.7)$$

$$X_3 = a_{11}R_3 + a_{12}G_3 + a_{13}B_3$$

where R_i, G_i and B_i represent the scanned values of the three points chosen. By inserting the values of X_i, R_i, G_i and B_i the coefficients can be calculated quite easily. Clearly, such a matrix must be correct for the points from which it was calculated, and if the scanner provides a good linear transformation of XYZ, then the matrix calculated from three points will equal that calculated from a best-fit solution to all 288 points in the target.

8.4.2 Least-squares fitting to a first-order model

As stated earlier, the data is a linear transformation of the tristimulus values of the target if a set of coefficients can be found which satisfy the following equations for all colours:

$$X = a_{11}R + a_{12}G + a_{13}B$$

$$X = a_{11}R + a_{12}G + a_{13}B \tag{8.8}$$

$$X = a_{11}R + a_{12}G + a_{13}B$$

where R, G and B are the scanned values for each patch of the target. However, in general the transformation will not be perfect and it is necessary to calculate and evaluate the best approximation. The best-fit function for each of the three equations above for the 288 colours in the target is found by least-squares regression, as described in Chapter 6.

8.4.3 Least-squares fitting to a higher-order model

As stated earlier, a simple first-order set of linear equations may not be adequate to characterize the scanner data. The procedure above can be extended to higher-order polynomial functions in order to establish the best-fit transformation.

If a higher-order polynomial is required it is worth investigating the linearity of the system to determine whether another domain than tristimulus values is appropriate for the calculation. Many scanners, for example, contain narrowband filters which are attempting to evaluate the dye concentration in the original. Others process the data to be more appropriate to an output device with a correction for gamma correction or density conversion. For either of these reasons a lower-order polynomial (or one with smaller coefficients for the higher-order terms) will be obtained by redefining the equations above for a different colour space. Either the RGB data is modified to 'undo' the correction made to it for output or, if dye concentrations are being determined, the equations should be modified in terms of logs of the tristimulus values. Analysis of the grey scales and colour scales of the ISO test target will be helpful in defining the correction required. Such correction will be referred to hereafter as linearization.

The reason why this conversion is desirable is that higher-order polynomials can introduce local maxima and minima in the transformation. Clearly these are not desirable and the higher the order the greater the risk. Hence finding the domains for the data which best approximate a first-order conversion is most desirable.

Such corrections are also desirable in that they may provide a colour space which is more perceptually uniform. Taking logs, for example, provides greater uniformity than for the tristimulus values alone. This helps to overcome the objection above that the use of simple tristimulus values may introduce an unwanted weighting to the least-squares fit.

In order to define the order of the polynomial which should be used, some measure of the error arising from any particular solution needs to be defined. Two measures which can be useful are the correlation coefficient and the colour differences determined between the values measured on the target and those calculated by the chosen polynomial from the scanner data.

To establish the correlation coefficient the following equation should be used:

$$r^2 = \frac{\sum (x(i) - \bar{x})(y(i) - \bar{y})}{288 s(x) s(y)} \tag{8.9}$$

where $x(i)$ and $y(i)$ are the individual calculated and measured values respectively for each of X, Y and Z, x and y are the mean calculated and measured values, and $s(x)$ and $s(y)$ are the standard deviations of the calculated and measured values. Clearly, there will be a separate coefficient for each of X, Y and Z.

The problem with this measure is the non-uniformity of the X, Y, Z colour space. Thus it is also useful to calculate the residual error in terms of colour difference for each patch of the target. Using either CIELAB or CIELUV an average error of around 3 will normally be acceptable in practice.

For analysing and reporting errors, specification of hue angle, chroma and lightness is recommended as well as the colour difference. So long as it is specified, either CIELAB or CIELUV may be used, although the former is preferred. The data may then be plotted for both predicted and actual values. The vector between them provides a measure of error. As stated earlier, there are a number of ways of plotting the data; any two can be specified for a constant level of the third, giving three possible plots.

8.4.4 Non-linear transformation

The algorithms in common use in Graphic Arts scanners have evolved over a number of years. In general they consist of a series of functions which have different coefficients depending upon the region of colour space in which the sample pixel falls. These algorithms have evolved in enabling the conversion from scanner RGB values to CMYK ink amounts but can easily be modified for this application, possibly in conjunction with a polynomial. Generally, the coefficients have been derived empirically, but there is no reason why they should not be defined numerically from data obtained by using the ISO colour chart.

8.4.5 Dye modelling algorithms

Most colour film manufacturers have developed mathematical models for predicting the colour produced on a film when specific amounts of each dye are present. Such models can be used for predicting the colour of a sample from scanned values, providing the scanner and film type are calibrated as a system. Dye modelling is also discussed in Chapter 10.

The basis of such systems is relatively simple and based on the work of Yule [2]. We recognize that photographic dyes exhibit little light scattering and therefore obey the Lambert–Beer model, whereby concentration is linearly related to the log of the light absorbed (i.e. density), fairly well. The extinction coefficient varies with wavelength and so the relationship is strictly valid only for monochromatic light, but providing a scanner has relatively narrow spectral bandpasses the linearity with respect to scanner density will be acceptable.

Clearly, by inverting the matrix obtained from the coefficients in Equation (10.18) in Chapter 6, and from prior calibration using a tool such as the ISO target which contains various concentrations of each dye, it is possible to compute the concentration of each dye present in a pixel. Knowing the spectral characteristics of the dyes (which can be measured from the ISO target if not readily available), it is possible to use the additive behaviour of the dye densities to compute the spectral characteristics of the pixel and hence the tristimulus values.

In practice there are some small departures from additivity, particularly at high dye concentrations and with finite bandwidth filters, which means that some approach to coefficient optimization, such as least-squares fitting over a range of concentrations, is preferable. Nevertheless, the results achieved can be quite acceptable, particularly for narrowband filter scanners. However, the computation is far more intensive than for polynomials and is of value only if the behaviour of the polynomials is known to be erratic.

8.4.6 Look-up table with interpolation

If the three methods above are deemed too inaccurate, the last resort is simply to measure the tristimulus values and scanner values of a large number of colours and create a look-up table between them. Any scanned pixel can then be converted into tristimulus values via the look-up table and interpolation used for intermediate points which do not fall in the table itself.

To obtain sufficient accuracy one of two (or both) approaches must be taken. Firstly, a large number of colours must be measured. In my experience 200 (i.e. all combinations of six levels in each channel) is an absolute minimum and 4000 (16 levels of each channel) is more typical. However, the number can be minimized by linearizing the data into and out of the look-up table by use of single-dimensional look-up tables for each channel or by use of non-linear interpolation. Such linearization is essential for a look-up table based on six levels in each channel.

The attractions of this approach are its very high level of accuracy if the colours are properly selected and either the number of colours chosen is large or the linearization is carefully carried out in order to minimize the number. The disadvantages are very clear. The number of measurements required is very high and for transparencies this is particularly troublesome.

8.5 SUMMARY

This chapter has discussed how scanners may be evaluated, calibrated and characterized by use of the ISO scanner calibration target. Results obtained from scanning and outputting it may be assessed visually to assist in defining a colour transformation empirically, or the data obtained from scanning may be correlated to that measured on the sample to set up a transformation mathematically. The same data can then be used as a measure of the accuracy of the transformation, signal-to-noise ratio and scanner repeatability.

Various methods have been described for defining the transformation from scanner *RGB* data to tristimulus values, including least-squares fitting, non-linear processing, dye modelling algorithms and look-up tables with interpolation.

REFERENCES

1. Maier, T. O. and Rinehart, C. E. (1990) Design Criteria for an Input Color Scanner Evaluation Test Object. *J. Photo. Sci.*, **38**, 169–172.
2. Yule, J. A. C. (1938) The Theory of Subtractive Colour Photography; I. The Conditions for Perfect Color Rendering. *J. Opt. Soc. Am.*, **28**, 419–430.

9

Color processing for digital photography

Jack Holm, Ingeborg Tastl, Larry Hanlon and Paul Hubel

9.1 INTRODUCTION

From a technical perspective, a digital photography solution requires the implementation of digital color reproduction processes that produce pleasing pictures. The entire imaging chain from capture to the end uses of an image must be considered, as each step can impact the final result. End-to-end imaging system design, as opposed to concentration on just a single component of the system, is therefore fundamentally important. In addition, the unique attributes of the digital paradigm, such as instantaneous image review and the sharing of images on a global scale, place challenging demands upon the design of the color reproduction processes. Instantaneous review allows the user effortlessly to compare the original scene to the reproduction presented on the camera's display, and the global electronic sharing of images means that other reproductions of the image will be presented on devices that will not be known *a priori*. Since small, portable digital cameras can simultaneously be an independent imaging appliance and a part of a complex imaging system, digital photography products from different manufacturers must all perform seamlessly as part of an overall system. This integration is essential if digital

Colour Engineering, edited by P.J. Green and L.W. MacDonald.
© 2002 John Wiley & Sons Ltd.

photography is to fulfill its potential of exceeding the performance and flexibility of traditional photographic systems. Consequently, the specification of system architectures, the definition of standards, and the design of specific solutions for particular amateur and professional applications are topics in which a great deal of thought has been invested in the past decade.

A significant part of the systems discussion has centered on defining architectures for the generation and sharing of color image data. While there is a user need for products from the different manufacturers to perform fluently together, there is also an economic need for companies to differentiate their products. Previous work in the international standards community has put in place a good foundation enabling manufacturers to address both needs simultaneously [1–44]. This gives digital camera designers some important reference points when making design decisions. We will take this body of work as a given, referencing it as needed, and only discuss areas currently under development and higher-level system issues in detail. In the rest of this chapter, we will concentrate on issues specific to the design and evaluation of color reproduction solutions for digital photography, and how they are implemented in digital cameras.

The designers of a digital camera face a multitude of choices and tradeoffs through which a path must be navigated to reach the target market. These decisions often pit imaging performance against camera size, weight, manufacturing and testing complexity, and so on, all of which affect product cost. The objective is to achieve an appropriate balance among the product attributes. Furthermore, an in-depth understanding of the primary factors affecting imaging outcomes is crucial. From the traditional descriptors of image quality – tone reproduction, color reproduction, sharpness, and noise – we will for the purpose of this chapter pay less attention to sharpness than to the other attributes. This is because sharpness is most easily decoupled from the others. While there is a sharpness–noise perceptibility tradeoff in image processing, it is a relatively simple tradeoff. The interactions of tone reproduction, color reproduction, and noise are more intimately connected.

It is our hope that this chapter will serve as a valuable reference for engineers involved in the design of digital cameras.

9.2 BACKGROUND

From an artistic point of view, photography is often about 'capturing a moment'. From a technical point of view, the moment an image is captured is also the crucial moment. The physical situation when the camera shutter is released determines the potential quality of resulting photographs. The spectral and geometric characteristics of the objects in the scene, the light sources illuminating it, the optical path to the image sensor, and the color separation filters used with the sensor all impact the way in which the scene colors are measured or 'seen' by the camera sensor. The electronic characteristics of the camera capture subsystem, including the image sensor well depth and sensitivity, the readout noise sources, the analog to digital conversion process, and the noise injected from the rest of the camera electronics, establish both a noise floor and an operating region for the digital image capture process. The conditions at the moment of capture, such as temperature and the total photon catch utilized, determine where in the camera's operating region the image falls. These conditions at 'the moment' establish the fundamental signal and noise structure of the image, and consequently the amount of information captured. Since information cannot be created later on, subsequent image manipulations can improve certain attributes only at the expense of others. So, the conditions at the moment of capture establish the potential quality of the image. The primary task of the capture subsystem is to provide data that best samples a variety of scene types under various camera operating conditions. The task for the image processing algorithms is to transform this data into pleasing photographic renderings.

9.2.1 Color processing approaches: uncalibrated versus calibrated image data

Consumer digital still cameras have traditionally employed image processing methods and techniques that are very similar to those used in camcorders [38, 78]. While this approach has the advantage of leveraging known solutions to a new class of products, it also brings performance limitations because of design tradeoffs previously made to optimize solutions for video. It is well known that still photography is more demanding of intrinsic image quality than video, because the images are not moving, the predominant medium for output is a reflection print, and because of long-established customer expectations.

Cameras employing techniques similar to those used in camcorders, and many other digital imaging systems, treat digital image data like a 'bag of bits' to which various image processing algorithms are applied to improve the picture quality. Without a link to the physical significance of the bits, these algorithms are in a sense forced to operate blindly. Information related to the camera used to capture the specific image (like sensitivity curves of the sensors, specifics about the lens, etc.) is not directly used within these algorithms. While many clever techniques have been developed, the range of photographic situations from which truly pleasing results can automatically be obtained is significantly more limited than the range of photographic situations commonly encountered.

In professional digital cameras to date, the prevailing paradigm has been that the captured image data is just a starting point from which a human makes modifications to produce the desired pictorial result. While some of these cameras offer a choice of preset tone reproduction curves, it is up to the user to choose curves appropriate to the given photographic situation. This workflow results from the demanding nature of professional photography and the fragile nature of automatic processing algorithms. As in the consumer case, the missing link to the physical meaning of the data has constrained the quality of the results. Professional digital photography image manipulation software packages tend to be quite complex, and the subjective nature of the operations performed has resulted in the establishment of a new specialization – the digital image editor.

An alternative approach to digital photographic color reproduction is embodied in the concept of scene estimation followed by color rendering. It represents a major paradigm shift in that the image processing is no longer being applied to relatively uncalibrated image data, but rather to data produced from a calibrated capture subsystem. Knowledge of the actual physical significance of the image data enables using 'physically aware' algorithms, which in turn makes possible the use of more aggressive, adaptive, and non-convergent algorithms.

A major issue in processing uncalibrated image data is the unknown history of often irreversible algorithms. When processing uncalibrated image data, it is highly desirable to use only algorithms that converge on a stable endpoint, because it is impossible to predict how many times an algorithm may be applied. However, experience in pictorial color reproduction points to the advantage of using non-convergent algorithms to produce preferred results. Implementation has been hindered by the lack of image data with known physical meaning. When such data is available, manually directed color rendering is simplified and improved, and automatic color rendering that explicitly accounts for scene key and dynamic range when processing each image becomes possible [61, 63, 65].

In practice, most image data is neither completely calibrated or uncalibrated [70]. In camcorders, 'gamma correction' is applied based on assumptions about the camera and display characteristics. Also, even the most carefully calibrated systems will have some calibration error. This error quickly becomes significant in mass-produced products because of unavoidable device variability. Therefore, while it is interesting to contrast the extremes of calibrated and uncalibrated image data, most systems operate somewhere between these extremes. The question is to what degree the state of calibration can be relied on. Since calibration is a matter of degree, both calibrated and uncalibrated images can be processed in basically the same workflow. It is possible to make tradeoffs as appropriate in a common context. The resulting different imaging outcomes are a consequence of design objectives and decisions applied to the definition of the camera processing pipeline. We will illustrate this by working through descriptions of possible camera processing pipelines.

9.2.2 Dealing with calibrated and uncalibrated image data in a common architecture

Fundamental differences exist between the more traditional relatively uncalibrated approach to image processing, and the more analytical calibrated scene estimation plus color rendering approach. However, it is possible to look at these approaches in a context where they are merged. For example, when the image data output by a sensor is only gamma corrected and output, these steps can be viewed as scene analysis plus color rendering, where the color rendering happens to be the combination of the optical flare present, the sensor and electronics nonlinearities, and the gamma correction applied. Such 'color rendering' may be sub-optimal, but by adopting the more complex architecture as the context, it is possible to present and compare different processing solutions in an organized way. The caveat is that, once a decision is made to abandon calibrated image data, the processing operations which may be subsequently applied are much more restricted in what they can do. If they are not mild, and convergent when applied repeatedly, dramatic failure cases will result. On the other hand, the system tracking requirements for maintaining calibrated image data are fairly rigorous. Color encodings for image data exchange need to be carefully and completely specified in order to assure unambiguous interpretation of the image data [36, 38, 41, 43, 44, 71]. In some applications, additional image metadata is required [42].

In many cases, there will be little difference in the actual physical processing pipelines for the two philosophical approaches. Sensor data is generally processed using some sort of preliminary linearization and normalization analog circuitry or digital look-up table (LUT), followed by a matrix operation, followed by the application of another LUT. In the uncalibrated case, the first operation performs white balancing and dark subtraction, followed by the matrix conversion to a standard color space, followed by a gamma correction curve. In a rigorous scene analysis plus color rendering case, the first operation will perform a more comprehensive linearization which will also address optical flare and sensor nonlinearities, the matrix operation is essentially the same, and the last operation contains a more thoughtfully derived and sometimes adaptive color rendering curve concatenated with the exchange encoding nonlinearity.

For the remainder of this chapter, the architecture shown in Figure 9.1 will be assumed. In Figure 9.1, the square boxes represent the image data states which specify the physical characteristics of some instantiation of the image. The ovals specify processing oper- ation(s) or associations. The digital photography processing path follows the arrows, although some processing choices will involve less rigor as mentioned above, and stages may be concatenated. The terminology used in Figure 9.1 is emerging as standard in expert forums, and is necessary for understanding this architecture. For convenience, many relevant terms and definitions (from CIE Publication 17.4 [40], ISO 12231 [15], and ISO 22028-1 [36]) are provided at the end of this chapter.

9.2.3 The meaning of device-independent color as applied to digital photography

Since the title of this book includes the claim 'achieving device-independent color', it is necessary to clearly describe what device-independent color means with respect to photographic reproductions.

For many decades, photographic systems have been designed to tune the desired appear- ance of a photographic reproduction to the reproduction medium and viewing conditions.

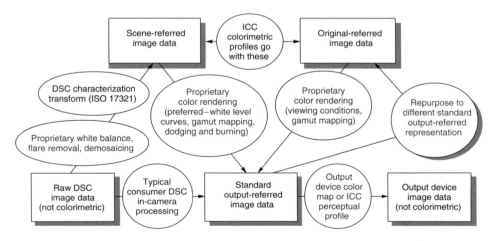

Figure 9.1   Image data encoding opportunities and intermediate transforms [88].

It is also well established that photographic reproductions frequently do not attempt to recreate the appearance of the scenes photographed. One might then question the value of the device-independent color concept to photography, since there is no intent to achieve identical reproductions regardless of the medium.

Historically, photographic scientists and engineers have not worried too much about device-independent color, because the number of media used were limited and could be grouped into a small number of categories. As a baseline, there were reflection prints and transparencies. Print and transparency media are available in black-and-white and color, from a variety of manufacturers, with a few different methods for production, and with a few different applications. For example, transparencies can be made using reversal processing or by printing from negatives, and may be viewed on light boxes, by still projection, or as motion pictures. Furthermore, because of the constraints of chemical photographic processes, dye set choices are limited and tend to be similar across products, even from different manufacturers. The photographic industry also has a long history of reliance on well-conceived and robust standards for measurement and product evaluation.

In general, photographic system engineering has focused on delivering well-designed closed-loop systems of limited flexibility. Major market needs were addressed by systems with very high potential for quality, when employed by expert users. Note that motion pictures from the 1940s still have pleasing color 60 years later (when restored for fading, or when new prints are made from archival black-and-white negatives). Also, in many cases major photographic manufacturers produced limited volume specialty products which provided solutions for niche (but experienced) users. However, the profitability of niche products is small to nonexistent. These products were made available by manufacturers largely as a community service and to build brand reputation. Over the past two decades, more careful evaluation of the bottom-line value of such practices has led to dramatic reductions in specialty offerings.

While the number of chemical-based photographic options has been declining, the number of digital options has been rapidly expanding. Many applications which required specialty products can now be accomplished using digital technology. Instead of duping a slide on duplicating film, one can scan the slide and write out another on a film recorder. Perhaps the most profound technological advantage afforded by digital photography is the softcopy display. The photographic quality ink-jet printer is close behind. Many also foresee great opportunity in the electronic sharing of digital images.

Digital technologies offer great promise for the future of photography, but they bring a new set of problems which shake the foundations of mature chemical photographic system design. It is no longer possible to rely on the constraints which allowed the closed-system approach. On the other hand, the maturity of photographic systems makes it difficult for emerging technologies to compete. Customer expectations are high, and simplistic approaches to digital photography often fall short.

Originally, the concept of device-independent color arose because of the frustrations of some digital imaging engineers as they tried to understand the complexities and limitations of the traditional closed-system approaches to photographic engineering. The densitometric measurement devices that were used would produce different measurements for color patches that appeared to be the same color, and could produce the same measurements for different colors. This was no surprise to those who understood the basis for densitometry, but it does point out the fact that densitometric instruments are designed to measure specific colorants, and then be applied using specific mathematical treatments which account for the spectral characteristics of the colorants. Development of such specialized spectral responses and mathematical treatments for arbitrary colorants is cumbersome.

Fortunately, it is possible to choose a special set of spectral responses when performing measurements which avoid the colorant dependency when comparing measurements to visual impressions. These spectral responses approximate the spectral responses of the human visual system. We all know this special branch of optical radiation measurement as colorimetry. Colorimetric measurements were deemed device independent because if samples from different devices, or on different media, measure the same, they will appear the same.

At this point, all but the most patient reader will be wondering about the motivations for this tangential discussion. There are two:

1. Virtually all digital cameras *do not* employ colorimetric capture. This means the data the cameras produce is by nature not device independent. Therefore, the scope of the problems to be solved goes beyond colorimetry. In processing digital camera image data, it is necessary to delve back into the complexities of device-dependent measurements and spectral characteristics.
2. In the enthusiasm that followed discovery of the joys of colorimetry, some budding digital photography engineers made an unjustified leap in their interpretation of the meaning of device independence. It seemed logical that if an image could be measured in a device-independent way, then it should be able to be reproduced in a device-independent way. While this approach does hold for reproductions on the same medium, it does not apply in general to image reproduction.

Therefore, the processing approaches described in this chapter will assume the goal is *not* (in general) to produce device-independent color images, but rather device-optimized color images. The optimization of an image for a particular medium will typically depend on the image content and statistical characteristics, the medium on which the image will be reproduced, the viewing conditions under which the image will be viewed, and the application for which the image will be used.

9.3 OPEN SYSTEMS ARCHITECTURE AND WORKFLOW OPTIONS

If one assumes a goal of image- and media-specific optimization, it becomes necessary to develop a system image processing architecture which allows this goal to be achieved in an open fashion. Open systems are essential if the full utility and flexibility of digital imaging is to be realized.

The design of an open systems architecture and component workflows which allow the necessary optimization while maintaining openness is nontrivial. Many experts have worked on this problem for a number of years. Three basic workflow approaches have emerged as viable [81]. They are:

1. *The intermediate reproduction description.* With this option, the image data captured is processed as if it will be reproduced on some standardized real or virtual output medium, and image data describing this reproduction is exchanged. Subsequent reproductions are produced by reprocessing the intermediate reproduction description to optimize it for the actual output medium. This option is not unlike using conventional photography to produce the original photograph, and then scanning it into a digital system. In this case, the original photograph is the intermediate reproduction description.

2. *Profiles on the fly.* With this option, a transform specifying the appropriate processing to some output representation is included in the image file and carried along with the image data captured. The predominant format for specifying such transforms is provided in ICC.1 [41]; such transforms are called ICC profiles. An ICC profile can specify transformations of the image data to a variety of representations. In digital photographic applications, the ICC perceptual intent is most commonly used. It specifies a transformation to or from a virtual high quality reflection print intermediate reproduction description. The expression 'profiles on the fly' comes from the fact that, for optimal quality, profiles may need to be image specific and therefore created as the images are created.

3. *The smart color processing engine.* With this option, a color processing engine receives the image data in its 'raw' state (along with the necessary metadata to allow correct interpretation) and performs all the processing required to produce the desired final image at the time of output. This option provides the greatest flexibility, and ultimately has the highest quality potential, but the processing algorithms required for robust implementations are still the subject of intensive research and are frequently protected as trade secrets or intellectual property. Furthermore, for this option to be widely successful, raw image data would have to be exchanged, and every device receiving the data would need to include or have access to robust smart color processing engines. We are a long way from seeing smart color processing engines as a general solution. It may even be unlikely that this approach will ever go much beyond professional imaging, as the quality limitations of approaches 1 and 2 are considered by most to be insignificant for consumer applications.

Most digital cameras use option 1, although option 2 is supported by the JPEG 2000 standard which has just been completed [25]. If JPEG 2000 is widely adopted for use in digital cameras, option 2 may encroach on the current dominance of option 1 for digital photography applications. As noted earlier, many professional digital photographers would say that option 3 is their current preference, but with an Adobe Photoshop™ expert as the 'smart color processing engine'.

9.3.1 Digital camera image processing steps related to the three different workflow options

Regardless of the workflow option used, the scene analysis plus color rendering image processing architecture requires calibrated image data. Part of this calibration is explicit knowledge of the image state. Defined image states are raw, scene-referred, original-referred, standard output-referred, specific output-referred, and output device (see Figure 9.1 and the definitions at the end of the chapter). Of these states, the scene-referred, original-referred, standard output-referred, and specific output-referred image states are encodings of image colorimetry and therefore can be rigorously interpreted to describe a specific image appearance. For the raw and output device states the image data are device values and are therefore not 'device independent' even in the colorimetric sense.

One way to obtain a practical understanding of the different image states is to consider which processing steps are applied to go from one state to another. Table 9.1 lists the digital camera image processing steps and the associated image states. Plate 3 illustrates what a digital still camera (DSC) image will look like as one steps through these processing operations.

It is generally most appropriate to consider an uncalibrated image for exchange to be standard output-referred to a typical video CRT display. The assumption is that whatever processing has been applied will attempt to produce a suitable image on such a display. However, when working with uncalibrated image data, it is important to keep in mind that the quality of the image described may be limited.

9.3.2 Device-independent color via exchange of standard output-referred image data and application of color rendering

If the chosen workflow is to achieve 'device-independent color' through the exchange of standard output-referred image data, one choice remains which is specific to digital cameras. This choice is whether the camera (or camera mode used) adopts a fixed or adaptive color rendering model. For historical reasons, we will call fixed color rendering models 'slide' models and adaptive color rendering models 'photofinishing' models.

The slide model and photographic color rendering

The slide model assumes one specific color rendering transform is applied within the image processing pipeline. Therefore, if the exposure provided to the image sensor is reduced the image will get darker, and if the exposure is increased the image will get lighter by amounts that are determined only by the exposure received and the characteristics of the fixed transformation. This behavior is similar to that of a film transparency. Consequently, the color rendering transforms used with this choice should approximate those of transparency film, except with modifications to account for the more common standard output-referred encodings being CRT display or reflection print based. For convenience, such a reproduction model will be called a 'photographic reproduction model'. Plate 4(a) and (c) illustrates the application of a photographic reproduction model to scene-referred image data (although in an unfavorable circumstance).

The advantages of the photographic reproduction model are:

1. The behavior of digital imaging systems using this model is similar to that of transparency systems. These systems are capable of very high image quality with control of the scene characteristics and careful exposure selection.
2. The operation of photographic reproduction model digital imaging systems is very similar to that of film transparency systems. Many professional photographers are highly experienced with film transparency systems.

Table 9.1 Digital camera image processing steps and the associated image states

---

Starting points for image processing

- Raw DSC image data
- Scene-referred image data
- Original-referred image data
- Standard output-referred image data.

Ending points for image processing

- Scene- or original-referred image data
- Standard output-referred image data
- Output-referred image data for a specific device
- Output device data.

Raw DSC to scene-referred processing steps

(order may vary in some instances)

- Linearize (if necessary)
- Subtract dark current (if necessary)
- Subtract flare (may be optional, but needs to be indicated)
- Determine scene adopted white (estimate adapted white)
- White balance and clip to illuminant neutral ceiling
  (this is the last step where the image data is raw CFA data for color filter array devices)
- Demosaic (if CFA data; after demosaicing the image data is typically raw device RGB)
- Transform to device-independent scene-referred color encoding (now the image data is scene-referred)
- Optional white luminance scaling.

Scene-referred to output-referred processing steps

- Scene-referred color encoding represents scene colorimetry (or possibly focal plane; need to indicate which). *Note.* Scene-referred images may be edited to produce a new 'scene'.
- Color rendering is performed to produce colorimetry appropriate for a particular real or virtual target output device (luminance range and color gamut) and viewing conditions.
- Output-referred colorimetry is encoded for exchange. If a standard rendering target is used, the image data encoded will be standard output-referred. May be edited to produce new output-referred image data.

Original-referred to output-referred processing steps

(The source of the original-referred image data may be standard or other output-referred image data.)

- Original-referred color encoding represents the colorimetry of the original captured (no capture device flare). *Note.* Original-referred images may be edited to produce a new original.
- Color rendering is performed to produce colorimetry appropriate for a particular real or virtual target output device (luminance range and color gamut) and viewing conditions.
- Output-referred colorimetry is encoded for exchange.

Output-referred to output device processing steps

- Determine whether output-referred image colorimetry encoded needs to be color rendered for intended actual output device.
- If color rendering is needed, perform it to produce image colorimetry for the actual output device.
- Map desired actual output colorimetry to device values which will result in the desired colorimetry on the output device.

The disadvantages of the photographic reproduction model are:

1. The lightness of the image produced is directly coupled to the exposure used to capture the image (and the camera gain setting). Exposure errors result in images that are too dark or too light. This situation has been long accepted in slide and instant print photography to the extent that lightness adjustments are frequently called 'exposure' adjustments. However, it is sometimes desirable to adjust exposure (for example, reducing it to increase highlight detail) without the intent of similarly changing the lightness of every part of the image. Also, a benefit of digital photography is the option to easily evaluate and adjust the image data after capture. This evaluation can be more accurate than exposure metering because it is performed on the actual image data.
2. Scenes are highly variable, and some scenes are not ideally suited to the fixed color rendering transformation applied when using this model. Professional photographers tend to avoid problematic scenes, and sometimes use 'darkroom' manipulation (digital or otherwise) when this is not possible. Less experienced photographers are familiar with 'subject failure'. Digital photography offers the opportunity to significantly reduce subject failure through adaptive color rendering.

Video color rendering

In some slide model applications, simplified color rendering is applied. Two of these simplifications are sufficiently common to deserve special names. The first is to apply a simple system gamma function with a gamma value between 1.1 and 1.4 as a fixed color rendering transformation, similar to video systems. Such a reproduction model will be called a video reproduction model. This model is historically the most common in digital cameras because of their video origins, and is still predominant. Plate 4(e) illustrates the application of a video reproduction model to scene-referred image data.

The advantages of the video reproduction model are:

1. It is commonly used to the point where some tolerance for its defects is established.
2. Color rendering is good for scenes with luminance ratios between 30:1 and 60:1.
3. Simple, intuitive user understanding of camera behavior.
4. Easy camera design and parity with video capture in combination video/still cameras.

The disadvantages of the video reproduction model are:

1. Exposure errors result in images which are too dark or too light.
2. Only about one-quarter of natural scenes have luminance ratios between 30:1 and 60:1, so three-quarters will be poorly color rendered. Highlight clipping is the most common objectionable artifact.

In many current cameras, highlight clipping is avoided by using a soft rolloff or 'knee' as opposed to an abrupt clipping. This practice has long been common in professional video, and results in a modified color rendering transform which looks similar to that of a photographic reproduction model.

Colorimetric reproduction

The second common type of simplified slide model will be called a colorimetric model, because with this model the goal is to reproduce the scene colorimetry as accurately as possible on the output device. Plate 4(g) illustrates the application of a colorimetric reproduction model to scene-referred image data.

Colorimetric reproduction models generally do not produce optimally pleasing images because there is no midtone contrast boost or color saturation boost, as are generally preferred for pictorial imaging (as is evidenced by the fact that such boosts have been employed in photography and television for decades). On the other hand, scene luminance ratios up to the limits of the output medium can be accommodated. Most photographic media have luminance ratio capabilities around 160:1, which is similar to the luminance ratio of a statistically average natural scene. This means that with this reproduction model only about half of the scenes captured will exhibit highlight clipping.

Colorimetric reproduction models are frequently used for the archiving of artwork, and for catalog photography where it is desirable to convey the color of a product as accurately as possible. Very few consumer digital cameras apply a colorimetric reproduction model, and those that do require the user to specifically select this mode (for example, the manual mode on the HP C912).

The photofinishing model and preferred reproduction

We will call the other baseline color rendering model the photofinishing model, because in use it parallels the negative-print system in conventional photography. With this model, the color rendering applied adjusts itself to produce a pleasing image in a way that is somewhat decoupled from the exposure received by the image sensor. This model requires analysis of the image data after capture to determine the appropriate color rendering. In much the same way as a photofinishing minilab analyzes photographic negatives, a preferred reproduction model analyzes scene-referred image data to create what it thinks is likely the best reproduction. Exposure errors are compensated for, and the digital processing individually optimizes each image. This 'image level customization' enables high quality results that are comparable to images processed by professional photofinishing laboratories.

The specifics of preferred reproduction model algorithms are typically proprietary trade secrets or covered by intellectual property protection. This is also the case with film minilab 'scene-balance' and exposure algorithms. Nevertheless, some results have been published, so these algorithms are not entirely mysterious. The preferred reproduction model described by Holm [61, 65, 98, 99] was used to create the standard output-referred representations shown in Plates 3E, 4(b), 4(d), 4(f) and 4(h). These plates show the advantages of adaptive color rendering when applied to scene-referred image data with different characteristics.

Color rendering, ISO speed ratings, and exposure determination

ISO speed ratings and exposure determination for digital photography are intimately connected to the color rendering applied. If one strongly under- or over-exposes an image, the color quality will be dramatically affected. In analyzing this relationship, it is necessary to distinguish between an ISO speed rating and an exposure index (EI). An ISO speed rating indicates the exposure necessary to achieve a particular image quality with a particular image capture system. The EI is a value input to an exposure metering system to tell it the exposure desired, so the exposure time and f-number settings can be selected (although the dial where the EI is set is typically labeled 'ISO') [16].

Most digital cameras can be used over a range of EIs. With slide model cameras, setting the EI (whether done automatically or manually) also sets the fixed color rendering transform. With photofinishing models, it is possible to let the adaptive color rendering algorithm determine the appropriate transform independent of the EI setting. In this case the (internal) transformation to scene-referred image data may be fixed, and the EI is used only to determine the camera exposure settings.

In developing ISO 12232, this situation was anticipated and methods for determining digital camera ISO speed ratings and ISO speed latitudes are specified. Basically, the ISO speed rating is the highest EI which allows the digital camera to produce the highest quality still images of which it is capable (on continuous-tone reflection prints printed at 170 pixels per inch). The ISO speed latitude is the range of EIs which can be used to obtain 'acceptable' images. Acceptability was determined through psychophysical experimentation, and is limited by highlight clipping for the ISO speed latitude lower limit, and by noise at the ISO speed latitude upper limit [62].

It is worth mentioning that, depending on the type of color rendering applied, the EI may go outside the ISO speed latitude. For example, the ISO speed latitude lower limit (saturation speed) assumes clipping begins at an exposure 7.8 times the scene arithmetic mean exposure. The video reproduction model starts clipping at about four times the arithmetic mean exposure. If the video reproduction model is used, EI values of almost half the speed latitude lower limit can be used without increasing the highlight clipping beyond the (typically rather large) amount produced by the video reproduction model. In fact, if the video reproduction model is used and the EI is set on the ISO speed latitude lower limit, the resulting images will tend to be too dark, because the scene arithmetic mean has been placed too far down on the tone curve. It is necessary to consider the color rendering that will be applied when selecting the EI a digital camera will use.

Utilizing standard output-referred image data

So far, we have been discussing the workflow option where the digital camera produces standard output-referred image data for exchange. This option is the overwhelming choice, and because of advantages previously noted will likely remain the dominant workflow for consumer imaging. However, it is important to keep in mind that the production of standard output-referred image data is not a complete color processing solution. It will be the final step if the output device directly utilizes this type of image data, such as sRGB image data viewed on a CRT display. In other cases, the actual output device will require a different kind of image data, such as the halftoned CMYK image data utilized by an ink-jet printer. In the second case, additional color rendering from the standard output-referred form to the specific output colorimetry is required. This type of color rendering is called tone and gamut mapping, and is required in addition to any color separation or halftoning processing that may be needed.

The job of tone and gamut mapping is to take original-referred, standard output-referred, or output-referred image data and map the colorimetry of the source image to colorimetry which will produce an optimally pleasing image on the actual reproduction medium. Tone and gamut mapping algorithms will be simple if the starting colorimetry is close to the colorimetry of the actual device. If large differences between the source image and destination medium (and viewing conditions) require the gamut mapping to make major alterations to the colorimetry, the gamut mapping may be quite complex, and may need to be image specific.

A common myth concerning gamut mapping is that the source gamut is always larger than the destination gamut. This is not necessarily true. It is equally important for a gamut mapping algorithm to be able to expand the source gamut to fill some larger destination gamut. In many cases, the differences between the source and destination gamuts will result in the destination gamut being smaller in some areas and larger in others. It is true that if a colorimetric reproduction model is being applied, the gamut mapping can only apply a clipping process. However, for most applications, preferred reproduction models are used and scene tone scales and gamuts are compressed and expanded as needed into the intermediate reproduction description luminance range and gamut. Then the intermediate reproduction description is compressed or expanded as needed to produce the final reproduction. The important consideration is how well suited the intermediate reproduction description is for the final reproduction medium.

In the limit where the intermediate reproduction description matches the final reproduction colorimetry, there is no need to perform color rendering for output. In this case, the quality achievable through the intermediate reproduction description approach equals that of the deferred color rendering approach, because all the color rendering is applied at once to go from the scene-referred image data directly to the final output-referred colorimetry. As this situation is approached, the need for complex (and perhaps image specific) color rendering algorithms, which may not be available in many output devices, is eliminated.

Selection of a standard output-referred rendering target

In the previous section the importance of careful selection of an intermediate reproduction description was stressed, and the requirements alluded to. These requirements are to choose an intermediate reproduction with characteristics which:

1. are as similar as possible to those of the expected higher quality output media,
2. are similar to those of output devices which will be most frequently encountered, and
3. are sufficiently similar to those of devices with limited tone and gamut mapping capabilities to allow images communicated to be widely utilized by such devices.

These requirements are somewhat conflicting and force application-dependent trade-offs to be made. These trade-offs are reflected in the standard output-referred color encoding specifications which have been developed. For example, the near-exact correspondence between the sRGB virtual medium and the widely used CRT display, and the similarity of its luminance range and color gamut to many print media, make it an excellent choice for high image quality exchange with maximum transportability. It is easily usable by almost any device, and at the same time can be used as the intermediate description in the production of high quality images.

In high-end graphic arts and photography applications, widespread usability is less important, and ultimate quality is at a premium. In these applications, the color accuracy limitations caused by gamut mapping from the sRGB color gamut to that of a photographic print may begin to be noticeable. For these applications, 'extended' color encodings have been developed with rendering targets more similar to a photographic print. These encodings allow for an increase in the accuracy of some high-chroma colors in prints at the expense of a slight decrease in the quality of the CRT displayed image. In some cases, such color encodings will also not be widely readable.

9.3.3 Device-independent color via deferred processing

The other workflow option is to defer some or all of the color processing to the time of output. This workflow has the advantage of allowing the color rendering to be performed all at once to produce the actual final output. The major disadvantage of this workflow is that the processing capability must be available at the time of output. Generally, it is not practical to offer this capability in all devices in an 'image appliance' environment, which precludes the use of deferred processing in mainstream consumer imaging. Another disadvantage is that frequently decisions about the artistic intent of the image must be made when performing the rendering. If the color processing is deferred, it may be difficult to communicate the desired artistic intent with the image data.

Advantages to deferred processing include:

1. The ability to alter the artistic intent with minimal quality degradation.
2. The ability to apply more advanced color rendering algorithms to archived image data as algorithms are developed.
3. Elimination of any image quality degradation that might result from going through the intermediate reproduction description.

Utilizing scene-referred image data

If the starting point for the color rendering is scene-referred image data, the color processing must apply the color rendering. This workflow allows users to take full advantage of 1 and 3 above, and to take advantage of 2 as applied to color rendering algorithms.

Utilizing scene-referred image data with an attached ICC profile

Probably the most practical path to robust exchange of scene-referred image data is by attaching an ICC profile. The ICC profile can be used by devices that support ICC color management to process the image data to a standard output-referred representation, while maintaining the advantages of a scene-referred encoding. The ICC profile can also indicate an artistic intent.

Utilizing raw image data

If the starting point for the color rendering is raw image data, the color processing must deal with both the transformation from raw device data to colorimetric (device-independent) data, and also perform the color rendering. This workflow requires more processing capability than utilizing scene-referred image data, particularly in the case of color filter array (CFA) data, but has the additional advantage that improvements in demosaicing, adapted white estimation, and transformations from raw to scene-referred data can be applied to the archived raw data as they are developed. However, a potential hazard in archiving raw image data is that camera characterization information is required for correct interpretation of the data. This information is typically included in the image file as metadata, but if somehow it is lost the image quality achievable by processing the raw data will be reduced, and human intervention in the processing almost certainly required. Unfortunately, ICC profiles do not provide a means for interpreting raw image data, because they do not address demosaicing and other necessary raw processing operations.

9.4 DIGITAL CAMERA IMAGE PROCESSING CHOICES: COMPONENT SELECTION, CHARACTERIZATION METHODS, AND PROCESSING ALGORITHMS

Having outlined the architecture, workflows, and steps used in the processing of digital photography image data, we can now focus on specific choices to be made when designing digital cameras and the image processing contained therein.

9.4.1 Capture of raw image data

The acquisition of high quality digital camera data starts with the choice of optics, image sensor, and color filters. While we will defer detailed discussion of optical issues, it is important to match the lens design to the angular acceptance characteristics and quantum efficiency of the pixels; to jointly consider the lens modulation transfer function, optical blur filters, and sensor pixel size; and to properly choose IR blocking filters that adequately suppress photons of wavelength greater than the desired red cutoff.

The study of image sensor attributes and their relation to imaging outcomes is an extensive topic in itself. Here we will summarize the key considerations for color reproduction. The fundamental sensor attributes underpinning image quality are pixel area, quantum efficiency as a function of wavelength, dark current, and read-out noise. Innovative design and silicon process improvements can alter the balance of performance characteristics from one device to another, but to date their effects have been relatively small. Pixel area determines photon collection area and charge capacity, which affect signal to noise, dynamic range, and ISO speed ratings. Quantum efficiency, dark current, and read-out noise are the other major factors affecting ISO speed ratings. The read-out noise sets the noise floor of the sensor and therefore, in conjunction with the charge capacity, the maximum dynamic range available in a single capture. At the present state of consumer camera CCD technology, read-out noise is typically in the range of $10-15$ electrons, and peak quantum efficiencies are typically in the range of $25-35\%$. Therefore, a fundamental realization is that, just as with film, the path to the highest quality images leads in the direction of sensors of larger area.

Since pixel size is a design variable, and a highly cost-sensitive one, the challenge for consumer cameras is to use most effectively the collection area and well depth available. The most significant factors affecting this utilization are the white balancing and the color matrixing which must be applied to transform raw image data to scene-referred image data. If the channels of a sensor are out of balance for a particular scene illumination source, it will be necessary to use channel multipliers significantly larger than one to perform the white balancing. In this case noise is amplified, and the effective saturation ceiling is lowered because even the channels which are not amplified have to be clipped so that a white object in the scene will have neutral (equal) digital values. This constraint is particularly troublesome as the channel multipliers required for different scene illumination sources can be quite different.

The amount of noise amplification by the color matrixing is determined by the amplitudes of negative cross-terms in the matrix (and the corresponding amplitudes of the diagonal coefficients), which are in part determined by the similarity of the camera color channel response functions to the color matching functions derived from the encoding primaries. Generally, these color matching functions will have negative lobes. This situation results in a tradeoff between matrix noise amplification and color analysis accuracy. Materials constraints in the design of the color channel responses can also negatively impact both noise amplification and color analysis simultaneously. At the very least, color channel responses should be selected to avoid this situation to the extent possible. Different response choices can be tested mathematically, given an analytical method for determining the channel multipliers and transformation matrix [74, 75].

Generally, once optimization has been performed to focus on the tradeoff, color accuracy is often sacrificed for reduced noise. This is because the reduction in color accuracy is not too noticeable in photographic applications (where the scene colors are modified by color rendering and gamut mapping anyway). In fact, it is even possible to achieve an advantage. For example, if the ITU-R BT.709 primaries [37] are used as the encoding primaries (such as is the case for sRGB), it is possible to optimize color channel responses to reduce cross-terms at the expense of color analysis accuracy. In the limit, the cross-terms will go to zero, the diagonal coefficients will go to one, and the effect on color analysis accuracy will be to map the scene colors into the encoding gamut.

9.4.2 Linearization to focal plane irradiance

Two factors must be dealt with to linearize the raw image data with respect to focal plane irradiance: dark current subtraction and correction of sensor nonlinearities. In the extreme case of nonlinearity, 'bad pixels' are also removed from the image, generally by interpolation of neighboring pixels of the same color channel.

Dark current subtraction is best achieved through dark frame subtraction. One or more dark frames are captured under the actual picture-taking conditions (exposure time and sensor temperature), and the frame or average of multiple frames is subtracted from the image. This process removes not only the mean level dark current, but also the fixed pattern point variation in the dark current, particularly when an average dark frame is used.

A simpler alternative to dark frame subtraction is optical black subtraction. In this case, pixels at the edges of the sensor are covered with metal to prevent photons from reaching them. These pixels are averaged to determine a mean level dark current which is then subtracted from the raw image data. This approach is more commonly used because it is simpler. It produces good results in sensors with low and uniform intrinsic dark current where temperatures are not too high and exposure times are not too long. In cases where there is gradual variation in the dark current across the sensor, the optical black averages can be moving averages.

Most CCD sensors used in digital cameras tend to be fairly linear, so correction of nonlinearities is not generally needed. Also, nonlinearities that are present can be image and operating condition specific, so correction can be difficult. In most cases nonlinearities are avoided or corrected at a gross level. It is also possible for the support electronics to introduce nonlinearities. ISO 14524 [23] provides methods for measuring the opto-electronic conversion function (OECF) of a digital camera. The focal plane measurement methods described are appropriate for measuring the linearity of the image sensor and support electronics combination. These methods can be used to check for linearity so that nonlinearity problems can be diagnosed and corrected.

9.4.3 Flare removal

The next processing step is to remove signal produced by the optical flare light. Flare light is primarily composed of light scattered by the camera optics and light reflected around inside the camera. Consequently the amount of flare light varies over the focal plane. Methods for the estimation and removal of flare light tend to be proprietary and covered by patents.

In uncalibrated processing approaches, flare light is frequently ignored because it is not a major cause of image degradation. Scene analysis plus color rendering approaches require estimation and removal of flare light, but it is possible to do an adequate job of this for pictorial applications by estimating a mean amount of flare light for the entire image and just subtracting this mean amount. The mean amount will depend on the camera design, but is typically in the neighborhood of 4% of the mean signal received by a color channel. Mean flare light values must be computed from and subtracted from linearized raw data. One should also be conservative in subtracting mean level flare light, because the actual flare light will be less than the mean in the darker areas of the image. If the true mean is subtracted, it could clip the image data in the darker areas.

Measurements of the flare light in a particular camera can be made by measuring both the focal plane and camera OECFs according to ISO 14524, and comparing the results. The flare light will be the difference.

9.4.4 Adopted white selection

Regardless of the workflow or specific processing algorithms used in a digital photography system, the adopted white selection and balancing are the fundamental procedures that define the frame of reference for color rendering [76]. There are two important and separable choices to be made in this regard: selection of the adopted white chromaticity, and selection of the adopted white luminance. In both cases, the goal is almost always to estimate the chromaticity and luminance of the adopted white as perceived by the human visual system, and then use the estimates obtained as the adopted white for the image. However, since digital cameras are usually not colorimetric, there will be some error in the estimation, and the adopted white chromaticities selected will be expressed as camera (not colorimetric) 'chromaticity' and 'luminance' coordinates. The terms 'adopted white' and 'adapted white' are differentiated to allow discrimination between the adopted white which is used by the imaging system and therefore definitely known, and the adapted white which is used by the human visual system and therefore can only be estimated (with varying degrees of accuracy depending on the scene).

Estimating the adapted white chromaticity

Color constancy is the ability of the human visual system to compensate for the color of illumination that exists in the scene: the color perception of a surface can remain constant under a variety of lights even when large shifts in measured radiance occur. This property can be described in imaging using the concept of 'adapted white' (as defined in the terminology section at the end of this chapter). One way to think of this conjecture is that the human visual system somehow comes up with a balance point that dictates the perception of white and neutral parts of a scene. Unfortunately, the visual mechanisms of color constancy are not yet understood; modeling and explaining this effect remains one of the most interesting areas of color research [48, 83, 89].

Any photographic system must balance the image to compensate for this 'adapted white' perception. The balance point chosen, as used in color appearance models [57, 68, 69], is referred to as the 'adopted white'. Although preference is used in color rendering to gauge properties such as tone, contrasts, and color saturation, it plays a much smaller role (if any) in the determination of the adopted white. Indeed, having the adopted white equal the adapted white almost always results in higher image quality – unfortunately, this condition remains rather elusive to automatic algorithms. The difficulty of white point determination and balancing, however, leaves room for proprietary algorithms and the potential of competitive advantage.

Professional analog and digital photography systems benefit from the abilities of a skilled photographer to manually correct for the color of illumination. In conventional photography, the professional photographer will measure the color of the light source in a scene using a color meter and then, based on the type of film being used, will select one or more filters from a collection of 'cc' or 'color correction' filters to be placed in front of the camera lens. The spectral transmission of the filter converts the neutral point of the scene to that designed into the film sensitivities. The results from this are tested by placing a gray card in the scene, exposing and developing a photograph, and then testing the gray card area of the scene for the desired color balance. In professional digital photography, the standard procedure is first to photograph the scene with the addition of a gray card, then probe this test image using the camera's control software at the point in the scene where the gray card lies, and use this image data to determine the adopted white point. As long as the lighting is not changed, the rest of the session can be shot and processed using this balance point.

Conventional consumer photography skirts this issue by assuming that most photographs are taken outdoors in sunlight or indoors using the camera's flash that is balanced for daylight. When these conditions do not arise, the photographs usually have a noticeable color cast (yellow-orange color casts for tungsten scenes are surprisingly common) that can be partially corrected at the printing stage – although this is usually left up to the consumer to complain and ask for reprinting. Newer photofinishing systems are now coming online that use more sophisticated methods to set the image white point, but there is little published information on the procedures they use.

Consumer digital photography, on the other hand, relies heavily on automatic algorithms to adjust for the wide variety of light source colors experienced. These automatic algorithms can be categorized based on whether they are intended for calibrated or uncalibrated image data. With uncalibrated algorithms, the adopted white is calculated using only the image data and not related back to the scene. These methods all use some kind of broad assumptions about the nature of scenes. The most basic method assumes that the average data values in an image (or sub-image region) should be set to gray, and all other colors are shifted in proportion. For this method to reliably give good results, all scenes must average to gray. The most famous example of the failure of this 'gray world' algorithm is the 'baby on the red rug' scene which when corrected using the gray world assumption shifts the red towards gray and the baby towards green (appearing sickly). Another popular technique, often employed in video processing, is to set either the total signal maximum of the scene to white or the maximum of each color channel to white, scaling the rest of the image proportionally. Although there is more evidence from psychophysics to support this 'max RGB' method [45], it is still easy to find failure conditions that lead to poor image quality in still images [77]. Furthermore, the limited dynamic range capabilities of many digital cameras force highlights to be clipped, which impedes the use of this method. Many improvements such as adding image semantics (e.g. blue sky detection and face recognition) have been added to these techniques to enable successful implementation in photographic products, but these methods are usually prohibitively computationally intensive and still do not completely eliminate failures.

More recently, several methods have appeared that depend on *a priori* knowledge of the capture system [46, 47, 49–51, 53, 54, 58, 60, 64]. In the 'color by correlation' method employed in the new HP and Pentax digital cameras [67, 73, 79, 97], the spectral sensitivities and linearity of each camera are used to define the color space in which statistical comparisons are made. This method uses a correlation matrix memory to compare the color signals that result from a captured image with a database of color signals derived from surface statistics, plausible light sources, and the particular camera's spectral sensitivities. Using this technique, the adopted white used by the camera has to be both physically possible and statistically probable, given the signals obtained from the optical system. Although this technique does give estimates of the visual system's adapted white points that are more reliable than uncalibrated approaches when applied to images of scenes with a wide variety of illumination conditions, it requires that the capture system be rigorously characterized and stable. It is also worth noting that 'gray world' and 'max RGB' become specific, simple instances of 'color by correlation' when applied to calibrated image data.

In view of the difficulty of adapted white chromaticity estimation and its profound impact on the color quality of images, it is likely that in the future algorithms will take advantage of whatever information is available. Scene estimation based algorithms will outperform uncalibrated algorithms. Image capture information and analysis of scene features will be used to the extent they are beneficial and practical. Finally, for critical pictorial imaging applications, human intervention will continue to be needed to either avoid or correct for corner cases. In this regard, the ability to process the raw image data after capture is a significant advantage of digital photography. If a professional photographer misjudges the adapted white when photographing on transparency film, the shooting session is usually a total loss. Currently, manual digital correction is accomplished in various applications using uncalibrated approaches. In the future, it will be possible to perform corrections based on calibrated image data, which will simplify and improve white balancing after capture.

Estimating adapted white luminance

It is also necessary to come up with an adequate estimate of the scene adapted white luminance to produce high quality images. This estimation is closely tied to the color rendering. If fixed (slide model) color rendering is applied, the selection of the adopted white luminance will determine the overall lightness of the images produced. If adaptive (photofinishing model) color rendering is used, the adopted white luminance selection will affect the overall lightness to a lesser degree, and determine where detail is lost in the highlights.

The overall lightness of an image is arguably the most important image quality attribute. The truism that the most important image quality attribute in any particular image is the one which is most lacking still holds, but in the absence of a particular problem, tone reproduction is often cited as the first item on the image quality shopping list [66]. Overall lightness is tone reproduction to the first order. Despite the importance of image lightness, no one has yet developed algorithms that adjust it perfectly for every conceivable scene. Relatively simple methods do an adequate job for most scenes in noncritical applications. The difficulties relate to gaps in the understanding of human visual perception, and to the fact that different viewers will tend to prefer somewhat different image lightnesses for the same scene.

The 'gray world' assumption noted previously can be applied to determine image lightness. With this method, the arithmetic mean scene luminance is mapped to some midtone luminance on a reproduction medium. The scene adopted white luminance is then just the reproduction medium white mapped back to the scene. Alternatively, some more perceptually uniform scene luminance average such as the geometric mean or mean L^* can be mapped to a target reproduction midtone luminance. The target midtone luminance will vary depending on the scene luminance average selected for mapping, the reproduction medium luminance ratio capabilities, and the reproduction viewing conditions. Some recommendations and typical values are provided in the tone reproduction section of the *Focal Encyclopedia of Photography* and in other work [52, 56].

It is not clear that the use of more perceptually uniform averages provides a significant advantage in determining a scene midtone luminance for mapping. This could be because the human visual system performs a more complex analysis of the scene when determining perceived lightness, so no simple average can reliably achieve the same result. For example, with some adaptive color rendering algorithms [61, 65] the scene dynamic range and key affect the lightness mapping. Recent studies show that, at least with respect to scene key, this is also the case for human vision [93].

Because of this situation, when encoding scene-referred image data it is generally desirable to maintain all the data captured. This means that no lightness adjustment will be performed, and the saturation level of the camera will be mapped to the largest digital code value of the encoding. This practice takes full advantage of the encoding precision, and defers the final selection of the adopted white luminance and resulting overall image lightness to the color rendering stage. However, specifications of scene-referred encoding standards for image exchange are in the relatively early stages of development [36, 42]. The amount of highlight headroom desirable for scene-referred encodings tends to be much larger than typical digital cameras can achieve because noise limits how far from the saturation ceiling the image exposures can be placed. This forces a choice:

1. Digital cameras can dump image data to scene-referred encodings without normalizing the white point luminance (as mentioned above). In this case, the selection of the adopted white luminance is either the result of the exposure selection (in the case of fixed or slide model color rendering) or deferred to adaptive color rendering (in the case of the photofinishing model). This choice is the simplest to implement and provides for the most compact storage of image data (particularly if the encoding bit depth is selected to match the camera's capabilities).

2. Digital cameras can select some scene adopted white luminance based on the image data captured and normalize it to some designated adopted white luminance for the color encoding. In this case, highlight data may be clipped if the actual image data highlights go above the color encoding headroom, or precision may be lost if it is necessary to scale the image data down. If the encoding has sufficient headroom and precision, the loss on normalization may be insignificant, although image file sizes may be unnecessarily large (particularly if uncompressed image data is exchanged). The advantage of this approach is that fixed color rendering transforms can be applied to generic scene-referred image data. The disadvantage is that an algorithm to estimate the scene adopted white is needed to prepare the scene-referred image data for encoding. It is possible to let the camera exposure determination system select the scene adopted white luminance, and map the image data to the scene-referred encoding in a fixed way. In this case, a particular digital camera will always utilize a fixed amount of the encoding headroom, depending on the relation of the camera headroom to the encoding headroom.

In current practice, scene-referred image data is rarely exchanged so this choice is dealt with inside the camera in an internally consistent way. In the future, scene-referred encoding specifications will need to be clear about how the adopted white luminance should be handled.

9.4.5 White balancing

After selection of the scene adopted white (in camera coordinates), it is necessary to perform the white balancing. This is typically done by multiplying the color channel

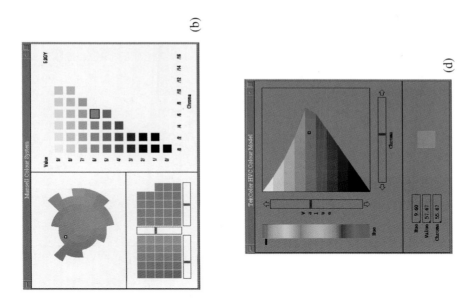

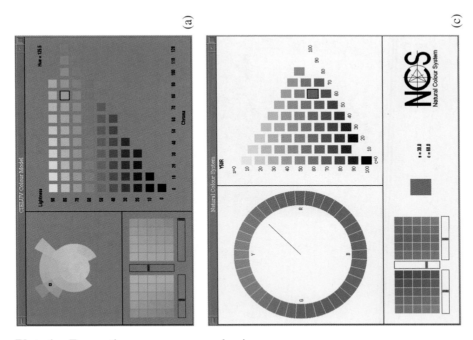

Plate 1. For caption see pages xv and xvi.

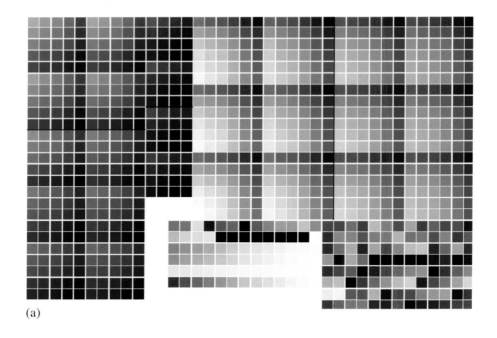

(a)

(b)

(c)

Plate 2. For caption see pages xv and xvi.

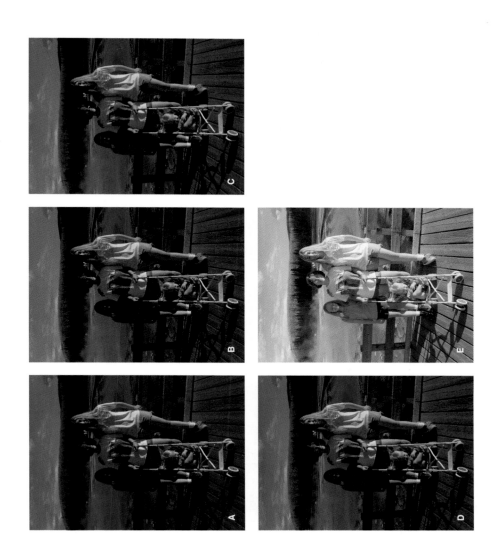

Plate 3. For caption see pages xv and xvi.

(a)

(b)

(c)

(d)

(e)

(f)

(g)

(h)

Plate 4. For caption see pages xv and xvi.

(a)

(b)

Plate 5. For caption see pages xv and xvi.

(a)

(b)

(c)

(d)

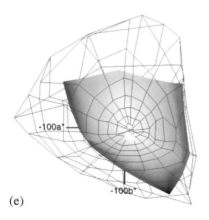

(e)

Plate 6. For caption see pages xv and xvi.

(a) sRGB   SWOP

(b) sRGB   SWOP

Plate 7.   For caption see pages xv and xvi.

(a) CMYK sRGB

CMYK sRGB

(b) CMYK sRGB

(c) sRGB SWOP

Plate 8. For caption see pages xv and xvi.

data by factors to result in equal RGB digital code values for neutrals. In the case of adopted white luminance normalization, these factors will also scale the adopted white luminance to the designated values for the scene-referred encoding. The effect of this common practice is to perform the white balancing in camera device space.

It would also be possible to perform the demosaicing and transformation to scene-referred colorimetry estimates prior to white balancing. Then, a chromatic adaptation transform could be used to transform the colorimetry estimates from the scene adopted white to the standard output-referred encoding (or actual reproduction) adopted white. From a color science standpoint this may seem to be more logical, but this approach is not generally used. This is because of the need for a color characterization transformation from device coordinates to colorimetry estimates. The essential question is: How can one most reliably obtain the desired scene-referred colorimetry estimates, where the estimates need to be relative to some particular scene-referred encoding adopted white that may be different from the actual scene adopted white?

It turns out that the method which gives the best results is first to white-balance the device values, and then apply a transform that goes from the white-balanced device values directly to scene-referred colorimetry estimates relative to some standard encoding adopted white. This practice allows the errors in the colorimetry estimates to be evaluated in the color space where they will be used. Since it is desirable to minimize errors in a nonlinear, perceptually uniform color space, application of a linear chromatic adaptation transform after errors are minimized will amplify the errors unevenly so the color characterization transformation is no longer optimal.

To calculate scene adopted white relative colorimetry estimates, it is then necessary to apply a chromatic adaptation transform from the encoding adopted white relative colorimetry back to the scene adopted white relative colorimetry. This transform will also cause uneven amplification of the errors. Research has shown that color characterization transforms are specific to the origination and destination adopted whites [64, 87]. Because of this specificity, general color characterization transforms followed by chromatic adaptation transforms produce inferior results. On the other hand, because of the specificity, it is possible to simultaneously optimize the colorimetry estimation and chromatic adaptation in a single transform, thereby reducing the combined errors.

Another advantage of performing the white balance first is that it allows one to constrain color characterization transforms to be 'white point preserving'. This constraint forces scene neutrals (relative to the adopted white) as analyzed by the camera to remain neutral. Because of the human visual system's sensitivity to small color deviations from neutrality, and because of the frequent use of preferential chroma boosts in color rendering, the white point preservation constraint can improve final image quality. It also has other benefits that make the derivation of color characterization transformations more robust and simple.

9.4.6 Demosaicing

Most consumer digital still cameras in today's market are single-CCD cameras, due to cost, size of the camera and difficulties in the optics for three-CCD configurations. In a single-CCD camera, color filters are layered over each pixel element of the CCD array in a mosaic pattern called a color filter array (CFA) pattern. The CFA pattern can vary, but the most commonly used one is the Bayer pattern [95].

Capturing only one color channel with each CCD element means that the two missing color channels have to be estimated from the existing information in order to get a full RGB image. This process is generally referred to as demosaicing. It is well known that the reconstructed image can contain aliasing artifacts that might manifest themselves as chromatic mottle or splotches in areas of the image with high spatial frequency content. The magnitude of the artifacts as well as the overall image quality of the final image depend on the optics of the camera, the sampling frequency of the sensor array, the CFA pattern, the demosaicing algorithm, and the scene captured.

Traditionally, the goal of the demosaicing process is to reconstruct an image that comes as close as possible to a full resolution image, but recently some scientists have been questioning that goal and aim for maximum perceptual quality instead [90]. Right now, we don't have a metric for perceptual demosaicing quality, thus the only way to quantify the results of different algorithms is by performing psychophysical tests. This holds true in particular as certain algorithms have the intrinsic property of increasing sharpness which might result in a reconstructed image that is preferred even over the full resolution image.

In general, demosaicing algorithms can be categorized into three different groups:

1. Simple, straightforward interpolation algorithms using surrounding cells of the same color channel.
2. Advanced interpolation algorithms that use surrounding cells of all three channels and take spatial information about the image data into account.
3. Reconstruction techniques that use information about the capturing device itself and that make certain assumptions about the captured scene.

The general concept of each group will be discussed in the following.

The most straightforward implementation of the first group is a so-called 'nearest neighbor' interpolation [92]. The Bayer CFA is divided into 2×2 blocks containing a green pixel in the upper left and lower right, a red pixel in the upper right and a blue pixel in the lower left. Now all red channels of the pixels of the block obtain the value from the red CFA element. The same holds true for the blue channel. The missing green channels obtain the averaged value of the two existing green elements of the CFA. The information of one element of the CFA spreads to the two or three neighbors depending on the actual channel, hence the name 'nearest neighbor'. The resolution of the final image isn't particularly good in comparison with a full resolution image and it might suffer from color fringing. Naturally, this specific interpolation algorithm can be substituted by other interpolation techniques like bilinear interpolation [90], but all techniques of this group use only CFA elements of the same color to estimate the missing elements for a particular color channel.

More advanced techniques, belonging to the second group, take spatial information of the image itself into account. In the simplest form this would mean interpolating the green pixels in the direction of the smallest graduation from the two options provided by the Bayer pattern. Another idea is to use CFA elements of more than one color to calculate the missing color channel information of one pixel [97]. The reason this approach tends to be superior is that the responses of the sensor in one channel can provide predictive information for the responses in other channels. This holds true in particular for natural images. From a mathematical point of view this phenomenon can be explained by the fact that the three color channels are not statistically independent, which also makes sense when we think about the relationships of the sensitivity curves of the three channels.

Conceptually, the algorithm first interpolates the green channel by bilinear interpolation and then subtracts the whole plane from the red plane. Subsequently, the (R−G) elements are interpolated and as a final step the G plane is subtracted form the interpolated cells. The same holds true for the blue channel. (R−G) and (B−G) might be thought of as the chromatic part of the image. Images resulting from this demosaicing technique will have higher resolution and a reduced number of colored edges. The argument for this method is that the sampling rate of the green CFA elements is twice as high as the one for red and blue, thus using the green CFA elements naturally increases the resolution. Colored edges are reduced because all three channels change if the green channel changes. Thus, neutrals stay neutral at the edges. Freeman [96] reduced the occurrence of small chromatic splotches even further by also applying a median filter to the (R−G) and (B−G) plane.

Obviously, there are many more ways to use the spatial information of an image to improve the quality of the interpolation process. The general idea is to perform interpolations along edges rather than across edges. Yet another improvement can be achieved by exchanging the linear interpolation in RGB camera space for a nonlinear interpolation that corresponds better with human perception. Conceptually, the same results can be achieved by a linear interpolation in a different color space whose metric corresponds better to the human visual system. This idea as well as an efficient implementation has been discussed in Tao *et al.* [82].

Tao *et al.* also discuss the selection of a more advanced interpolation filter in correspondence with the filtering and/or sampling process performed by both the lens and the CCD. The lens performs a low-pass filtering which is dependent on the modulation transfer function of the lens, while a CCD performs a spatial integration, which can be seen as a subsampling process. The interpolation filter used during the demosaicing process has to take into account the filtering and subsampling process in order to retain as much high frequency components as possible while avoiding aliasing effects. Using two distinct wavelet filters in combination with the above-mentioned, more perceptually uniform interpolation yields superior results, with improved sharpness and chromatic component accuracy.

The third group of demosaicing algorithms has been less widely discussed in the digital camera and color science literature. The basic idea is to see the initial problem as a reconstruction problem rather than an interpolation problem. The specifics of the capturing system (spectral sensitivities, spatial frequency response, and noise characteristics) are measured, and the information used for the reconstruction of missing pixels. In addition, appropriate assumptions about the spatial and spectral correlation statistics of the captured scene can further improve the output. From a mathematical point of view, the camera characterization information and statistical assumptions are used to determine a 3D tensor convolution operator which is applied to the raw image data. Demosaicing algorithms which fall into this category are sometimes referred to as Bayesian recon-struction algorithms [55, 59, 84]. Naturally this process results in higher computational complexity, but can bring the resulting image very close to a full resolution image if the camera characterization measurements and statistical assumptions are accurate.

From a theoretical point of view, the choice of a specific demosaicing algorithm becomes less critical as the digital camera pixels are viewed at smaller and smaller sizes. The increasing numbers of pixels in digital cameras therefore reduce the importance of demosaicing, if one assumed the resulting images will continue to be viewed on CRT displays without enlargement, or printed at a 'snapshot' size. On the other hand, a digital camera user might want the higher pixel count to take part of the full image and blow it up, or print the image at a large size. In both cases, the result of the demosaicing process regains its importance.

In conclusion, no matter what algorithm is used, it is very critical to avoid artifacts. Having a slightly blurred or less sharp image seems to be less of an issue with users than 'popping up pixels'. Using know-how of the human visual system to a larger extent seems to be another way forward. And naturally there is still work to be done in the development of a metric to evaluate the results of different demosaicing algorithms. In the meantime, we have to rely on observers and psychophysical experiments.

9.4.7 Determination of transformations to scene-referred colorimetry estimates

As mentioned before, the color reproduction processing within a digital camera can be separated, at least conceptually, into two stages: a scene analysis stage and a color rendering stage. Recently, examples have emerged where this separation is explicitly implemented [34, 36, 42, 93, 91]. These implementations result in an increase in flexibil-ity and utility due to the possibility of specifying rendering algorithms in a generic way. These algorithms can then be used with a variety of different scene capturing devices. The interface between the two steps is a scene-referred representation, which, unlike the raw data, is independent of the actual capturing device. Another benefit of this separa-tion arises from the fact that the majority of the proprietary techniques in digital camera image processing relate either to adapted white estimation, or to color rendering. It is therefore possible to approach the problem of obtaining scene-referred colorimetry esti-mates, given an adopted white, in open forums. This is in fact one of the objectives of ISO TC42 JWG20.

In the discussions in JWG 20, it has become clear that there is a shortage of research results in this area. Typically, the scene analysis problem is solved by photographing a test chart in the actual scene illumination, and then determining a transformation which results in the digital code values produced for the test chart patches being mapped to the colori-metric values of the chart patches. At first glance, this seems a reasonable approach, and indeed it is. However, the preceding description leaves a number of particulars unspeci-fied. These particulars can have a significant effect on the transformation determined, and therefore on a cameras scene analysis behavior. Some of these particulars are:

1. Are the test chart patch spectral radiance characteristics adequately representative of the scene; i.e. can both the patch and scene spectral radiance distributions be decomposed into similar basis functions?

2. Were the measurements of the patch spectral radiances performed using the scene illumination (both the spectral and geometric characteristics of the illumination)?
3. If the spectral sensitivities of the camera being 'characterized' are not a linear transform of color matching functions, and the scene spectral characteristics cannot be decomposed into a number of basis functions less than or equal to the number of camera analysis channels, it is not possible to determine an exact characterization transform. In this case, no matter how precise the image data and patch measurements are, there will be errors in the estimates of the patch colorimetry obtained from the image data. The magnitude of these errors for each patch will depend on the color space in which the errors are evaluated. The question is therefore: Which color space is preferred for minimization of these errors?

The answers to questions 1 and 2 above can be obtained in a straightforward manner. It is clear what the best practices are. These details are mentioned not because they present novel scientific questions, but because they are frequently overlooked. However, the literature does not offer a definitive answer to question 3.

The basic issue presented in question 3 concerns the minimization of colorimetric analysis errors that cannot be eliminated. Ideally, these errors should be minimized in a way that produces colorimetry estimates which are perceptually the closest to the scene colorimetry. Unfortunately, the only way to truly evaluate 'perceptual closeness' is through psychophysical experimentation. Therefore, the experimental approach indicated for research in this area is to evaluate the correlation of subjective appearance to the minimization of various error metrics.

In the WG20 discussions, concern has been expressed that the use of subjective methods for this evaluation will unavoidably confound preference considerations in the selection of the best error minimization criterion. For example, since human viewers tend to be critical of skin color fidelity, error metrics may favor the accurate analysis of skin colors over the accurate analysis of some other colors. While it is true that subjective methods will tend to produce a 'preferred error metric', the authors do not feel this is a disadvantage. So long as the transformations determined do in fact minimize colorimetric analysis error, it is not clear that certain preferential biases in the error minimization are a problem.

A related issue is the interaction between analysis errors and subsequent color rendering. One would like to avoid colorimetric analysis errors that might be unduly amplified or otherwise made objectionable by the color rendering processing. However, this desire potentially recouples color analysis and color rendering. The current opinion of the authors is that, except for enforcing correct analysis of neutrals, there is little need to consider the color rendering algorithm to be used when minimizing scene analysis errors. This stance, if widely adopted, could reinforce the decoupling between color analysis and rendering by discouraging color rendering algorithms which are especially sensitive to particular color analysis errors. It may be wise to avoid such color rendering algorithms in any case because the color analysis errors are affected not only by the error minimization criteria, but also by the scene illumination and camera spectral sensitivities.

In determining an error minimization criterion, there are two basic variables: the stimuli or patches for which the analysis error is minimized, and the color space in which the error minimization is performed. A variation on the patch selection is the use of increased weights for specific patches, resulting in higher importance of errors in certain color areas.

Another useful consideration is the class of scene to be analyzed. In the development of ISO 17321 [34, 35] the following three classes have been identified.

Class A analysis (known and limited 'scene' spectral radiance behavior)

Analysis is of this class when the 'scene' spectral radiance distributions can be decomposed into basis functions not greater in number than the number of independent camera spectral analysis channels. The most common instance of class A analysis is the capture of photographic and digital or photomechanically printed images. In the production of such images, three independent colorants are typically used (cyan, magenta, and yellow). These colorants are well suited to analysis by typical RGB scanners and cameras. In printing processes, a black colorant is also typically used, and sometimes light cyan and magenta colorants are used, but these colorants are not spectrally independent of the CMY colorants to a significant degree. If colorants which are independent of the primary CMY colorants are used, such as in 'hi-fi' color, then the resulting printed images will not be suitable for class A analysis, unless the number of independent camera channels is also increased.

A major advantage of class A analysis is that it is possible to determine a unique transform which will produce exact colorimetry estimates (except for noise and calibration measurement errors) of the original captured. This is possible because of the limited number of colorants. The camera responses are directly related to the amount of each colorant present, and the colorimetry of the original is also directly related to the amounts of each colorant present, so it is possible to determine a definitive transformation from camera responses to original colorimetry. Class A analysis is typically supported using test targets constructed according to the specifications in ISO 12641 [21].

Because of layering and other interactions between colorants, class A analysis characterization transforms can be complicated, and are represented using $3 \times n$ matrices, polynomials, and 3D look-up tables in addition to the 3×3 matrices more common to the other classes of analysis. The more complex transforms are justified because of the knowledge that a unique, exact transform exists. They are made practical by the fact that with class A analysis a well-defined and fixed reference white is usually available. If no fixed reference white exists, the application of more complex transforms is problematic because in accounting for different colorant behavior at different densities, it is necessary to know the colorant densities. If no fixed reference white exists, it may be difficult to determine absolutely the colorant densities.

Class B analysis (statistically expected scene spectral radiance behavior)

With this class of analysis, the assumption is that all spectral radiance distributions present in the scene can be decomposed into a relatively small number of basis functions (usually between 6 and 12). Frequently this decomposition is further divided into two parts: assuming a known illumination spectral power distribution, and assuming that the scene's spectral reflectance distributions can be decomposed into x basis functions. In practice, further statistical assumptions, such as the probabilities of occurrence of each basis function, may be used to overcome the need for the x color channels necessary to determine the weighting coefficients for x basis functions. However, if such assumptions are made, they may reduce the accuracy of the scene spectral reconstruction to the extent that they are incorrect. Also, since the smoother spectral correlation statistics are more likely to correspond to less saturated colors, such statistical assumptions will tend to bias analyzed image colors toward desaturation. Further bias toward desaturation can be introduced if the basis functions used do not fully represent the sharp transitions in the scene spectral radiance statistics. Because of the higher probability of occurrence of smooth basis functions, when 'drawing the line' at the number of basis functions to be used, functions with sharper transitions will more likely be dropped.

A bias towards desaturation in scene color analysis is counter to typical viewer preferences. In critical applications, the number of analysis channels used should be sufficient to accurately determine the presence of each basis spectral radiance distribution, thereby moving to class A analysis. When this is not practical, some experts argue it is better to make no assumptions about the likelihood of each basis spectral radiance distribution. If no assumptions are made, the analysis errors over an ensemble of images will increase, but the bias toward desaturation caused by the statistical assumptions will be avoided (although the bias resulting from the basis function selection will remain). Other experts argue that it is better to fully utilize statistical assumptions to minimize analysis errors, and then incorporate additional chroma boost in the color rendering to compensate for the analysis bias. This issue is significant because it affects the degree to which scene analysis and color rendering can be decoupled.

In summary, errors in Class B analysis typically arise from the fact that the camera doesn't have sufficient color channels to enable the accurate calculation of all the weighting coefficients, and/or the number of basis functions is not sufficient to accurately reconstruct the radiance and reflectance distributions of the scene. Another important issue is the fact that using the basis functions for the specification of the transformation results in a limitation of the achievable colors after applying the transformation. In other words, spectral colors in a scene won't be mapped into spectral colors. Thus, the basis functions used constrain the gamut of achievable scene-referred colors.

Class C analysis (unknown and variable scene spectral radiance behavior)

This class of scene analysis is more general than class B and is sometimes referred to as the maximum ignorance case – assuming that we know nothing about the spectral correlation statistics of the actual scene. Naturally, the question arises whether we should be that general. From a theoretical point of view, it is advantageous to use knowledge about a scene; the results should be better. On the other hand, if the assumptions made are wrong, we are in greater trouble than assuming nothing. To obtain true knowledge of general scene spectral radiance behaviors, it is necessary to measure vast numbers of scene spectral radiance distributions *in situ* with an accuracy that is very difficult to achieve. Also, the questions remain of how well we can reconstruct radiance and reflectance distributions of a natural scene with a limited set of basis functions, and how well the statistical probabilities of occurrence of the basis functions represent their actual occurrence in a specific scene.

In practice, the relative advantages of class B and class C analysis may result in the best choice being application specific. For example, class B analysis may be preferred when capturing artwork created using known colorants using known illumination. Class B analysis may also be preferred when there are other sources of uncertainty, such as in demosaicing. Class C analysis may be a better choice for determining general color transforms for arbitrary scenes that are decoupled from any color rendering considerations. Its 'maximum ignorance' nature may also make it preferable for things like color analysis error metrics. If there are no preconceived assumptions about the scene, it is impossible to engineer a good metric value rather than accurate analysis. On the other hand, if the assumptions about scenes are truly representative, a metric of this type could correlate well with camera analysis performance.

Scene analysis error metrics

Now that we have established a context for discussion, let us go back to the original question of how to determine transformations to produce scene colorimetry estimates from digital camera image data. This question boils down to the selection of an analysis error metric for the evaluation of candidate transformations. As mentioned previously, the metric choices relate to the test stimuli or patches used, and the color space in which the analysis errors are evaluated.

So far, the metrics of Table 9.2 have been investigated using psychophysical experimentation. More specific information about the different metrics and why they were chosen can be obtained from Holm *et al.* [87].

Two details that are important for the general understanding are that there are two different groups: CIE based metrics (CIE XYZ rms error, CIE $L^*u^*v^*$ ΔE, CIE $L^*a^*b^*$ ΔE, CIE $\Delta E94$) and RGB based metrics (ITU-R BT.709, so called 'prime colors' [72], and RIMM primary based metrics). For the second group we distinguish between illumination-independent Class C analysis versus illumination-weighted Class C analysis. In the second case the spectral error at a particular wavelength is weighted by the actual illumination spectral power at that wavelength.

Evaluation of the performance of different metrics

The color science community hasn't agreed on one specific method to evaluate the different error metrics, but the authors strongly believe that the right way to do it is to perform psychophysical experiments. In designing such experiments different approaches are possible, but in general it is necessary to capture images of a variety of natural scenes, send them through the steps described before in this chapter, apply the transformation matrices based on the different error metrics, perform some type of color rendering, and evaluate the results subjectively.

One experimental approach is to use the method of paired comparison, presenting pairs of images of the same scene transformed with different matrices, and ask observers to choose the image which they believe has a higher color accuracy (the colors are closer to the colors in a real scene). Afterwards, statistical methods are used to achieve interval scales of the different methods. Obviously, the goal is to use several different cameras,

Table 9.2 Error minimization criteria tested by Holm *et al.* [87]

Class B analysis

Macbeth Color Checker spectral reflectance correlation statistics, illumination dependent transformations

- CIE XYZ error minimization (or any linear transformation thereof)
- CIE $L^*u^*v^*$ ΔE minimization
- CIE $L^*a^*b^*$ ΔE minimization
- CIE $\Delta E94$ minimization
- Nonlinear ITU-R BT.709 RGB primary-based error minimization.

Class C analysis

Maximum ignorance, illumination independent

- CIE XYZ error minimization
- CIE $L^*u^*v^*\Delta E$ minimization
- CIE $L^*a^*b^*\Delta E$ minimization
- CIE $\Delta E94$ minimization.

Class C analysis

Maximum ignorance, illumination spectral power distribution weighted

- Linear ITU-R BT.709 RGB primary-based error minimization
- Nonlinear ITU-R BT.709 RGB primary-based error minimization
- Linear prime colors RGB primary-based error minimization
- Nonlinear prime colors RGB primary-based error minimization
- Linear RIMM RGB primary-based error minimization
- Nonlinear RIMM RGB primary-based error minimization.

---

a variety of scenes and a high enough number of observers, especially if the variance of the results is high.

Initial conclusions from one set of psychophysical experiments

In Holm *et al.*, the authors used two different cameras, 10 different scenes and about 10 observers to perform a ranking of those methods that weren't obviously very bad as identified by experienced observers. We won't go into details of the results here, but some trends are:

1. Transformations based on the CIE XYZ error metric and all three CIE ΔE criteria tested were found to be unacceptable for class C analysis. This is very interesting, in particular as the CIE XYZ metric tends to be used quite often. Also, error minimization in any non-illumination-weighted linear color space based on CIE color matching functions (such as a linear RGB) is mathematically equivalent to CIE XYZ minimization. A new illumination-weighted error metric is needed for determining class C transformations.

2. Nonlinear, illumination-weighted 'prime color' RGB primary-based metrics seem to perform reasonably well for class C analysis, although there is some indication that the choice of RGB metric for optimal subjective results may be camera dependent.

3. Of the Class B analysis error metrics tested using Macbeth Color Checker™ patches, CIE $L^*u^*v^*$ and CIE $L^*a^*b^*\Delta E$ error minimization scored highest. However, nonlinear illumination-weighed prime color RGB error minimization has not yet been tested for determining class B analysis transforms.
4. The differences in results for class B and class C analysis were below the level of statistical significance; however, the respective methods which scored highest for class B and class C analysis have yet to be tested against each other.

Further investigation into scene analysis error metrics

The first set of psychophysical experiments showed that using different error minimization metrics to determine the matrix transforming raw DSC image data into scene-referred image data definitely makes a difference in the final results. Furthermore, the widespread use of the CIE *XYZ* error metric unnecessarily inhibits the use of class C analysis. The question of which metric performs best has not yet been answered, particularly for class C analysis. Further research will be performed, and while we might not be able to find one metric that always performs best for each camera when capturing each scene, we are confident a metric can be found that performs reasonably well for most scenes when captured using a particular type of camera.

The person designing the color reproduction process within a digital camera should at least be aware that there are different options for the determination of this transformation and that there is a way to evaluate the different options.

A generic approach for determining class C analysis transforms

In closing this topic, we would like to provide a brief overview of how to determine illumination-weighted linear or nonlinear RGB error minimization metrics for arbitrary RGB primaries. This information is prerequisite to the investigation of class C analysis, and is not readily available elsewhere.

1. Select the RGB primaries which define the error minimization color space and determine their tristimulus values.
2. Normalize the RGB primary tristimulus values by the factors necessary so that mixing the RGB primaries together produces the desired adopted white chromaticity.
3. Construct a matrix transformation from CIE XYZ tristimulus values to the normalized RGB primary tristimulus values. This is done by arranging the normalized tristimulus values of the RGB primaries as the rows of a 3×3 matrix, and then inverting the matrix.
4. Select the error minimization color space nonlinearity. The specification of the error minimization color space is now complete.
5. Select the test patches (including the illumination source) to be used for error minimization. Note that spectral patches as produced by a monochromator can be used.
6. Measure the patch tristimulus values, and calculate aim values in the error minimization color space by converting the tristimulus values to the color space. This is accomplished by first applying the matrix transformation, and then the nonlinearity.

Absolute tristimulus values, which convey information about both the chromaticities and the luminances of the test patches, must be used.

7. Determine the camera responses to the test patches. In the case of reflective or transmissive patches, the patches should be illuminated using the same illumination as when measuring the patch tristimulus values. In the case of spectral patches, it may be desirable to choose patch luminances corresponding to those of a particular illuminant.

8. Convert the camera data values to the error minimization color space by linearizing the camera data, applying a matrix **M**, and then applying the error minimization color space nonlinearity. The matrix **M** should be the one that produces the smallest root-mean-square error between the transformed camera values and the aim values. Depending on the color space nonlinearity, the coefficients of matrix **M** may have to be determined iteratively. Also, weights may be applied to the patch errors. In the case of spectral patches, one approach is to make the weights dependent on the illumination spectral power at each wavelength.

9.4.8 Color rendering to standard output-referred representations

As mentioned previously, the second major stage in color processing is the color rendering to an output-referred image representation. This stage is considered to be proprietary, although some example implementations such as photographic, video, and colorimetric rendering are widely known. For exchange interoperability, the output reference chosen should be a standard output reference such as an sRGB display in the sRGB viewing conditions [38, 44], or the ICC perceptual intent 'ideal reflection print' in the ISO 3664 viewing condition P2, which is also used for ROMM RGB [41, 43].

The job of the color rendering algorithm is to produce a colorimetric description of the desired image on the reference output medium in the specified viewing conditions. In most cases, the desired image is a pleasing or 'preferred' image. To produce such an image, the color rendering algorithm must deal with appearance, preference, and gamut mapping issues. Appearance issues relate to the fact that equivalent colorimetric values do not convey equivalent appearance under different viewing conditions. To complicate matters further, transformations that convert scene colorimetry to equivalent appearance colorimetry on a reproduction can vary for different regions in a scene if the viewing conditions associated with the regions are different. In current photographic systems (chemical and digital), this issue is usually addressed by manual manipulation of images. However, research results on this topic are presented on occasion [45, 80, 85, 86, 94]. In the future, it is likely that proprietary solutions will be offered.

The preference and gamut mapping aspects of color rendering are closely linked, although the objectives are different. The linkages are frequently due to tradeoff situations. 'Bright' colors and dark shadows may be preferred in an image, but not available on a particular medium. When creating a preferred reproduction, it is usually necessary first to ascertain the capabilities of the reproduction medium, and then to create the most preferred reproduction possible on that medium. Other tradeoffs in color rendering can deal with non-color-related issues. For example, color saturation or shadow detail is sometimes traded off for improved noise suppression.

In discussing color rendering algorithms, it is convenient to separate them into three categories: fixed algorithms that apply the same transformation to all scenes; globally adaptive algorithms that customize the transform to the scene, but then apply the same transform to the entire scene; and locally adaptive algorithms that may vary the transform within a scene. It is also useful to indicate the rendering target of the algorithm, although in many cases the algorithms are parameterized to allow them to go to a variety of output references.

9.4.9 Encoding, compression, and file formats for exchange

After color rendering, image data is ready for encoding, compression (if desired), and exchange. The encoding step is relatively straightforward, so long as the colorimetry resulting from the color rendering processing is appropriate for the encoding. For example, if scene-referred image data is color rendered to an sRGB display, all that remains is to translate that image colorimetry into the corresponding sRGB digital code values. In this case, the translation will involve a matrix transformation to the sRGB (ITU-R BT.709) primaries (if the image colorimetry is not already represented in terms of these primaries), followed by application of the sRGB nonlinearity.

In applying the encoding transformation, it is important to consider the 'normalization' of the color rendered image data. Some color rendering processes will produce output medium normalized image colorimetry scaled so that the encoding zero maps to the output medium black, and the encoding white maps to the output medium white. Other color rendering processes will produce actual colorimetric values for the output medium. In performing the color rendering, it is important to know what the real output medium black and white are, and then to normalize as specified for the encoding. This normalization step is sometimes not spelled out. For example, in the sRGB case, there is a more explicit description of these issues in PIMA 7667 [44] than in IEC 61966-2-1 [38].

Compression and file formats

The vast majority of image files exchanged are compressed. This is because image files are large, because image data possesses a significant amount of redundancy, and also because by taking advantage of characteristics of the human visual system, it is possible to dramatically reduce file sizes without much apparent loss of visual image quality. Generally, image compression can be separated into four steps: the application of a decorrelation transform, the application of a spatial transform such as a discrete cosine or wavelet transform, a re-quantization, and entropy-based compression. Further discussion of most of these details is beyond the scope of this chapter, but they are listed because one of these steps, the decorrelation transform, is fundamentally related to color processing.

The most common file format for digital camera image exchange is EXIF [18, 20]. The EXIF file format is a combination of JPEG compression [10–14] and ISO 12234 metadata [18, 19]. EXIF specifies the use of a color transformation from sRGB to a luminance–chrominance (YCC) representation (but for decorrelation, subsampling, and quantization purposes) prior to compression. An interesting aspect of this transformation is that the YCC image data encoded potentially has a larger color gamut than sRGB. This makes it possible to color render to colors outside the sRGB CRT display gamut, while still using the standard EXIF file format. However, when interpreting the image data, it is frequently desirable to know that this has been done. For this and other reasons, a revision of the EXIF specification is currently under development. In the future, it is possible that the JPEG 2000 image compression and file format may become predominant. The JPEG 2000 standards [25–27] already distinguish the use of sRGB and 'sRGB YCC'.

9.4.10 Color rendering for output

The final step in producing a digital photograph is to color render for output. This color rendering has two components: gamut mapping from the encoded standard reproduction description to the actual output device gamut, and rendering the actual output image colorimetry to output device values (including halftoning if necessary). In some cases, for example sRGB image data to be viewed on an sRGB display, the color rendering for output does not do anything. In other cases, the encoded image colorimetry may not be well suited to the device gamut, and sophisticated gamut mapping may be necessary to produce optimal output image quality. Such gamut mapping may compress or expand the image colorimetry as needed. In this case, less sophisticated devices may employ simpler gamut mapping techniques (such as clipping) to enable viewing of the image at somewhat reduced quality levels. However, frequently this quality loss is not a concern because of inherent quality limitations in the device image viewing medium.

Color rendering for output resides in the output device or device driver, not in the digital camera. It is mentioned here because it is an important part of color processing for digital photography.

REFERENCES

1. ISO 5-1:1984, *Photography – Density measurements – Part 1: Terms, symbols and notations.*
2. ISO 5-2:2001, *Photography – Density measurements – Part 2: Geometric conditions for transmission density.*
3. ISO 5-3:1995, *Photography – Density measurements – Part 3: Spectral conditions.*
4. ISO 5-4:1995, *Photography – Density measurements – Part 4: Geometric conditions for reflection density.*
5. ISO 3028:1984, *Photography – Camera flash illuminants – Determination of ISO Spectral Distribution Index (ISO/SDI).*
6. ISO 3664:2000, *Viewing conditions – Graphic technology and photography.*
7. ISO 6728:1983, *Photography – Camera lenses – Determination of ISO color contribution index (ISO CCI).*
8. ISO 7589:1984, *Photography – Illuminants for sensitometry – Specifications for daylight and incandescent tungsten.*
9. ISO 8478:1996, *Photography – Camera lenses – Measurement of ISO spectral transmittance.*
10. ISO/IEC 10918-1:1994, *Information Technology – Digital compression and coding of continuous-tone still images: Requirements and guidelines.*
11. ISO/IEC 10918-2:1995, *Information Technology – Digital compression and coding of continuous-tone still images: Compliance testing.*

12. ISO/IEC 10918-3:1997, *Information Technology – Digital compression and coding of continuous-tone still images: Extensions.*
13. ISO/IEC 10918-3:1997/Amd 1:1999, *Provision to allow registration of new compression types and versions in the SPIFF header.*
14. ISO/IEC 10918-4:1999, *Information Technology – Digital compression and coding of continuous-tone still images: Registration of JPEG profiles, SPIFF profiles, SPIFF tags, SPIFF color spaces, APPn markers, SPIFF compression types and Registration Authorities (REGAUT).*
15. ISO/DIS 12231, *Photography – Electronic still picture cameras – Terminology.*
16. ISO 12232:1998, *Photography – Electronic still picture cameras – Determination of ISO speed.*
17. ISO 12233:2000, *Photography – Electronic still picture cameras – Resolution measurements.*
18. ISO/FDIS 12234-1, *Photography – Electronic still picture cameras – Removable memory – Part 1: Basic removable memory reference model.*
19. ISO/FDIS 12234-2, *Photography – Electronic still picture imaging – Removable memory – Part 2: Image data format – TIFF/EP.*
20. ISO/DIS 12234-3, *Photography – Electronic still picture imaging – Part 3: Design rule for camera file system (DCF).*
21. ISO 12641:1997, *Graphic Technology – Prepress digital data exchange – Color targets for input scanner calibration.*
22. ISO 13655:1996, *Graphic Technology – Spectral measurement and colorimetric computation for graphic arts images.*
23. ISO 14524:1999, *Photography – Electronic Still Picture Cameras – Methods for measuring opto-electronic conversion functions (OECFs).*
24. ISO/DIS 14807, *Photography – Transmission and reflection densitometers – Method for determining performance.*
25. ISO/IEC 15444-1:2001, *Information Technology – JPEG 2000 image coding system.*
26. ISO/IEC 15444-1:2001/FDAM2, *Inclusion of additional color space.*
27. ISO/IEC/FDIS 15444-2, *Information Technology – JPEG 2000 image coding system: Extensions.*
28. ISO/DIS 15739, *Photography – Electronic still picture imaging – Noise measurements.*
29. ISO/DIS 15790, *Graphic technology and photography – reflection and transmission metrology – Certified reference materials – Documentation and procedures for use, including determination of combined standard uncertainty.*
30. ISO/CD 16067-1, *Photography – Electronic scanners for photographic images – Spatial resolution measurements – Part 1: Scanners for reflective media.*
31. ISO/NP 16067-2, *Photography – Electronic scanners for photographic images – Spatial resolution measurements – Part 2: Film scanners.*
32. ISO/NP 16067-3, *Photography – Electronic scanners for photographic images – Dynamic range measurements.*
33. Proposal for ISO/NP (28 April 1997), *Graphic technology and photography – Standard monitor RGB specifications.*
34. ISO/WD 17321.4 (1 October 1999), *Graphic technology and photography – Color characterization of digital still cameras (DSCs) using color targets and spectral illumination.*
35. ISO/WD 17321-1, *Graphic technology and photography – Color characterization of digital still cameras (DSCs) – Part 1: stimuli, metrology and test procedures.*
36. ISO/WD 22028-1, *Extended color encodings for digital still image storage, manipulation and interchange – Part 1: Architecture and requirements.*
37. ITU-R BT.709-3:1998, *Parameter values for the HDTV standards for production and international programme exchange.*
38. IEC 61966-2-1:1999, *Multimedia systems and equipment – Color measurement and management – Part 2-1: Color management – Default RGB color space – sRGB.*
39. CIE Publication 15.2, *Colorimetry.*
40. CIE Publication 17.4, *International Lighting Vocabulary.*
41. ICC.1:2001-11, *File Format for Color Profiles.*

42. PIMA/WD 7466, *Photography – Electronic still picture imaging – Reference Input Medium Metric RGB Color Encoding: RIMM RGB.*

43. PIMA 7666:2001, *Photography – Electronic still picture imaging – Reference Output Medium Metric RGB Color Encoding: ROMM RGB.*

44. PIMA 7667:2001, *Photography – Electronic still picture imaging – Extended sRGB color encoding – e-sRGB.*

45. Land, E. H. (1986) Recent advances in retinex theory. *Vision Research*, **26**, 7–21.

46. Maloney, L. T. and Wandell, B. A. (1986) Color constancy: a method for recovering surface spectral reflectance. *J. Opt. Soc. Am. A*, **3**, 29–33.

47. Wandell, B. A. (1987) The synthesis and analysis of color images. *IEEE Trans. Patt. Anal. and Mach. Intell. PAMI-9*, 2–13.

48. Funt, B. V. and Drew, M. S. (1988) Color constancy computation in Near-Mondrian scenes using a finite dimensional linear model. *IEEE Computer Vision and Pattern Recognition Proceedings*, pp. 544–549, Ann Arbor, MI.

49. Tominaga, S. and Wandell, B. A. (1989) Standard surface-reflectance model and illuminant estimation. *J. Opt. Soc. Am. A*, **6**, 576–584.

50. Forsyth, D. (1990) A novel algorithm for color constancy. *IJVC*, **5**, 5–36.

51. D'Zmura, M. (1992) Color constancy: surface color from changing illumination. *J. Opt. Soc. Am. A*, **9**, 490–493.

52. Holm, J. (1993) Tone reproduction. In *Focal Encyclopedia of Photography*, 3rd edn. pp. 791–801. Boston & London: Focal Press.

53. D'Zmura, M. and Iverson, G. (1993) Color constancy I: basic theory of two-stage linear recovery of spectral descriptions for lights and surfaces. *J. Opt. Soc. Am. A*, **10**, 2148–2165.

54. Brainard, D. H. and Freeman, W. T. (1994) Bayesian method for recovering surface and illuminant properties from photosensor responses. *Human Vision, Visual Processing, and Digital Display V*, pp. 364–376. Bellingham, WA.

55. Brainard, D. (1994) Bayesian method for reconstructing color images from trichromatic samples. *Proceedings, 47th Annual Convention of the Society for Imaging Science & Technology*, pp. 375–380. Rochester, NY.

56. Holm, J. (1994) Exposure-speed relations and tone reproduction. *Proceedings, 47th Annual Convention of the Society for Imaging Science & Technology*, pp. 641–648. Rochester, NY.

57. Hunt, R. W. G. (1995) *The Reproduction of Color*, 5th edn. Fountain Press.

58. Finlayson, G. D. (1995) Color constancy in diagonal chromaticity space. *IEEE Computer Society: Proceedings of the Fifth International Conference on Computer Vision*, pp. 218–223. Cambridge, MA.

59. Brainard, D. and Sherman, D. (1995) Reconstructing images from trichromatic samples: from basic research to practical applications. *Proceedings, IS&T/SID Third Color Imaging Conference: Color Science, Systems, and Applications*, pp. 4–10. Scottsdale, AZ.

60. Finlayson, G. D. (1996) Color in perspective. *IEEE transactions on Pattern Analysis and Machine Intelligence*, **18**(10), 1034–1038.

61. Holm, J. (1996) Photographic tone and color reproduction goals. *Proceedings, CIE Expert Symposium '96: Color Standards for Image Technology*, pp. 51–56. Vienna, Austria.

62. Holm, J. (1996) The Photographic Sensitivity of Electronic Still Cameras. *Journal of Photographic Science & Technology, Japan, 70th Anniversary Edition*, **59**(1), 117–131.

63. Holm, J. (1996) Factors to consider in pictorial digital image processing. *Proceedings, 49th Annual Conference of the Society for Imaging Science & Technology*, pp. 298–304. Minneapolis, MN.

64. Funt, B. V., Cardei, V. and Barnard, K. (1996) Learning color constancy. *Proceedings, IS&T/SID Fourth Color Imaging Conference: Color Science, Systems, and Applications*, pp. 58–60. Scottsdale, AZ.

65. Holm, J. (1996) A strategy for pictorial digital image processing (PDIP). *Proceedings, IS&T/SID Fourth Color Imaging Conference: Color Science, Systems, and Applications*, pp. 194–201. Scottsdale, AZ.

66. Hunt, R. W. G. (1996) Why is black and white so important in color. *Proceedings, IS&T/SID Fourth Color Imaging Conference: Color Science, Systems, and Applications*, pp. 54–57. Scottsdale, AZ.

67. Finlayson, G. D., Hordley, S. and Hubel, P. M. (1997) Color by correlation. *Proceedings, IS&T/SID Fifth Color Imaging Conference: Color Science, Systems, and Applications*, pp. 6–11. Scottsdale, AZ.

68. Hunt, R. W. G. and Luo, M. R. (1998) The Structure of the CIE 1997 Color Appearance Model (CIECAM97s). *Color Research and Application*, **23**, 138–146.

69. Fairchild, M. D. (1998) *Color Appearance Models*. Reading, MA: Addison-Wesley.

70. Adams, J., Parulski, K. and Spaulding, K. (1998) Color Processing in Digital Cameras. *IEEE Micro*, **18**(6), 20–31.

71. Holm, J. (1998) Issues relating to the transformation of sensor data into standard color spaces. In *Recent Progress in Color Management and Color Communications*, R. Buckley (ed.), pp. 193–197. IS&T Press.

72. Brill, M. H., Finlayson, G. D., Hubel, P. M. and Thornton, W. A. (1998) Prime colors and color imaging. *Proceedings, IS&T/SID Sixth Color Imaging Conference: Color Science, Systems, and Applications*, pp. 33–41. Scottsdale, AZ.

73. Hubel, P. M., Holm, J. and Finlayson, G. D. (1999) Illuminant estimation and color correction. In *Color Imaging: Vision and Technology*, L. MacDonald (ed.), pp. 73–95. Wiley.

74. Baer, R. L., Holland, W. D., Holm, J. and Vora, P. (1999) A comparison of primary and complementary color filters for CCD-based digital photography. *Proceedings of SPIE, Volume 3650: Sensors, Cameras, and Applications for Digital Photography*, pp. 16–25. San Jose, CA.

75. Baer, R. L. and Holm, J. (1999) A model for calculating the potential ISO speeds of digital still cameras based upon CCD characteristics. *Proceedings, 2nd IS&T PICS Conference*, pp. 35–38. Savannah, GA.

76. Holm, J. (1999) Adjusting for the scene adopted white. *Proceedings, IS&T PICS Conference*, pp. 158–162. Savannah, GA.

77. Hubel, P. M. (1999) Color image quality in digital cameras. *Proceedings, IS&T PICS Conference*, pp. 153–157. Savannah, GA.

78. Saito, O., *et al.* (1999) Signal processing system for high resolution digital camera. *Proceedings, IS&T PICS Conference*, pp. 76–79. Savannah, GA.

79. Finlayson, G. D., Hordley, S. and Hubel, P. M. (1999) Unifying color constancy. *Proceedings, IS&T/SID Seventh Color Imaging Conference: Color Science, Systems, and Applications*, pp. 120–126. Scottsdale, AZ.

80. McCann, J. (1999) Lessons learned from mondrians applied to real images and color gamuts. *Proceedings, IS&T/SID Seventh Color Imaging Conference: Color Science, Systems, and Applications*, pp. 1–8. Scottsdale, AZ.

81. Holm, J. (1999) Integrating new color image processing techniques with color management. *Proceedings, IS&T/SID Seventh Color Imaging Conference: Color Science, Systems, and Applications*, pp. 80–86. Scottsdale, AZ.

82. Tao, B., Tastl, I., Cooper, T., Blasgen, M. and Edwards, E. (1999) Demosaicing using human visual properties and wavelet interpolation filtering. *Proceedings, IS&T/SID Seventh Color Imaging Conference: Color Science, Systems, and Applications*, pp. 252–256. Scottsdale, AZ.

83. Hubel, P. M. (2000) The perception of color at dawn and dusk. *J. Im. Sci. Tech.*, **44**(4), 371–375.

84. Taubman, D. (2000) Generalized wiener reconstruction of images from color sensor data using a scale invariant prior. *Proceedings of IEEE International Conference of Image Processing*, **3**, 801–804. Vancouver, Canada.

85. Moroney, N. (2000) Local color correction using non-linear masking. *Proceedings, IS&T/SID Eighth Color Imaging Conference: Color Science and Engineering, Systems, Technologies, Applications*, pp. 108–111. Scottsdale, AZ.

86. Funt, B., Ciurea, F. and McCann, J. (2000) Retinex in matlab. *Proceedings, IS&T/SID Eighth Color Imaging Conference: Color Science and Engineering, Systems, Technologies, Applications*, pp. 112–121. Scottsdale, AZ.

87. Holm, J., Tastl, I. and Hordley, S. (2000) Evaluation of DSC (digital still camera) scene analysis error metrics – Part 1. *Proceedings, IS&T/SID Eighth Color Imaging Conference: Color Science and Engineering, Systems, Technologies, Applications*, pp. 279–287. Scottsdale, AZ.
88. Holm, J., Edwards, E. and Parulski, K. (2000) Extended color encoding requirements for photographic applications. *Proceedings, CIE Expert Symposium on Color Encodings*, in press, Scottsdale, AZ.
89. Süsstrunk, S., Holm, J. and Finlayson, G. D. (2001) Chromatic adaptation performance of different RGB sensors. *Proceedings of SPIE, Volume 4300: Device Independent Color, Color Hardcopy, and Graphic Arts VI*, pp. 172–183. San Jose, CA.
90. Longere, P., Zhang, X., DelaHunt, P. and Brainard, D. (2001) Perceptual assessment of demosaicing algorithm performance. Submitted to *Proceedings of IEEE*, in press.
91. Hanlon, L. R., Baer, R. L., Holm, J. and Hubel, P. (2001) Generating scene-referred data in a digital still camera. *Proceedings, 4th IS&T PICS Conference*, pp. 271–276. Montreal, Canada.
92. Palum, R. (2001) Image sampling with the Bayer color filter array. *Proceedings, 4th IS&T PICS Conference*, pp. 239–245. Montreal, Canada.
93. Moroney, N. (2001) Background and the perception of lightness. *Proceedings of the AIC*, in press, Rochester, NY.
94. Moroney, N. (2001) Chroma scaling and crispening. *Proceedings, IS&T/SID Ninth Color Imaging Conference: Color Science and Engineering, Systems, Technologies, Applications*, in press, Scottsdale, AZ.
95. Bayer, B. (1975) Color Imaging Array. US Patent 3,971,065.
96. Freeman, W. (1988) Method and Apparatus for Reconstructing Missing Color Samples. US Patent 4,774,565.
97. Hubel, P. M. and Finlayson, G. D. (2000) Whitepoint Determination Using Correlation Matrix Memory. US Patent 6,038,339.
98. Holm, J. (2001) Strategy for Pictorial Digital Image Processing. US Patent 6,249,315.
99. Holm, J. (2001) Pictorial Digital Image Processing Incorporating Image and Output Device Modifications. Patent Application US2001/0009590.

APPENDIX 1: TERMINOLOGY

The following terms and definitions are selected from CIE Publication 17.4 [39], ISO 12231 [15], and ISO 22028 [36].

adapted white – absolute colorimetric co-ordinates of the spectral radiance distribution that an observer who is adapted to the environment would consider to be a perfectly reflecting diffuse white; i.e. color stimulus that an observer would judge to be perfectly achromatic and to have a luminance factor of unity.

adopted white – for a particular area of a scene as seen by an image capture or measurement device, the absolute spectral radiance distribution (and colorimetric co-ordinates thereof) that is considered by an appearance, reproduction, or preferred reproduction model to be a perfectly reflecting diffuse white.

Note. No assumptions should be made concerning the relation between the adapted or adopted white and measurements of near-perfectly reflecting diffusers in a scene, because measurements of such diffusers will depend on the illumination and viewing geometry, and other elements in the scene that may affect perception. It is easy to arrange conditions where a near-perfectly reflecting diffuser will appear to be gray or colored.

color encoding – a quantized numerical representation of an image in a color space, including any associated data required to interpret the color appearance of the image.

Note. The image colorimetry encoded is referred to a specific manifestation of the image. For example, if scene colorimetry is encoded the encoding is classed as scene-referred.

color rendering – mapping of image data representing the colorimetric co-ordinates of the elements of a scene or original to image data representing the colorimetric co-ordinates of the elements of a reproduction.

color space – geometric representation of colors in space, usually of three dimensions (CIE Publication 17.4, 845-03-25).

Note. CIE based color spaces are traceable to CIE colorimetry either by a mathematical transform or by a look-up table.

gamut mapping – mapping of the colorimetric co-ordinates of the elements of a source image to colorimetric co-ordinates of the elements of a reproduction to compensate for differences in the source and output medium color gamut capability.

Note. The term gamut mapping is somewhat more restrictive than the term color rendering because gamut mapping is performed on colorimetry which has already been adjusted to compensate for viewing condition differences and viewer preferences, although these processing operations are frequently combined in reproduction and preferred reproduction models.

ICC PCS – International Color Consortium profile connection space, with two standard color spaces for representing the colorimetry of an image that is either original-referred or standard output-referred depending on the rendering intent of the profile.

ICC profile – International Color Consortium's file format, used to store transforms from one color encoding to another, e.g. from device color coordinates to profile connection space, as part of a color management system.

original-referred image data – input-referred image data which represents the colorimetric co-ordinates (or an approximation thereof) of a two-dimensional image produced by scanning artwork, photographic transparencies or prints, or photomechanical or other reproductions.

Note 1. Original-referred image data should relate to the colorimetric co-ordinates of the original as measured according to ISO 13655, and should not include any additional veiling glare or other flare.

Note 2. The characteristics of original-referred image data that most generally distinguish it from scene-referred image data are that it is referred to a two-dimensional surface, and the illumination incident on the two-dimensional surface is assumed to be uniform (or the image data is corrected for any non-uniformity in the illumination).

Note 3. There are classes of originals that produce original-referred image data with different characteristics. Examples include various types of artwork, photographic prints, photographic transparencies, etc. In selecting a color rendering algorithm, it is usually necessary to know the class of the original in selecting the color rendering to be applied. For example, a colorimetric intent is generally applied to artwork, while different perceptual algorithms are applied to produce photographic prints from transparencies, or newsprint reproductions from photographic prints. In some cases the assumed viewing conditions are also different between the original classes, such as between photographic prints and transparencies, and must also be considered.

Note 4. In a few cases, it may be desirable to introduce slight colorimetric errors in the production of original-referred image data, for example to make the gamut of the original more closely fit the color space, or because of the way the image data was captured (such as a Status A densitometry-based scanner).

output-referred image data – encoding of image colorimetry which has undergone color rendering appropriate for a specified real or virtual output device and viewing conditions.

Note. The output-referred image data is then referred to the specified output device and viewing conditions. A single scene can be color rendered to a variety of output-referred representations depending on the anticipated output viewing conditions, media limitations, and/or artistic intents.

preferred reproduction model – mathematical model that produces transformations which are applied to scene- or original-referred image data to produce image data describing a pleasing reproduction.

Note. Preferred reproduction models are different from reproduction models in that the pleasing reproduction need not be an attempt to reproduce the appearance of the original. In fact, what is considered pleasing may depend on viewer preferences. The transformations produced by a preferred reproduction model are generally dependent on the characteristics of the scene or original and the output medium.

raw DSC image data – image data produced by or internal to a DSC that has not been processed, except for A/D conversion and the following optional steps: linearization, dark current/frame subtraction, shading and sensitivity (flat field) correction, flare removal, white balancing (e.g. so the adopted white produces equal RGB values or no chrominance), missing color filter array pixel reconstruction (without color transformations).

reproduction model – mathematical model that produces transformations which are applied to scene- or original-referred image data to produce image data describing a reproduction which is as close as possible to being an appearance match to the original.

Note. Transformations produced by reproduction models will generally depend on the luminance ratio and color gamut of the scene or original and the output medium.

scene-referred image data – input-referred image data which represents the colorimetric co-ordinates of the elements of a scene.

standard output-referred image data – output-referred image data which is referred to a standardized real or virtual output device and viewing conditions.

Note 1. Image data intended for open interchange is most commonly standard output-referred. This is because with standard output-referred image data it is generally sufficient to specify the standard output to which the image data is referred to interpret the color appearance described by the image data.

Note 2. Standard output-referred image data may become the original for a subsequent reproduction process. For example, sRGB output-referred image data is frequently considered to be the original for the color rendering performed by an sRGB printer.

tone mapping – adjusting the tone reproduction of an image to compensate for differences in the input and output medium luminance ratio capability.

10

Characterizing hard copy printers

Phil Green

10.1 INTRODUCTION

Hard copy output systems include those of the traditional mechanical variety, on which an image carrier is mounted on the press and produces a run of identical copies; or of the electronic variety, in which the image is formed according to data transferred to it. They employ a wide range of techniques for transferring colorant to a substrate, and can be described in terms of their resolving power, the density of colorant applied, the method of varying tone, and the surface properties of the colorant and substrate (such as gloss and fluorescence).

Mechanical systems include those printing by relief, sub-relief, planographic and stencil processes. The relief processes (letterpress and flexography) are distinguished by relatively poor resolving power in comparison to offset lithography. Sub-relief (gravure) printing uniquely has the capability of imaging a variable density of colorant, all other mechanical processes being effectively binary in their transfer of colorant to substrate.

While some devices are restricted to printing on paper and board substrates, the range of media which are imaged also includes other materials such as plastics, metals and textiles. Where the target of the imaging process cannot itself be marked by the imaging process, a transfer material can be imaged and then applied to the surface of the target.

In the traditional four-colour printing processes, the chromaticity of the colorants used and the densities at which they are printed are subject to published standards [24–28], although there are quite wide tolerances in both formulation and printing. The gamuts yielded by these processes depend to a considerable extent on the surface characteristics of the printed substrate, and in the case of low-grade materials such as newsprint can be quite restricted in comparison to other media such as transparencies and displays. A number of techniques are available to enhance the gamuts of printed processes [17]. These include:

- Using inks with higher chroma or printing at higher densities
- Using a larger number of primaries, adding additional colorants with hue angles intermediate to cyan, magenta and yellow, together with higher-density black inks
- Increasing the colorant density by marking it more than once at the same location.

Other factors affecting the gamut of the printing processes include:

- The rheology and stability of the ink, which will influence printed density, wet-on-wet trapping and permanence
- The use of resins and other components which control surface properties such as gloss
- The properties of the substrate, including its chromaticity, absorption, texture and gloss
- The viewing environment, especially the chromaticity and luminance level of the source under which prints are viewed.

There are several methods of determining the gamut of a hard copy printer, including direct measurement of a printed target [18]. Such methods are described in detail in Chapter 12.

Colour imaging systems are conventionally based on three primaries, this being the minimum number needed to reconstruct a full range of hues for viewing by the trichromatic human visual system. In a sampling system each red, green or blue filter records the intensity information in a single waveband, while in a printing system each of the cyan, magenta and yellow inks similarly controls the absorption of incident illumination in a single waveband (although a black is also used to ensure an adequate density range). In some situations in sampling or printing there are advantages in increasing the number of absorption bands.

Multi-ink systems make it possible to obtain a better spectral match to the original and reduce metamerism, thus improving the likelihood that the match between original and reproduction will hold under different illuminants [42, 66]. If the match is non-metameric and thus invariant with respect to illuminant, there is no need either for standardization of viewing conditions or for colour appearance models to predict the appearance under different viewing conditions [5]. The increase in colour gamut using a multi-ink set has been described in several papers [42, 48].

Multi-ink systems can utilize either a fixed set of primaries, capable of being standardized, or a variable set which are adapted to the spectral or gamut characteristics of the original. A variable set of primaries clearly requires an adaptive transformation into colorant space.

10.1.1 Halftoning

When an image is reproduced on a digital output device, unless the number of intensity levels and the addressable resolution of the device match those of the source image, some form of dithering will be required in order to render all the intensity levels of the image. Regardless of the marking technology employed, digital halftoning imposes the constraint of a regular grid of discrete pixels, whose spacing corresponds to the 'addressable resolution' of the device. Halftoning involves the selection of a spatial region containing some multiple of these device pixels which is to visually integrate to a given tonal value, and marking the device pixels within this region in a way that achieves the requested tonal value (the region is often referred to as the halftone cell). The cell may be composed of any number of device pixels. A cell consisting of a single device pixel which can be modulated with 256 discrete intensity levels (as in dye sublimation printers) produces the same range of tones as a cell of the same size which is marked by a 16×16 array of bi-level pixels (such as an imagesetter or platesetter). For a given addressable resolution, the number of grey levels can be increased by simply increasing the cell size, but there is a risk that doing so will make the visual impact of the halftone structure objectionable. Careful selection of the halftone algorithm is required on low-resolution devices in order to avoid such effects.

It should be noted that not all halftoning algorithms are based on a rectangular cell, and of those that are, some are applied to an array of cells (or 'supercell') [16] in order to achieve better control of screen angle or tonal value.

The main problems that arise in halftoning are:

1. Visually apparent patterns or noise
2. Contouring from an insufficient number of grey levels
3. Interference patterns when halftones of different colours are overprinted.

The proportion of device pixels which are imaged in a given halftone cell is referred to as the dot area, and corresponds (within the available precision) to the halftone value requested. The halftone cell will appear to have a higher dot area owing to the effects of ink spread and internal light scattering, and this tonal value increase will vary with the surface properties of the media.

Final print dot area is also affected by tonal value transfer during platemaking. The main variables here are plate exposure amount and plate type – i.e. exposed from a negative or positive film intermediate, or directly from data.

Halftoning methods in high-resolution mechanical reproduction processes usually produce a periodic halftone screen, which has the advantage of being relatively simple to compute when compared to non-periodic halftoning algorithms. In lower resolution electronic marking technologies, periodic dot patterns produce less pleasing results, so other methods of generating dot patterns are used. In selecting a halftoning algorithm it is important to consider the physical and mechanical effects of the marking technology and the substrate: for example, in some processes there is a lower limit on the range of spot sizes that can be imaged, the choice of substrate will affect the tonal value transfer characteristics of the process, and the registration precision of the marking system will limit the accuracy with which dot placement can be controlled.

Halftoning algorithms can be classified into ordered dither methods (including clustered and dispersed dither and threshold arrays) and error diffusion (including stochastic and frequency modulated algorithms). Halftoning methods are not discussed in detail in this chapter, and readers are referred to texts where the topic has been comprehensively reviewed [41, 16, 63].

10.2 PRINTER MODELS

In order to exchange data between colour imaging devices through the medium of a device-independent colour space, it is necessary to define the relationship between the device colour space and a CIE colour space. For some hard copy media it is possible to define this relationship by modelling the physical properties of the media, as in the case of the Kubelka–Munk model, although in practice the relationship is more often defined through a numerical model determined from measurement of a sample of the colours produced by the device.

The frequency of the need to model a printing device varies widely. If the device can be recalibrated when, for example, operating conditions change or different colorant batches are used, then the device model may continue to be valid. For digital printers this recalibration may not be possible and it may be necessary to redefine the device model for each significant change that occurs. Commercial printing equipment based on processes such as lithography and flexography is subject to additional variables in operation (such as changes in ink rheology as temperature or water content fluctuate), and in practice operates within a relatively wide tolerance band. For a printer model to have any value, it is essential that colorants are colorimetrically standardized and printed to fixed and, where appropriate, standardized densities.

In commercial colour reproduction, the device model may be determined for either the printing system or the proofing system. It is sometimes desirable to apply a device model to a group of devices, such as:

1. A number of similar devices (such as printers that share the same design of marking technology or the same colorants).
2. Presses in the same industry sector whose printing characteristics can be made to conform to a process control standard.
3. A proofing system whose characteristics a production run is intended to match.

Many of the printer model evaluations in the literature were performed for electronic hard copy printers. In one study a number of different models and training sets were compared for characterizing transparency media, proofs and offset newspaper printing, and some of the results are given below in the relevant sections. In this comparison, proofs of an ISO 12641 test target were measured and a two-stage model developed to transform the measurement data to the corresponding tristimulus values for the transparency media. Prints of an ISO 12640 test target were made according to ISO 12647:3 [28] on a cold-set offset newspaper press. Some of the results have been reported previously [19].

10.2.1 Block dye model

The simplest printer model is the 'ideal ink' or 'block dye' model. If a set of CMY colorants have the block absorptions shown in Figure 10.1, then the colorant amounts

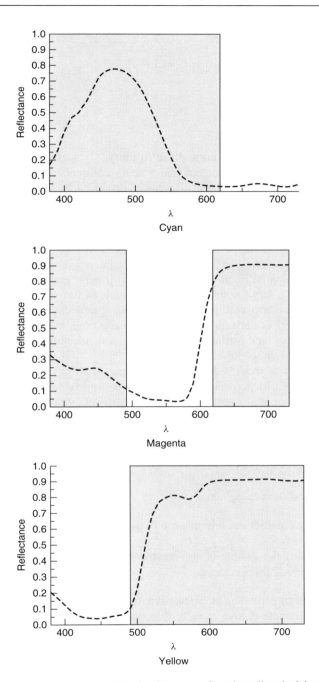

Figure 10.1 Block dye model showing ideal reflectances for a broadband trichromatic ink system, together with spectral reflectance curves of a set of real printing inks.

required to print a stimulus similarly constituted of block RGB emissions is given by:

$$C = 1 - R$$

$$M = 1 - G \tag{10.1}$$

$$Y = 1 - B$$

where C, M, Y, R, G and B are normalized to unity.

Of course, the spectral absorptions of real primary colorants do not correspond to those of the theoretical block dyes, and the reflectances produced by varying colorant densities and overprints cannot be determined by a simple subtractive model. Moreover, it would be necessary to define the RGB stimulus in terms of a set of colour matching functions in order to create a colorimetric model.

The relationship between the amount of a single colorant produced by a hard copy printer and the corresponding reflectance can be relatively simple to model. However, the relationship between colorant amounts and colorimetric quantities is inevitably more complex, owing to the large number of variable such as the spectral absorbance of the colorants used, the method of dithering employed, optical factors such as first surface reflections and internal scattering, and colorant interactions which lead to variations in the colour of secondary and tertiary overprints. As a result simple linear models cannot be used with any expectation of accuracy.

The different types of models described in Chapter 6 all have applications to printing systems. The principles of numerical models and look-up tables having been introduced in that chapter, here we will consider the physical models that have been developed for hard copy devices, and the application of numerical models and look-up tables.

10.3 PHYSICAL MODELS

Physical printer models include:

1. Those that aim to predict the relationship between reflectance and dot area or colorant strength.
2. Those that predict the colour of different colorant combinations, in terms of either colorimetry or spectral reflectance.

The term colour mixing models is often used for physical models of printing systems, although some are single-colour models that do not describe the interactions between colours.

The parameters of a physical model must often be determined empirically for a given process and substrate. Most colour mixing models can be extended or made more accurate by spectral extensions (evaluating the model at discrete spectral intervals) or cellular extensions (partitioning colour space into cubic cells and evaluating the model within each cell) [60, 51, 45].

10.3.1 Density

It is often convenient to use density as a measure of colorant amount, since it is more perceptually uniform than reflectance and also more linearly related to colorant concentration. Optical density D is calculated from reflectance as described in Chapter 1:

$$D = -\log_{10} R \qquad (10.2)$$

The value of R is found by weighting the reflectance factor at each wavelength by a suitable function. This function could be the $V(\lambda)$ function for visual density, or a function based on Status T densities for coloured inks.

10.3.2 Dot area models

Dot area models that define the relationship between dot area and either reflectance or density include the Murray–Davies, Yule–Nielsen and Clapper–Yule models described below. Models that define a relationship between colorimetry and ink density (Neugebauer equations, masking equations) are described later in the chapter.

Murray–Davies

If the reflectance measurement is normalized so that unprinted paper is given a value 1 and solid ink the value 0, then the fractional area covered by the dots of a halftone, A, and the reflectance factor of the halftone, R, are related by:

$$R = 1 - A \qquad (10.3)$$

If the reflectance measurement is not normalized to unprinted paper and solid ink, then:

$$R = R_\mathrm{w} - A(R_\mathrm{w} - R_\mathrm{s}) \qquad (10.4)$$

where R_w is the reflectance of the unprinted surface, and R_s is the reflectance of the solid. In this case the dot area is found by:

$$A = \frac{R_\mathrm{w} - R}{R_\mathrm{w} - R_\mathrm{s}} \qquad (10.5)$$

In the density domain, equation (10.5) becomes:

$$A = \frac{10^{-D_\mathrm{w}} - 10^{-D_\mathrm{t}}}{10^{-D_\mathrm{w}} - 10^{-D_\mathrm{s}}} \qquad (10.6)$$

where D_w is the density of unprinted paper, D_s is the density of the solid area and D_t is the density of the tint for which the dot area is to be calculated.

If D is normalized to paper white, this can be simplified to equation (10.7), which is the usual form of the Murray–Davies equation:

$$A = \frac{1 - 10^{-D_t}}{1 - 10^{-D_s}} \qquad (10.7)$$

Yule–Nielsen

The density of a tint area is affected by internal light scattering, and as a result the apparent dot area is a function of both the geometric dot size and the optical properties of the substrate. The relationship between the apparent dot area, the geometric dot size reproduced on the media, and the fractional dot area requested (known as the 'actual dot area') can be defined by tonal value transfer functions.

If a tonal value transfer function from actual dot area to apparent dot area for a given media is known, then one can easily be predicted from the other. However, if the transfer function is not known and the geometric or actual dot area is required from the measurement of apparent area, it is necessary to include a correction for the physical effects of light scattering and mechanical dot growth. One method of doing this is to add an exponent n to the reflectance [70]:

$$A = 1 - R^{1/n} \qquad (10.8)$$

where n is usually given a value between 1.0 (for a glossy substrate) and 2.0 (for a perfect diffuser). Attempts have been made to define n from the physical properties of the paper [56], but in practice the value of n is somewhat arbitrary and can exceed 2.0.

In the density domain, the Yule–Nielsen equation is:

$$A = \frac{1 - 10^{-D_t/n}}{1 - 10^{-D_s/n}} \qquad (10.9)$$

Clapper–Yule

Multiple internal reflections in glossy prints are a possible source of errors in both the Murray–Davies and Yule–Nielsen models. The Clapper–Yule [8, 9] model introduces a correction for such errors by modelling the internal scattering, the transmittance through the ink film, and the reflectance from the surface:

$$R = K_s + \frac{x I r (1 - A + At)^2}{1 - r(1 - x)(1 - A + At^2)} \qquad (10.10)$$

where K_s is the first surface reflection from the print, t is the ink transmittance, I is the amount of light incident in the medium, and x is the fraction of light which is internally reflected.

Neugebauer

The reflectance from a region of a coloured halftone is a function of the relative dot areas of the primary colorants. As shown in Figure 10.2, the halftone dots and their overlaps give rise to eight possible colours, consisting of the primary colorants and their secondary and tertiary overprints. Demichel showed in 1924 [12], that if it is assumed that there is local randomness in the dot structure (which is approximated by the structure of a halftone screen with its different screen angles), then the reflectance from each of these eight possible colours can be considered as the product of its solid reflectance and its area coverage.

If it is assumed that these reflectances integrate together to form the total coloured reflectance from the substrate, and that this total reflectance is the additive combination of the amounts for each colour, then:

$$R = A_w R_w + A_c R_c + A_m R_m + A_y R_y + A_r R_r + A_g R_g + A_b R_b + A_k R_k \qquad (10.11)$$

where R is the total reflectance, A is the area coverage, and the subscripts w, c, m, y, r, g, b and k refer to the paper white, the three colorant primaries, their secondary overprints, and the tertiary overprint of all three primaries respectively.

This approach can be applied to reflectances for any set of primaries, including X, Y and Z tristimulus reflectances:

$$X = A_w X_w + A_c X_c + A_m X_m + A_y X_y + A_r X_r + A_g X_g + A_b X_b + A_k X_k$$

$$Y = A_w Y_w + A_c Y_c + A_m Y_m + A_y Y_y + A_r Y_r + A_g Y_g + A_b Y_b + A_k Y_k$$

$$Z = A_w Z_w + A_c Z_c + A_m Z_m + A_y Z_y + A_r Z_r + A_g Z_g + A_b Z_b + A_k Z_k \quad (10.12)$$

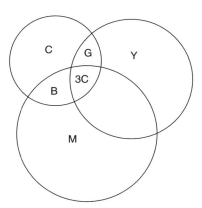

Figure 10.2 The Neugebauer model assumes that the reflectance of a spatial area is the additive combination of the reflectances of the primary colours and their overprints.

This form of equation (10.11) is known as the Neugebauer equations [53]. The Neugebauer model can be expressed as:

$$X_{\text{cmy}} = \sum_{i=1}^{8} w_i X_i$$

$$Y_{\text{cmy}} = \sum_{i=1}^{8} w_i Y_i$$

$$Z_{\text{cmy}} = \sum_{i=1}^{8} w_i Z_i \tag{10.13}$$

where X_{cmy} is the X tristimulus value of the colorant combination to be found, X_i is the set of X tristimulus values of the eight known colours, and the vector of weights w_i is given by Demichel's equation:

$$w_i = [(1 - c)(1 - m)(1 - y), c(1 - m)(1 - y), m(1 - c)(1 - y),$$

$$y(1 - c)(1 - m), cm(1 - y), cy(1 - m), my(1 - c), cmy] \tag{10.14}$$

These weights represent the fractional dot areas for unprinted substrate, single-colour, two-colour and three-colour combinations. The dot areas can be found from the reflectances using any of the dot area models described earlier. It can be seen from equations (10.13) and (10.14) that although the Neugebauer model is derived from a physical model, in its three-colour version it has the same form as a trilinear interpolation (as described in Chapter 6) in which the eight known colours form the vertices of a cube.

The model is equally applicable to colorant primaries other than cyan, magenta and yellow, and can be extended to include other colours. The model for a four-primary colorant set would be:

$$X_{\text{cmyk}} = \sum_{i=1}^{16} w_i X_i \tag{10.15}$$

where the vector w_i includes the weights in equation (10.14) above, together with their combinations with black. The Y and Z amounts are computed in the same way.

The Neugebauer equations can also be solved for colorimetric densities rather than tristimulus values [58, 31]. The original Neugebauer equations gave the results shown in Table 10.1 when applied to a newsprint test set.

Since the colorant primaries are the principal inputs into the model, the errors that occur using the Neugebauer equations are smallest in solid prints of the colorant primaries, and tend to be largest in neutrals.

The Neugebauer equations are not analytically invertible, so the inverse solution in order to predict colorant amounts from colorimetric values is not trivial [1, 30, 49, 50, 59]. One method is to use an iterative solution [57]. The difficulty of inversion increases for the modified Neugebauer equations described below.

Table 10.1 Performance of original Neugebauer equations on a newsprint test set

Test set	Mean	Median	Max.	95th percentile
323 CMY colours	4.29	4.56	8.70	7.69
592 CMYK colours	16.56	17.11	32.89	25.45

Modified and extended Neugebauer equations

The original Neugebauer equations are based on a linear transformation between a CIE colour space and the colorant space defined by an ink–paper interaction. They ignore non-randomness in halftone structure, non-uniformity in the inking of halftone dots, and light scattering within the paper. Their predictive accuracy can be improved through various corrections.

n-Modified Neugebauer equations

The *n*-modified versions incorporate an exponential correction for light scatter using, for example, either the Yule–Nielsen [70] or Yule–Colt [71] methods.

Vector-corrected Neugebauer equations

In the 'vector-corrected' Neugebauer equations [45] there is a correction factor based on the vector distance between measured colours and the colours predicted by the original or modified Neugebauer equations.

This approach is in effect a hybrid between modelling and a look-up table with interpolation, and requires a large number of measurements to be made to be effective.

Cellular extensions

An alternative approach to vector correction, similar in principle to the vector model, is to increase the number of grid points associated with measured data, thus sub-dividing the cube into sub-cubes [22], as illustrated in Figure 10.3. This reduces the errors arising from interpolation.

Spectral extensions

This method [64, 65] substitutes spectral reflectance data for colorimetric measurements:

$$R(\lambda) = \sum_{i=1}^{8} w_i R_i(\lambda) \qquad (10.16)$$

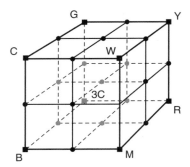

■ Nodes in original Neugebauer model

● Additional nodes in cellular version of the model

Figure 10.3 Sub-dividing the colorant space into sub-cubes, and thus including inputs for an increased number of colorant combinations, reduces the error of models like the Neugebauer model that make predictions based solely on the basis of the solid colorants and their overprints.

This approach is more computationally complex, since it requires measurements at a series of wavelength intervals which are each interpolated within the colour cube, and summation of the results to obtain tristimulus values.

The features of the n-modified and extended forms of the equations can be combined together.

Evaluation of different forms of the Neugebauer equations

A number of evaluations of the different forms of the Neugebauer equations have been reported. These include Heuberger *et al.* [23], who found that cellular extensions improved accuracy. Rolleston and Balasubramanian [60] reviewed the performance of several different forms of the Neugebauer equations, finding that the spectral n-modified combination performed best, followed by spectral cellular. Kang [37] compared both three-primary and four-primary versions of Neugebauer equations with both Yule–Nielsen and Clapper–Yule methods of calculating dot areas, and also with Beer–Bouguer and Kubelka–Munk colorant models (see below). The modified Neugebauer performed better than the original Neugebauer equations, but were outperformed by halftone-corrected versions of Beer–Bouguer and Kubelka–Munk.

10.3.3 Colorant models

Colorants models predict the reflectance or transmittance of varying thicknesses of colorants by modelling the subtractive effects of the colorant and the substrate.

Masking equations

Masking equations [68] have been used in colour reproduction systems to compute the densities of CMY required to reproduce an original colour by correcting for the 'unwanted absorptions' of the three inks. If it is assumed that the densities of printing inks are additive (i.e. that the D_r colorimetric density of a three-colour overprint is the sum of the D_r densities of the three separate ink or dye layers, and similarly for the D_g and D_b densities), we have:

$$
\begin{bmatrix} D_r \\ D_g \\ D_b \end{bmatrix} = \begin{bmatrix} C_r \\ C_g \\ C_b \end{bmatrix} + \begin{bmatrix} M_r \\ M_g \\ M_b \end{bmatrix} + \begin{bmatrix} Y_r \\ Y_g \\ Y_b \end{bmatrix}
\tag{10.17}
$$

where C_r is the red colorimetric density of the cyan ink.

Equation (10.17) can also be written:

$$
\begin{bmatrix} D_r \\ D_g \\ D_b \end{bmatrix} = \begin{bmatrix} C_r + \dfrac{M_r}{M_g} M_g + \dfrac{Y_r}{Y_b} Y_b \\[2mm] \dfrac{C_g}{C_r} C_r + M_g + \dfrac{Y_g}{Y_b} Y_b \\[2mm] \dfrac{C_b}{C_r} C_r + \dfrac{M_b}{M_g} M_g + Y_b \end{bmatrix}
\tag{10.18}
$$

The ratios C_g/C_r etc. in equation (10.18) give the amount of unwanted absorptions. If it is further assumed that these ratios are independent of the colorant amount (the proportionality rule), then these ratios are constants and can be determined by measurement. The required densities of C_r etc. are then given by:

$$
\begin{bmatrix} C_r \\ M_g \\ Y_b \end{bmatrix} = \mathbf{A} \begin{bmatrix} D_r \\ D_g \\ D_b \end{bmatrix}
\tag{10.19}
$$

where $\mathbf{A}$ is a 3×3 matrix of coefficients found by inverting the ratios C_g/C_r etc.

Clapper [10] showed that errors arising in this first-order model from the failure of the additivity and proportionality assumptions could be empirically corrected by regression to produce a second- or third-order model. This has the same form as the polynomial regression described in Chapter 6.

Although originally defined for dye layers of varying thicknesses, as found in photographic media, the higher-order masking equations also perform well for halftone processes if a suitable model is used to convert between density and dot area.

Beer–Bouguer

The physical laws of Beer [4] and Lambert [46] (or Bouguer [7]) can be used to define a relationship between transmittance and colorant amount in absorption filters, whereby transmittance through a medium is proportional to the thickness and the absorptivity of the medium [67, 40, 61]:

$$
A = \log\left(\frac{1}{T}\right)
\tag{10.20}
$$

where A is the absorbance and T the transmittance. A is also given by [40]:

$$A = \varepsilon c w \qquad (10.21)$$

where ε is a constant for the absorptivity of the medium, c is its concentration and w is the length of the light path (or thickness of the medium). Thus the spectral transmittance is given by:

$$T(\lambda) = \exp(-\varepsilon(\lambda)cw) \qquad (10.22)$$

where $\varepsilon(\lambda)$ is the spectral absorption coefficient of the colorant.

The assumption of proportionality and additivity implied by equation (10.20) holds at low to medium dye concentrations but fails at higher concentrations [67].

Kubelka–Munk

Where a colorant layer is imaged on to a reflective surface, the model needs to include the effect of the substrate and its reflectance, as well as the properties of the colorant itself.

The Kubelka–Munk equation [43, 44] models the absorption and scattering of the light path in an ink film, through terms for absorption and scattering coefficients, together with the reflectance of the substrate on which the colorant is imaged [62]. The quantities in the model are illustrated in Figure 10.4.

$$\frac{K}{S} = \frac{(1 - R)^2}{2R} \qquad (10.23)$$

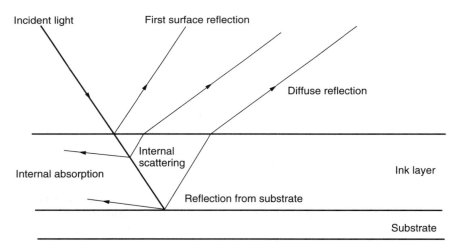

Figure 10.4   The reflectance of an ink film is a function of the incident light, the internal scattering and absorption within the ink layer, and the reflectance from the substrate.

where R is the reflection factor for a sample of infinite thickness, K is the absorption coefficient and S is the scattering coefficient. Reflectance R can then be found [54]:

$$R = 1 + \frac{K}{S} - \left(\left(1 + \frac{K}{S} \right)^2 - 1 \right)^{\frac{1}{2}} \tag{10.24}$$

The constants K and S are determined empirically. They can be found individually, although since it is the ratio of these constants rather than their absolute values that is required in equation (10.23), it is sufficient to determine the combined ratio K/S. These are known as two-constant and single-constant methods respectively.

If spectral reflectance $R(\lambda)$ is considered, then the function

$$\frac{K}{S}(\lambda) = \frac{(1 - R(\lambda))^2}{2R(\lambda)} \tag{10.25}$$

has a linear relationship with colorant concentration [72].

If the colorant thickness and the reflectance of the substrate are included in the model, it is possible to define $R(\lambda)$ as follows [40]:

$$R(\lambda) = \frac{1 - R_g(\lambda)[\alpha(\lambda) - \beta(\lambda)\coth(\beta(\lambda)S(\lambda)W]}{\alpha(\lambda) - R_g(\lambda) + \beta(\lambda)\coth(\beta(\lambda)S(\lambda)W} \tag{10.26}$$

where $R_g(\lambda)$ is the reflectance of the substrate, W is the colorant film thickness, and:

$$\alpha(\lambda) = 1 + \frac{K(\lambda)}{S(\lambda)}$$

$$\beta(\lambda) = [\alpha(\lambda)^2 - 1]^{\frac{1}{2}}$$

Extensions

Colour mixing models based on the physical properties of substrate and colorants can make broadly accurate predictions where the model parameters are known and can be optimized for a given reproduction situation. Several authors have extended the physical models described here to refine the model predictions or incorporate further physical properties [2, 3, 20, 64, 65, 13–15, 37].

Kang [37] proposed a modification to the Beer–Bouguer and Kubelka–Munk models for use with halftone printers. This takes the form of a spectral halftone correction factor $h(\lambda)$ which is computed by:

$$h(\lambda) = \frac{Q_m(\lambda)}{Q_c(\lambda)} \tag{10.27}$$

where Q is the parameter to be corrected, such as density, scatter coefficient S or absorption coefficient K; and the subscripts m and c denote the values measured from a halftone step wedge and computed by the model respectively.

Emmel and Hersch [14] extended the Kubelka–Munk model to incorporate a more complex description of light scattering and ink spreading. They showed that the dot area models described above can be seen as particular cases of this modified Kubelka–Munk model.

10.4 NUMERICAL MODELS AND LOOK-UP TABLES

Regression methods and look-up tables for characterization were described in Chapter 6. The choice of model is not always obvious, and it is worth testing alternative approaches in order to determine the optimum model for a particular application.

Johnson and others [34] compared original and n-modified Neugebauer with a numerical model using third-order polynomials, with the result that the third-order model performed best, followed by second-order and n-modified Neugebauer.

Herzog and Roger [21] compared a number of numerical approaches, and found that higher-order polynomials and spline methods all gave sufficient accuracy.

Numerical models are most frequently applied using trichromatic quantities, although spectral models have also been developed. One spectral method [11] requires only measurements made of tone scales of the primaries, avoiding the need to measure colorant combinations.

An important factor in the design of look-up tables is the number of measurements required. A full characterization can populate all the entries in the table and obviate the need for a printer model., albeit at the cost of a large number of measurements. As was noted in Chapter 6, in a typical test target the colours of many patches can often be predicted from their neighbouring patches, and the number of patches required to characterize a colour device can thus be reduced if the spacing of the patches is carefully chosen [28]. If the spacing is perceptually uniform, a three-dimensional grid with between 6 and 16 patches in each dimension should be adequate [32].

The ISO 12640 [25] target is designed to sample the colorant space of a four-colour printing process and provide sufficient information for characterization. If all the patches in this target are measured it is possible to adopt a full characterization approach and create a LUT directly from the data. The ISO 12640 target is illustrated in Plate 2(a).

Most printing processes operate within relatively broad tolerances, and as suggested in Chapter 6 a single printed target is unlikely to be representative of the performance of a process over time. To characterize a single device it is necessary to take a number of prints over a period and average the measurements. Alternatively one can measure prints made at a number of different sites to obtain a data set that represents a given process. Averaged data sets for a number of different printing processes can be found at the ICC website www.color.org.

10.5 BLACK PRINTER

Many printing devices include a black printer in addition to the chromatic primaries cyan, magenta and yellow. The main role this plays in colour images is to compensate for additivity failure in dark neutrals, and thus increase the contrast range. The black printer also enhances the perception of sharpness when unsharp masking is applied, and can substitute for a proportion of coloured ink through grey component replacement (GCR). In a printed page, black is also required for text and graphical elements.

For a three-primary device, there will be a unique combination of primaries associated with any colour that is to be matched. With a four-primary device the amount of black can vary continuously over a range of values for a given colour, and so there are many different colorant combinations that will result in a match. Hence there is no unique set of CMYK values corresponding to any given set of tristimulus values.

The amount of black is usually found by first computing the CMY amounts and then determining the black amount required. An advantage of computing the black printer independently of the colour transformation is that parameters such as GCR amount and total ink limits can be modified without having to adjust the colour transformation.

The black printer amount is frequently computed from the minimum colorant amount:

$$K = f(\min(C, M, Y)) \qquad (10.28)$$

where K is the amount of black and f is a function of the minimum colorant amount. It is common to set $K = 0$ where the minimum colorant amount is below a specified threshold.

Several methods of modelling the black printer have been described [6, 29, 33, 35, 55, 38]. One widely used method is to correct a model based on additivity [55, 29, 47]. A simple additivity failure model uses a multiple linear regression to derive a convergence point [69] k, where the functions D_{cmyk} against D_k converge in an additivity diagram, as shown in Figure 10.5. The function D_{cmyk} is the densities of the primary colorants combined with progressive increments of black, while the function D_k is increments of black alone. K tends to be fractionally greater than $D_{g_{max}}$ for uncoated substrates, but higher for coated ones.

The relationship between the required total density and the three-colour and black components is given by the following [29]:

$$\begin{bmatrix} Dr_{cmyk} \\ Dg_{cmyk} \\ Db_{cmyk} \end{bmatrix} = \begin{bmatrix} Dr_{cmy} \\ Dg_{cmy} \\ Db_{cmy} \end{bmatrix} + \begin{bmatrix} Dr_k \\ Dg_k \\ Db_k \end{bmatrix} - \begin{bmatrix} \dfrac{Dr_{cmy} Dr_k}{k_r} \\ \dfrac{Dg_{cmy} Dg_k}{k_g} \\ \dfrac{Db_{cmy} Db_k}{k_b} \end{bmatrix} \qquad (10.29)$$

where Dr_{cmyk}, Dg_{cmyk}, Db_{cmyk} are the colorimetric densities of the four-colour print, and Dr_{cmy}, Dg_{cmy}, Db_{cmy} the colorimetric densities of the three-colour component.

A modified additivity failure model of polynomial form determines a value for k by second- or third-order regression in place of a linear convergence [32, 35].

10.5.1 Spectral grey-component replacement

A four-colour spectral colour-mixing method [39] was developed using Kang's halftone correction approach described above, in which the spectrum of a given colour is considered

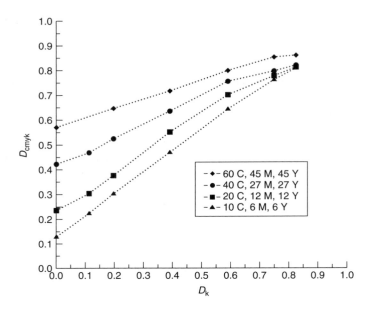

Figure 10.5 The convergence point on the additivity diagram indicates the amount of black required to extend the gamut of the trichromatic primaries.

as an additive combination of the density spectra of the three primaries and the black printer. It employs colour mixing models to model the spectral fit between three-colour and four-colour prints. The amount of the grey component to be removed from the colour predicted by the halftone-corrected mixing model is then determined, and each primary colour is adjusted until a good match between the peak wavelengths of the three-colour and four-colour spectra is obtained. Of the models tested, the halftone-corrected Kubelka-Munk model gave the closest fit, with an average ΔE_{ab}^* between CMY and CMYK prints of 64 colour patches of 3.11. This was followed by the halftone-corrected Beer–Bouguer and Yule–Nielsen models.

10.6 BLACK GENERATION ALGORITHM

A step-by-step method of computing the black printer using the additivity failure method, based on one described by Johnson [29, 35], is given below. The main steps are:

1. Convert XYZ to D_{rgb} colorimetric densities.
2. Convert D_{rgb} to D_{cmy} ink densities.
3. Calculate the black needed to extend the gamut.
4. Apply GCR. In summary, this involves the following steps:
    (i) Select the amount of the grey component that is to be replaced
    (ii) Calculate an enhancement factor to correct for additivity failure, then for each colour being mapped:

(iii) Grey balance D_{cmy} so that $D_c = D_m = D_y$ results in a neutral grey

(iv) Find the grey component of the colour

 (v) Calculate the total amount of black required

(vi) Reduce the grey component by an amount corresponding to the black to be substituted

(vii) Subtract this reduced grey component from (grey-balanced) D_c, D_m and D_y

(viii) Reverse the grey balancing

 (ix) Multiply D_c, D_m, D_y by the enhancement factor f to compensate for additivity failure.

5. Convert ink densities to dot percentages by look-up.

Process

1. Convert tristimulus values to colorimetric densities. This is done as described in Chapter 6:

$$
\begin{bmatrix} D_r \\ D_g \\ D_b \end{bmatrix} = \log \begin{bmatrix} \dfrac{X_o}{X} \\ \dfrac{Y_o}{Y} \\ \dfrac{Z_o}{Z} \end{bmatrix} \tag{10.30}
$$

where $[X_o, Y_o, Z_o]$ are the tristimulus values of the unprinted paper.

2. Convert D_{rgb} to D_{cmy} ink densities:

$$
\begin{bmatrix} D_c \\ D_m \\ D_y \end{bmatrix} = \mathbf{AB} \tag{10.31}
$$

where $\mathbf{A}$ is a matrix of coefficients found by regression and $\mathbf{B}$ is the matrix D_{rgb} found in equation (10.30), plus polynomial expansion terms, as described in Chapter 6. A third-order model with cross-products will often give good results.

3. Calculate the black needed to extend the gamut ($D_g K$):

(i) Determine the maximum three-colour density $D_{g\max}$ and additivity failure convergence K, the point where the functions D_{cmy} against D_k converge in an additivity diagram, as described above.

(ii) Calculate the gamut extension:

$$
D_g K = \frac{D_g - D_{g\max}}{1 - \dfrac{D_{g\max}}{K}}, \quad D_g > D_{g\max}
$$

$$
D_g K = 0, \qquad\qquad D_g \leqslant D_{g\max} \tag{10.32}
$$

4. Apply GCR to find cmyk.

(i) Select the desired value for *gcr* (the fractional amount of three-colour grey that is to be replaced by black) between 0 and 1.

(ii) Calculate the enhancement factor, f:

$$f = 10^{gcr^2} \tag{10.33}$$

(iii) Grey-balance the D_{cmy} densities. This can be done by scaling in the ratios of the tristimulus values of a neutral grey; an alternative method described here is to use the D_{cmy} values for neutral grey to perform the scaling.

(a) Calculate grey balance ratios $[D_{c_{grey}}, D_{m_{grey}}, D_{y_{grey}}]$:

$$\begin{bmatrix} D_{c_{grey}} \\ D_{m_{grey}} \\ D_{y_{grey}} \end{bmatrix} = \begin{bmatrix} D_{c3c} \\ D_{m3c} \\ D_{y3c} \end{bmatrix} \Big/ D_{m3c} \tag{10.34}$$

where $\begin{bmatrix} D_{c3c} \\ D_{m3c} \\ D_{y3c} \end{bmatrix}$ are the values of $\begin{bmatrix} D_c \\ D_m \\ D_y \end{bmatrix}$ for three-colour grey. $\begin{bmatrix} D_{c3c} \\ D_{m3c} \\ D_{y3c} \end{bmatrix}$ can be determined from a single colour where C, M, Y are printed in a ratio that forms a neutral grey (e.g. $C = 50$, $M = 40$, $Y = 40$), or from the mean values of $\begin{bmatrix} D_c \\ D_m \\ D_y \end{bmatrix}$ of a tone scale where C, M, Y in each patch are neutral grey.

(b) Grey-balance the ink densities and D_{max}:

$$\begin{bmatrix} D_{c'} \\ D_{m'} \\ D_{y'} \end{bmatrix} = \begin{bmatrix} D_c D_{c_{grey}} \\ D_m D_{m_{grey}} \\ D_y D_{y_{grey}} \end{bmatrix} \tag{10.35}$$

$$\begin{bmatrix} c'_{max} \\ m'_{max} \\ y'_{max} \end{bmatrix} = \begin{bmatrix} c_{max} D_{c_{grey}} \\ m_{max} D_{m_{grey}} \\ y_{max} D_{y_{grey}} \end{bmatrix} \tag{10.36}$$

(iv) Find the grey component for each colour by finding the minimum density and multiplying this by the grey balance ratio:

$$\begin{bmatrix} c_{grey} \\ m_{grey} \\ y_{grey} \end{bmatrix} = \min(D_{c'}, D_{m'}, D_{y'}) \begin{bmatrix} D_{c_{grey}} \\ D_{m_{grey}} \\ D_{y_{grey}} \end{bmatrix} \Big/ D_{grey} \tag{10.37}$$

where D_{grey} is the element of $\begin{bmatrix} D_{c_{grey}} \\ D_{m_{grey}} \\ D_{y_{grey}} \end{bmatrix}$ that corresponds to $\min(D_{c'}, D_{m'}, D_{y'})$ e.g. if $\min(D_{c'}, D_{m'}, D_{y'}) = D_{c'}$, then $D_{grey} = D_{c_{grey}}$.

(v) Compute total black:

$$k = D_g K + \min(D_{c'}, D_{m'}, D_{y'})gcr\left(\frac{1 - D_g K}{K}\right) \tag{10.38}$$

(vi) Reduce grey component:

$$\begin{bmatrix} c'_{grey} \\ m'_{grey} \\ y'_{grey} \end{bmatrix} = (1 - gcr)\begin{bmatrix} c_{grey} \\ m_{grey} \\ y_{grey} \end{bmatrix} \tag{10.39}$$

(vii) Subtract reduced grey component from D_{cmy}:

$$\begin{bmatrix} D_{c''} \\ D_{m''} \\ D_{y''} \end{bmatrix} = \begin{bmatrix} D_{c'} \\ D_{m'} \\ D_{y'} \end{bmatrix} - \begin{bmatrix} c'_{grey} \\ m'_{grey} \\ y'_{grey} \end{bmatrix} \tag{10.40}$$

(viii) Reverse cmy grey balancing:

$$\begin{bmatrix} c' \\ m' \\ y' \end{bmatrix} = \begin{bmatrix} D_{c''}/D_{c_{grey}} \\ D_{m''}/D_{m_{grey}} \\ D_{y''}/D_{y_{grey}} \end{bmatrix} \tag{10.41}$$

(ix) Enhance cmy to compensate for additivity failure:

$$\begin{bmatrix} c \\ m \\ y \end{bmatrix} = \begin{bmatrix} c'\left(1 + f\dfrac{c_{grey}}{c'_{max}}\right) \\ m'\left(1 + f\dfrac{m_{grey}}{m'_{max}}\right) \\ y'\left(1 + f\dfrac{y_{grey}}{y'_{max}}\right) \end{bmatrix} \tag{10.42}$$

5. Use the single-colour tone scales in cyan, magenta, yellow and black from the test target to establish a one-dimensional look-up table between CMYK dot percentages and the corresponding cmyk densities found using steps 1–3 above.

After establishing the model, the accuracy of the predicted CMYK dot percentages can be checked against the dot percentages in the original target, and further single channel corrections may be determined at this stage if required in order to improve the fit.

10.7 THE REVERSE MODEL

Computing the reverse transformation from CMYK to a CIE colour space is somewhat simpler than the forward model described above, and again polynomial and additivity failure models can be used [29, 32, 35, 47]. In an evaluation of four black printer algorithms [35], a second-order polynomial with cross-products of D_r, D_g and D_b produced the best results, with an average $\Delta E_{CMC(1:1)}$ error of 1.23 (Cromalin proof) and 1.72 (Iris print), followed by third-order polynomial, modified additivity failure and simple additivity failure models.

In a modified version of the second-order polynomial [18], prints of the ISO 12640 [25] colour patches on newsprint were made and the measurement data sorted into three groups: patches including black and CMY, CMY only, and black only. Coefficients were calculated by second-order polynomial regression for CMY-to-colorimetric density and black-to-colorimetric density. The black and CMY components of the CMY + K patches were then separately transformed to colorimetric density using these coefficients, and a further set of 24 coefficients calculated by polynomial regression to define the mixing of CMY and black to give the colorimetric densities of the four-colour patches.

This model was compared with a single-stage regression from four-colour CMYK directly to colorimetric density [18]. The poor performance of this latter approach can be seen in a comparison of the two models in predicting a test set of 264 colours from the ISO 12641 target. This comparison is shown in Table 10.2.

Table 10.2 Performance of different regression models in predicting a reverse colorimetric transformation

Method	Mean ΔE_{ab}^*	Median	Max.	95th percentile
Two-stage regression	2.83	2.63	9.60	5.42
Single-stage regression	17.23	10.04	98.11	47.77

10.8 RGB PRINTERS

Many desktop printers are GDI devices designed to work with RGB inputs rather than CMYK. The printer driver supplied by the manufacturer performs a conversion into colorant amounts, following rules devised by the manufacturer. These rules will depend on the marking technology and on the properties of the colorants and media; for example, many inkjet printers restrict the total ink that is permitted in one location in order to avoid ink bleeding into the substrate or forming thick layers that fail to dry. Thus the CMYK amounts needed to reproduce a given colour will be quite different from the amounts needed on another printer that employs a different marking technology, and when data is sent to the printer (whether in CMYK or RGB) it will have to be converted into appropriate ink amounts for that printer.

The algorithm that the driver uses to convert RGB inputs into CMYK amounts (and by extension into other ink sets such as six-colour CcMmYK) is not accessible, and thus the methods of characterization described above may not be suitable.

One approach is to treat the printer as a 'virtual CMYK' device [52], and characterize the relationship between XYZ and CMYK as for other printers, accepting that a further unknown but consistent 'black box' conversion will be performed on the resulting CMYK data. In many cases the inputs to the printer driver conversion to CMYK amounts are required to be in RGB, and thus an additional step of conversion from CMYK to RGB will be performed prior to calculating colorant amounts. In such cases the printer can be characterized in terms of XYZ to RGB. Whether the input data is RGB or CMYK, the printer driver will compute final colorant amounts, ink limits and black using its own internal algorithms.

A third alternative is to bypass the printer driver and use a PostScript interpreter to send CMYK amounts directly to the printer. If this is done it may be necessary to devise appropriate ink limits, black printer algorithms and any other rules that are necessary to give good results on the media.

REFERENCES

1. Archer, H. B. (1981) An inverse solution to the Neugebauer equations for the calculation of grey balance requirements. *Proc. IARIGAI 16*.
2. Arney, J. S. and Arney, C. D. (1996) Modelling the Yule-Nielsen effect. *J. Imaging Sci. Tech.*, **40**, 233–238.
3. Arney, J. S., Engledrum, P. G. and Zeng, H. (1995) An expanded Murray-Davies model of tone reproduction in halftone imaging. *J. Imaging Sci. Tech.*, **39**, 502–508.
4. Beer, A. (1852) Bestimmung der Absorption des rothen Lichts in farbigen Flüssigkeiten. *Ann. Phy. Chem.*, **86**, 2–78.
5. Berns, R. (1999) Challenges for colour science in multimedia imaging. In *Colour Imaging: Vision and Technology*, L. MacDonald and M. R. Luo (eds), pp. 99–127. Wiley, Chichester.
6. Birkenshaw, J. W. (1977) *The black printer*. Pira report PR/154.
7. Bouguer, P. (1729) *Essai d'Optique sur la gradation de la lumière*, Claude Tombert, Paris.
8. Clapper, F. R. and Yule, J. A. C. (1953) The effect of multiple internal reflections on the densities of half-tone prints on paper. *J. Opt. Soc. Am.*, **43**, 600–603.
9. Clapper, F. R. and Yule, J. A. C. (1955) Reproduction of halftone colored images. *Proc. TAGA*, 1–12.
10. Clapper, F. R. (1961) An empirical determination of halftone colour reproduction requirements. *Proc. TAGA Conf.*, 31–41.
11. de Capelle, J. P. and Van Meireson, B. (1997) A new method for characterizing output devices and its fit into ICC and HIFI color workflows. *Proc. IS&T/SID 5th Color Imaging Conf.*, 66–69.
12. Demichel, E. (1924). Procédé de la Société Française de Photographie, **26**(3), 17–21, 26–27.

13. Emmel, P. and Hersch, R. D. (2000) Exploring ink spreading. *Proc. 8th IS&T/SID Color Imaging Conf.*, 335–341.
14. Emmel, P. and Hersch, R. D. (1999) A model for colour prediction of halftoned samples incorporating light scattering and ink spreading. *Proc. IS&T/SID 7th Color Imaging Conf.*, 173–181.
15. Emmel, P. and Hersch, R. D. (2000) A unified model for color prediction of halftoned prints. *J. Imaging Sci. and Tech.*, **44**, 351–359.
16. Fink, P. (1992) *PostScript screening*, Adobe Press, Mountain View, CA.
17. Green, P. J. (1999) *Understanding digital color*, 2nd edn. GATFPress, Sewickley, PA.
18. Green, P. J. (2001) A test target for defining media gamut boundaries. *Proc. Conf. Col. Imaging: Device-Independent Color, Color Hardcopy and Graphic Arts*, **6**, 105–113.
19. Green, P. J. and Johnson, A. J. (2000) Issues of measurement and assessment in hard copy colour reproduction. *Proc. SPIE Conf. Col. Imaging: Device-Independent Color, Color Hardcopy and Graphic Arts*, **5**, 281–293.
20. Gustavson, S. (1997) Dot gain in colour halftones. PhD Thesis, Department of Electrical Engineering, Linköping University, Sweden.
21. Herzog, P. and Roger, T. (1998) Comparing different methods of printer characterization. *Proc. IS&T 51st PICS Conf.*, 23–30.
22. Heuberger, K. J. and Jing, Z. M. (1991) Colour transformations from RGB to CMY. *Proc. IARIGAI Conf.*
23. Heuberger, K. J., Jing, Z. M. and Persiev, S. (1992) Color transformations and look-up tables. *Proc. TAGA/ISCC Conf.*, **2**, 863–881.
24. ISO 12641 *Graphic technology – Prepress digital data exchange – colour targets for input scanner calibration.*
25. ISO 12640 *Graphic technology – Prepress digital data exchange – Input data for characterization of 4-colour process printing targets.*
26. ISO 12647-1:1996 *Graphic technology – Process control for the manufacture of halftone colour separations, proof and production prints Part 1: Parameters and measurement methods.*
27. ISO 12647-2 *Graphic technology – Process control for the manufacture of halftone colour separations, proof and production prints Part 2: Coldset offset lithography.*
28. ISO 12647-3 *Graphic technology – Process control for the manufacture of halftone colour separations, proof and production prints Part 3: Coldset offset lithography and letterpress on newsprint.*
29. Johnson, A. J. (1985) Polychromatic colour removal – evolution or revolution. *Proc. TAGA.*
30. Johnson, A. J. (1992) Techniques for reproducing images in different media: advantages and disadvantages of current methods. *Proc. TAGA*, **2**, 739–756.
31. Johnson, A. J. (1989) Defining optimum photomechanical colour reproduction. *Proc. TAGA*, 350.
32. Johnson, A. J. (1995) *Colour management in graphic arts and publishing*. Pira International, Leatherhead.
33. Johnson, A. J. (1996) Methods for characterizing colour printers. *Displays*, **16**, 193–202.
34. Johnson, A. J., Luo, M. R., Lo, M.-C. and Rhodes, P. A. *Aspects of colour management Part 1: Characterisation of three-colour imaging devices.* Unpublished paper.
35. Johnson, A. J., Luo, M. R., Lo, M.-C. and Rhodes, P. A. *Aspects of colour management Part 2: Characterisation of four-colour imaging devices and colour gamut compression.* Unpublished paper.
36. Johnson, A. J., Tritton, K. T., Eamer, M. and Sunderland, B. H. W. (1982) *Systematic lithographic colour reproduction.* Pira report PR171, Pira, England.
37. Kang, H. R. (1993) Comparison of colour mixing theories for use in electronic printing. *Proc. 1st IS&T/SID Color Imaging Conf.*, 78–82.
38. Kang, H. R. (1994) Applications of colour mixing models to electronic printing. *J. Electronic Imaging*, **3**, 276–287.
39. Kang, H. R. (1994) Gray component replacement using colour mixing models. *Proc. SPIE*, **2171**, 287–296.

40. Kang, H. R. (1996) *Color technology for electronic imaging devices*, SPIE Press, Bellingham, WA.
41. Kang, H. R. (1999) *Digital color halftoning*, SPIE Press, Bellingham, WA.
42. Kohler, T. and Berns, R. (1993) Reducing metamerism and increasing gamut using five or more colored inks. In *Proc. IS&T/SID 3rd Technical Symposium on Prepress*, Proofing and Printing pp. 24–28.
43. Kubelka, P. (1948) New contributions to the optics of intensely light-scattering materials. *J. Opt. Soc. Am.*, **38**, 449–457.
44. Kubelka, P. and Munk, F. (1931) Ein Beitrag zür Optik der Farbanstriche. *Zeitschrift für technische Physik*, **12**, 593–601.
45. Laihanen, P. (1987) Colour reproduction theory based on the principles of colour science. *Proc. IARIGAI Conf.*, **19**, 1–36.
46. Lambert, J. H. (1760) *Photometria sive de mensura et gradibus luminis, colorum et umbrae*, Eberhard Klett, Augsburg.
47. Lo, M. C. and Luo, M. R. (1994) Models for characterizing four-primary printing devices. *J. Phot. Sci.*, **42**, 94–96.
48. MacDonald, L. W., Deane, J. M. and Rughani, D. N. (1994) Extending the colour gamut of printed images. *J. Phot. Sci.*, **42**, 97–99.
49. Mahy, M. and Delabatista, P. (1996) Inversion of the Neugebauer equations. *Col. Res. Appl.*, **21**, 6.
50. Mahy, M. (1998) Insight into the solutions of the Neugebauer equations. *Proc. SPIE Conf. Col. Imaging: Device-Independent Color, Color Hardcopy and Graphic Arts*, **3300**, 76–85.
51. Mahy, M. (1998) Output characterization based on the localized Neugebauer model. *Proc. IS&T 51st PICS Conf.*, 30–34.
52. Morovic, J. and Luo, R. M. (1996) Characterizing desktop printers without full control of all colorants. *Proc. 3rd IS&T/SID Color Imaging Conf.*, 70–74.
53. Neugebauer, H. (1937) Die theoretischen Grundlagen des Mehrfarbendrucks. *Zeitschrift für wissenschaftliche Photographie, Photophysik und Photochemie*, **36**(4), 73–89.
54. Nobbs, J. H. (1997) Colour match prediction for pigmented materials. In *Colour Physics for Industry*, R. MacDonald (ed.), pp. 292–372. Soc. Dyers and Colourists, Bradford.
55. Otschik, G. (1981) *Untersuchengen zur Veranderung des Farbsatzaufbaues durch geanderte Farbausuge fur die Teilfarbe Schwarz*. FOGRA Report 1.203.
56. Pearson, M. (1980) n value for general conditions. *Proc. TAGA*, 415–425.
57. Pobboravsky, I. and Pearson, M. (1972) Computation of dot areas required to match a colorimetrically specified colour using the Neugebauer equations. *Proc. TAGA*, 65–77.
58. Pobboravsky, I. and Pearson, M. (1973) Colour transformations from psychophysical parameters to photomechanically printed dot areas. *Proc. TAGA*, 12–181.
59. Pobboravsky, I., Pearson, M. and Yule, J. A. C. (1975) Design of a printed spectrum using modified Neugebauer equations. *J. Opt. Soc. Am.*, **65**(3), 323.
60. Rolleston, R. and Balasubramanian, R. (1994) Accuracy of various types of Neugebauer models. *Proc. IS&T/SID Colour Imaging Conf.*, **1**, 32–37.
61. Sinclair, R. (1997) Light, light sources and light interactions. In *Colour Physics for Industry*, R. MacDonald (ed.), pp. 1–56. Soc. Dyers and Colourists, Bradford.
62. Smith, K. (1997) Colour-order systems, colour spaces, colour difference and colour scales. In *Colour Physics for Industry*, R. MacDonald (ed.), pp. 121–208. Soc. Dyers and Colourists, Bradford.
63. Ulichney, R. (1987) *Digital halftoning*, MIT Press, Cambridge, MA.
64. Viggiano, J. A. S. (1985) The color of halftone tints. *Proc. TAGA Conf.*
65. Viggiano, J. A. S. (1990) Modelling the color of multi-colored halftones. *Proc. TAGA Conf.*, 44–62.
66. Wyble, D. R. and Berns, R. S. (2000) A critical review of spectral models applied to binary colour printing. *Col. Res. & Appl.*, **25**, 4–19.
67. Wyszecki, G. and Stiles, W. S. (1982) *Color science: concepts and methods, quantitative data and formulae*, 2nd edition. Wiley, New York.

68. Yule, J. A. C. (1938) The theory of subtractive colour photography. I: The conditions for perfect rendering. *J. Opt. Soc. Am.*, **28**, 419–430.
69. Yule, J. A. C. (1967) *The principles of color reproduction*, Wiley, New York.
70. Yule, J. A. C. and Nielsen, W. J. (1951) The penetration of light into paper and its effect on halftone reproduction. *Proc. TAGA Conf.*, 65–75.
71. Yule, J. A. C. and Colt, R. S. (1951) Colorimetric investigations in multi-colour printing. *Proc. TAGA*, 3–77.

11

Color management and transformation through ICC profiles

Dawn Wallner

This chapter reviews the need for a common, open interface for transforming color data between different devices and media and considers the color management architecture implied by the specification of the ICC profile format and its reference color space. It then continues with a description of the ICC profile structure and its content, additional information for profile creators, and some examples of profile interfaces. The latest version of the specification can be found at www.color.org

11.1 THE NEED FOR COLOR MANAGEMENT

The market requirements for color management are to provide predictable and consistent color results with maximum throughput and minimum operating skill levels. It must be easy, moreover, to achieve these with systems comprising many different imaging peripherals from different suppliers and delivering colored images to multiple media. As

Colour Engineering, edited by P.J. Green and L.W. MacDonald.
© 2002 John Wiley & Sons Ltd.

publishing moves into an era of digital image archives, trans-continental data communications and multimedia presentation, workflows are becoming much more open-ended to the extent that the image originator will generally not know even in which medium an image will be reproduced, let alone its precise color rendering characteristics. The ability to capture an image once, store it digitally and reproduce it with consistent color many times on different media is becoming essential.

The new developments in color management systems have arisen during the past decade from co-operation between leading companies in the desktop computer industry and in the color reproduction industry The principles of device characterisation, color transformation, color appearance, gamut mapping, upon which these initiatives are based, are well-established and are documented elsewhere in this book. What is new is the level of agreement and standardisation across the industry, and the way in which the facilities for color management are being embedded into computer operating systems software, enabling future inter-operability of all software applications for desktop color imaging.

With the advent of open systems in desktop publishing, customers now expect to be able to purchase standard computer platforms, such as Apple Macintosh or PC, and to configure heterogeneous systems with components from many different vendors, tailored to meet the exact requirements of the business. The challenge for vendors and users alike is to provide the same color consistency and image quality as from closed systems, whilst enjoying the benefits of open systems architecture.

The objective of any color management system is to provide a means of managing and communicating color consistently throughout a system made up of disparate components. For some users this will mean adjusting several image capture devices (scanners, digital cameras, PhotoCD, etc.) so that any device will produce near-identical image representation. For other users it will mean processing color images to give near-identical results on several output devices (displays, hard-copy proofers, printing presses, etc.). To achieve these objectives three critical tasks must be performed:

(a) Provide calibration and characterisation of input devices, such as scanners, so that image data can be interpreted in terms of a colorimetric reference space.
(b) Provide calibration and characterisation of output devices, such as displays, proofers and printing processes, so that the appropriate device signals can be generated to produce any desired color (within gamut).
(c) Provide an efficient means for processing images along the chain from input device to output device, together with a convenient user interface for setting up and controlling the process.

Device calibration is the initial setup process that ensures that the device performs to some known color specification. Once the device has been calibrated, device characterisation establishes the relationship between the signals sent to the device and the color produced.

A common framework for efficient color management requires agreement on a data structure for the device model which includes the characterization together with other necessary parameters such as appearance modelling and gamut mapping and a definition of the intermediate device-independent color space. Implementation also requires a standard framework, preferably at the level of the computer operating system, which can handle different profiles and manage color transformations.

The ICC color management architecture consists of four main elements, which are:

The color management **framework** is an extension to the capabilities of the computer operating system which serves as a 'connect and dispatch' mechanism, enabling applications to access profiles and CMMs by routing application program calls to the most appropriate one of a set of 'plug-in' modules. In the Macintosh platform the framework is known as ColorSync; in Windows it is ICM.

The **profile** defines the device model by providing the relationship between the device Co-ordinates and those of the reference color space.

The **color management module** (CMM) connects together profiles to produce transformations between source and destination device color spaces. It also performs any interpolation that is required. The color management framework includes a default CMM, and also allows third party vendors to modify or extend the functionality of the CMM with their own plug-ins.

The **application** program can then make calls to the operating system to handle color transformations as required by the user.

11.2 THE PCS

The definition of a reference color space is key to the success of any color management system. Rather than provide an ad hoc conversion between every possible combination of input and output devices, it is far better to define a common CIE-based reference color space to which every input and output device can be related. Such a reference color space must define the colorimetry of both the reference medium and its assumed viewing condition. The reference color space specified by ICC is known as the Profile Connection Space (PCS), and is defined as 'the CIE colorimetry required to achieve the desired color appearance on ... an ideal reflection print viewed in a standard viewing booth'. [7] This corresponds to the ISO 3664 P2 viewing condition with a surround reflectance of 20%, an adapting illumination with a chromaticity of D50, and an illumination level of 500 lux.

PCS encodings can be either 16-bit XYZ, or 8- or 16-bit LAB, using the measurement conditions defined in ISO 13655. In the case of the perceptual rendering intent, the reference medium is defined as one having a large gamut and dynamic range which approximate the limits of current print technology. This is specified as a substrate whose white point has a neutral reflectance of 89% and whose darkest printable color has a neutral reflectance of 0.30911% (corresponding to a dynamic range of 287.9:1 and a density range of 2.4593). For colorimetric intents no limits on gamut or dynamic range are assumed.

While accepting that this is a somewhat 'print-centric' definition of the PCS, it is believed that this will provide good results in other applications such as video production, slide production, and presentation graphics. [7]

It is not generally intended that color image data will actually be stored as PCS data (although that is possible). Instead the PCS operates as a virtual space, defining the relationship between different device spaces so that a unique transformation can be created for any pair of devices or color spaces for which profiles are available.

If the device model in a profile is defined for a viewing condition whose chromaticity is different from D50 (as, for example, would often be the case for a display profile), the profile must include a chromatic adaptation transform so that a translation between PCS and device colorimetry can be performed by the CMM. Otherwise, the profile must ensure that PCS colorimetry is corrected to account for any change in color appearance caused by differences in the absolute luminance level between viewing condition of the PCS and that for which the device model was defined, which will lead to changes in perceived lightness and colorfulness. This will normally require the use of a color appearance model. Adjustments to the device model to include preferred reproduction strategies such as gamut mapping techniques and tone scale mappings should also be included in the profile.

This definition of the PCS implies that the color matching module simply provides a means of connecting two or more profiles together, and where necessary interpolating the LUT data and performing any required chromatic adaptation. Modelling device behaviour, color appearance, gamut mapping and so on are all done by the profile.

This definition of the role of the PCS, the profiles and the CMM together make up default behaviour of the system, which can be implemented at the level of the operating system with minimal computational overhead. It is possible for third party CMM vendors to implement more complex transformations, but the default behaviour ensures that profiles will operate on any system, regardless of the operating system or CMM used.

11.3 THE ICC PROFILE

The function of the ICC profile is to provide the data necessary to transform the colors of an image from the color characteristics of one device to those of another. To accomplish this, it employs a common intermediate ground on which these profiles can communicate. A simplified example is a profile containing the information to convert the color data from a specific scanner's characteristics into this well-defined common ground so that the printer's profile can convert the color from there to its own color characteristics. This common ground is a reference colour space, defined by the ICC as the profile connection space (PCS) described in Section 11.2 above, and using either CIE XYZ or CIELAB. The PCS is described in Section 11.2.2 below. The conversions necessary for the example above may be illustrated as follows:

RBG scanner profile provides the conversion from: scanner RGB $\rightarrow$ PCS
CMYK printer profile provides the conversion from: PCS $\rightarrow$ printer CMYK

But the profile can be even more powerful. Suppose the image producer wants to use a CRT monitor to preview the effects of a printer profile on an image prior to printing it. The printer profile may contain the information to accomplish this in another set of data. Note that because the colors will have gone through printer CMYK, only in-gamut colors will be shown on the monitor.

(RBG scanner profile)   scanner RGB → PCS
(CMYK printer profile)   PCS → printer CMYK → PCS
(RGB monitor profile)   PCS → monitor RGB

In addition, since the printer's color gamut may be much narrower then the scanner's color gamut, the printer profile can include data which will allow identification of out-of-gamut colors. It would then be up to the CMM in use to do something with this information by adjusting the out-of-gamut colors in some way. Chapters 12 and 13 in this section discuss many of these gamut issues.

As illustrated above, the profile frequently needs to include information to convert color data from the PCS back to the device color. This is a reverse transformation, which is required content in some profiles.

ICC profiles can aid in almost any color workflow, provided the profiles exist and the software managing the workflow supports color management using ICC profiles. The profile is generally needed to convert the color between any device creating, viewing, or printing the graphic or image to maintain some degree of color management.

11.3.1 ICC Profile Types

The profile user will want to be aware of the different types of ICC profiles defined by the specification that may be encountered in color management. Each profile type defines a slightly different set of required vs. optional tags so each can contain slightly different information (see the ICC specification for details). So far, only device profiles have been mentioned – these are further identified as 'Input', 'Display' and 'Output' profiles, which describes generally where the profile appears in the workflow.

The 'Colorspace' profile converts data from one colorspace to the PCS. Another profile is normally used with the colorspace profile to take the data to another defined color space for storage or transport or to an output device. For example, if you receive an image in some RGB colorspace and your software wants to manipulate the image in HSV space (this problem was actually presented to this writer), a set of colorspace profiles could accomplish the following:

(source profile)   input device's RGB → PCS →
(colorspace profile)   PCS → HSV
Image manipulation software outputs new image in HSV
(colorspace profile)   HSV → PCS →
(destination profile)   PCS → output device's color

The 'Device link' profile is the combination of the transforms of two or more profiles. The value of creating a device link profile is that it can be created once and used for numerous images in the same workflow. Some color management systems create the device link on the fly, use it once, and then destroy it as their normal mode of operation. Creating the device link and storing it can save a lot of processing. Once a device link profile is created, it cannot be further linked to other profiles (as defined by the ICC specification). The example above may involve two device link profiles, the combined source and colorspace device link and the combined colorspace and destination device link.

An interesting, but little used, profile type is the 'Abstract' profile. It can provide a means of making artistic alterations on images. The abstract profile can accompany an image, along with its source device profile, to produce certain effects on an image prior to it being printed. These effects could be incorporated into the image itself, but providing this capability allows for more flexibility. For example, one may want the source image intact, but have the option to reproduce it with the artist's unique rendition, say light hazy effects over the entire image. That hazy effect can be captured in the abstract profile, which is then inserted between the source and destination profiles to accomplish the necessary transforms. The abstract profile is defined by the ICC to stay in PCS space, resulting in a set of transforms such as:

$$
\begin{array}{ll}
\text{(source profile)} & \text{device space} \rightarrow \text{PCS} \rightarrow \\
\text{(abstract profile)} & \text{PCS} \rightarrow \text{PCS} \rightarrow \\
\text{(destination profile)} & \text{PCS} \rightarrow \text{device space}
\end{array}
$$

The 'Named color' profile deals with sample library colors or color which can be similarly identified with a name and CIE color values. This is especially useful in graphics where a patented logo's color must be exactly recreated.

11.3.2 Use of ICC Profiles

When a user selects an option that requires a color transformation a call is made via the operating system to the profile and CMM.

If an application supports color management, there should be information about that fact in the documentation and some way of choosing a profile for the devices to which or through which users are able to manage color data. The application would normally provide a list of available profiles using the description in the profile's text fields. So, for example, if you have a scanned print you wish to display on your monitor, you would choose the profile which describes your scanner type and settings as well as the media type scanned, then also choose the profile describing your monitor type and settings. Providing this information to the application will allow it to process your data through the profile to provide the best possible rendition of the original onto your monitor.

Also of interest to profile users are embedded profiles. Several file formats allow for embedding device profiles along with the image related to the device that created the image. These formats include PICT, EPS, TIFF, JFIF and GIF. In receiving or sending images in these file formats, the user will want to be aware of how to access or append this information. See the ICC specification's Annex for detailed information. The embedded profile serves as the device source profile to be used with your monitor or other destination device profile in effecting color management on the image. The applications that support color management will extract and use the embedded profile if it exists.

Printing an image entails the same scenario. One must choose the profile for the device which produced the data (perhaps a scanner, computer display, or digital camera) and the profile for the printer/media on which it is to be printed.

What if the result is not what the user expected? There are several reasons why this can occur. The profile may not represent the media you are using or viewing conditions under which the data were created, viewed, or printed. The number of combinations of these factors can result in a staggering number of required profiles. Companies are working on providing these profiles, however. Not having the correct profile will make a difference in the result, some differences being greater than others.

Even if the user has the correct profile, the devices in the workflow may require calibration and that calibration information must be reflected in the profile chosen for the device. Some applications provide this capability, monitor calibration being the most common. It provides the capability through software either to use a hardcopy color test target to visually match colors on your monitor, or to use a device such as a colorimeter to measure colors and convert them to a known color space. The result of the measurements can either create a new profile or update an existing profile to represent the device's current condition.

Scanner calibration is also provided by some software. This normally uses some sort of known target of color and grayscale data that the user scans and the software rectifies with the profile. With calibration, a new profile is created which contains the updated data from the original profile. The user should be aware that monitors need regular calibration, even when the user's viewing condition is held constant.

Printer calibration is more difficult and the software for accomplishing this is not as readily available for the casual user. Several manufacturers have very sophisticated software to aid in the calibration and creation of profiles, including printer profiles.

Some of the software referred to above can also accomplish profile creation, as well as profile updates, for monitors, scanners, and, less commonly, printers. This makes the profile user a profile creator without needing the knowledge of how to manipulate the data to create the color conversions to enter into a profile. The application handles all the required transformations.

11.4 THE ICC PROFILE STRUCTURE

The ICC profile is a data file consisting of information which may include:

- The company/software that created the profile
- The device and its settings/media for which the profile was created
- The color characteristics of the device
- The data necessary to convert the color out of the device's color space and settings into a PCS
- The data to convert from the PCS back to the device's colorspace (if required)
- Additional information to aid in printing, viewing or otherwise reproducing the image with the desired color.

The file format is a tag-based structure which allows a variable amount of data (tags) to be included. There is a fixed length header for each profile and a listing of the tag keywords, their byte offsets from the beginning of the profile, and their length. Following this is the actual tag data, identified by the tag identification (tag ID), length, and specific tag data. This structure allows the profile reader (CMM) to skip over tag data that it does not use in processing the profile. The definition, but not necessarily the contents, of each tag, comprises the ICC Profile Format Specification.

It would be laborious and repetitive to include an explanation of all the tags in the ICC profile. They are well explained in the specification. For quick reference, however, Table 11.1 lists the tags supported in the ICC profile at the time of this writing.

The profile user may want to skip the next section if they are not also interested in creating profiles.

11.4.1 ICC Profile Creation

The profile creator needs much more knowledge than one who only uses profiles in applications. Much of this knowledge is provided throughout this book. To create ICC format profiles, the ICC Profile Format Specification is necessary. The ICC website [7] host the latest version of this document, together with characterization data for standard printing processes, technical information on profiles and the ICC architecture, and other resources. One document, which has been helpful to some profile creators, is the book *Building ICC Profiles – the Mechanics and Engineering* by this author [1] which documents the programming mechanics of reading and writing profiles. This can be downloaded free of charge from the ICC website along with 'C' language code examples for implementation help.

Table 11.1 Summary of tags in ICC profile format

Tag name	General description
AToB0Tag	Multi-dimensional transformation structure: perceptual rendering intent
AToB1Tag	Multi-dimensional transformation structure: colorimetric rendering intent
AToB2Tag	Multi-dimensional transformation structure: saturation rendering intent
blueColorantTag	Relative XYZ values of blue phosphor or colorant
blueTRCTag	Blue channel tone reproduction curve
BToA0Tag	Multi-dimensional transformation structure: perceptual rendering intent
BToA1Tag	Multi-dimensional transformation structure: colorimetric rendering intent
BToA2Tag	Multi-dimensional transformation structure: saturation rendering intent
calibrationDateTimeTag	Profile calibration date and time
charTargetTag	Characterization target such as IT8/7.2
chromaticAdaptationTag	Multi-dimensional transform for non-D50 adaptation
chromaticityTag	Set of phosphor/colorant chromaticity
copyrightTag	7-bit ASCII profile copyright information

Table 11.1 (*continued*)

Tag name	General description
crdInfoTag	Names of companion CRDs to the profile
deviceMfgDescTag	Displayable description of device manufacturer
deviceModelDescTag	Displayable description of device model
deviceSettingsTag	Specifies the device settings for which the profile is valid
gamutTag	Out of gamut: 8-bit or 16-bit data
grayTRCTag	Gray tone reproduction curve
greenColorantTag	Relative XYZ values of green phosphor or colorant
greenTRCTag	Green channel tone reproduction curve
luminanceTag	Absolute luminance for emissive device
measurementTag	Alternative measurement specification information
mediaBlackPointTag	Media XYZ black point
mediaWhitePointTag	Media XYZ white point
outputResponseTag	Description of the desired device response
preview0Tag	Preview transformation: 8-bit or 16-bit data
preview1Tag	Preview transformation: 8-bit or 16-bit data
preview2Tag	Preview transformation: 8-bit or 16-bit data
profileDescriptionTag	Profile description for display
profileSequenceDescTag	Profile sequence description from source to destination
ps2CRD1Tag	PostScript Level 2 color rendering dictionary: colorimetric rendering intent
ps2CRD2Tag	PostScript Level 2 color rendering dictionary: saturation rendering intent
ps2CRD3Tag	PostScript Level 2 color rendering dictionary: ICC-absolute rendering intent
ps2CSATag	PostScript Level 2 color space array
ps2RenderingIntentTag	PostScript Level 2 Rendering Intent
redColorantTag	Relative XYZ values of red phosphor or colorant
redTRCTag	Red channel tone reproduction curve
screeningDescTag	Screening attributes description
screeningTag	Screening attributes such as frequency, angle and spot shape
technologyTag	Device technology information such as LCD, CRT, dye sublimation, etc.
ucrbgTag	Under color removal and black generation description
viewingCondDescTag	Viewing condition description
viewingConditionsTag	Viewing condition parameters

The following sections outline a range of topics which are or frequently overlooked or misunderstood by profile creators.

11.4.2 ICC device models

Some input devices may be able to model their color characteristics and color conversions to PCS with a single matrix, and one-dimensional look-up tables (1D

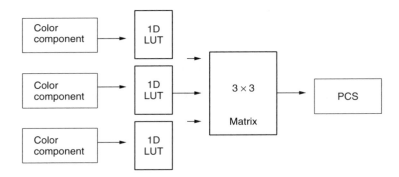

Figure 11.1 Shaper/matrix processing model for the device-to-PCS direction. Inversion of this model gives the PCS-to-device model.

LUTs). Figure 11.1 is representative of the transforms for this 'shaper/matrix' model. The three color components are processed through the device's tonal response curves (red/green/blueTRCTag), then through the 3 × 3 matrix derived from the red, green, and blue colorants (red/green/blueColorantTag). Only the CIEXYZ encoding of the PCS can be used with this model. The one-dimensional LUTs map individual color component values into other color component values either directly or via interpolation between the specific values provided in the LUT. The input device can be any of a number of devices like scanners, digital cameras, video cameras, etc. All these devices must have profile data specifying how to convert from their device-dependent color data to the PCS, but not all require the data to convert back. In some cases it just doesn't make sense to need the data back in, for example, scanner space because that may be a very uncommon workflow. But there are reasons to want to convert monitor data back to the monitor space for previewing printer results prior to printing.

Other devices or conditions (viewing conditions, adaptation models, or media) are more complex to model and will need to use the multi-dimensional color LUT (CLUT) tags provided in the profile format. Figure 11.2 illustrates the conversion from PCS to output

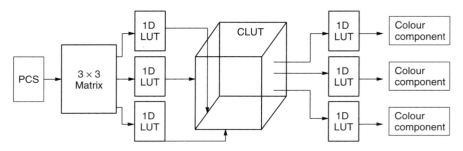

Figure 11.2 Matrix/tabulated function processing model for the PCS-to-device direction. The profile also includes the inverse device-to-PCS model.

device color using the 3×3 color conversion matrix (red/green/blueColorantTag) and the multi-dimensional CLUT (BtoA#Tag) composed of a set of one-dimensional input LUTs, a multi-dimensional LUT, and a set of 1D output LUTs. The CLUT is also more difficult to create and can result in a very large profile. The use of the multi-dimensional LUTs is required for some profile types.

Profile creators must ensure that their conversions are such that no confusion exists as to how a table is used. Monotonically increasing or decreasing tables are necessary to eliminate interpolation errors and ambiguities.

So why aren't all these transforms spelled out in the ICC specification? Allowing companies to have a 'value-added' ability in their color management systems has been an important goal for the ICC. It may make it harder for the profile creator to figure out exactly how to massage the data going into and out of a profile if they choose to write their own color management/profile creation software, but it also allows innovation and improvements to be realized in this inexact and frequently subjective field. At the time of this writing, the ICC is working on software which will provide a baseline for those who wish to create their own software to create and manipulate profiles. This reference implementation will be posted on the ICC website [7] when it is made publically available.

11.4.3 ICC rendering intents

One result of the complexity of providing color management, which attempts to satisfy everyone in all workflows, is the inclusion of the rendering intent flag. Its concept is similar to a quality flag that allows the user to choose speed over accuracy. In this case, quality is always desired, but with different intents for use. The intents are related to gamut compression techniques. The ICC supports two styles of controlled gamut compression – perceptual and saturation – in addition to the colorimetric option, which clips abruptly at the gamut boundary. A photographer, hoping to retain the color balance and perceptual effect of a photo from a scanner to a printer, would want a CMM to use the perceptual rendering intent. However, since the same devices may be used to create company logos or other graphics whose color must be exact, would be needed colorimetric intent. An image with large areas of uniform color, such as a pie-chart, might require the saturation intent. Ideally, these tags would not be necessary – the original color would always be recreated. However, this is frequently not possible due to the varying range of color gamuts of which devices are capable. Some compromise must be made. This involves gamut mapping, a topic covered elsewhere this book.

Rendering intents are discussed further in Chapters 13 and 16.

11.4.4 Required processing tags

The header, copyrightTag, and profileDescriptionTag are always required for a profile. The mediaWhitePointTag is required for all but the device link profile type, which requires the profileSequenceTag. The named color profile type also requires the namedColor2Tag.

The primary tags that are necessary for processing the color are listed below. Please see the ICC specification for which profile types require which tags.

red/green/blueTRCTag:	1D tonal response curves
grayTRCTag:	1D tonal response curve for grayscale data
red/green/blueColorantTag:	creates a 3×3 color conversion matrix

CLUT tags (# is 0 for perceptual rendering intent, 1 for colorimetric rendering intent, 2 for saturation rendering intent) are:

AtoB#Tag:	device to PCS color conversion CLUT
BtoA#Tag:	PCS to device color conversion CLUT
preview#Tag:	PCS to device space and back to PCS for previewing data
gamutTag:	PCS to 1D output representing an out-of-gamut indicator

It is not uncommon to find all the AtoB#Tags in the tag table pointing to the same data in the profile. The same may be true of the BtoA#Tags and preview#Tags.

11.4.5 Private tags

As mentioned previously, the ICC specification does not handle all of the capabilities that might be useful for color management. The concept of private tags in the ICC specification provides a convenient means to enable expansion and innovation.

Logically, the need and use of private tags for any basic color management would defeat the purpose of interoperability of profiles. The private tags that are in use are primarily informative, although there are some CMMs that can use the private tags for value-added quality results when using their own profiles within their own software. The goal, however, is that each CMM should strive to provide the best possible results using any profile which follows the ICC specification. This, of course, is dependent upon the quality of the data in the profile – something that is not guaranteed, but certainly intended. When writing software to read ICC profiles, the software must account for skipping over these private tags.

Private tags can be registered with the ICC. Their structure must follow the ICC tag structure so that other CMMs may pass over the data, safely ignoring these private tags. You may get a list of the registered tags from the ICC website. Occasionally, there may be enough information about a private tag in the registry so that your CMM may be able to make use of the information. While the use of private tags is discouraged, if the information is of sufficient general importance, it should be presented to the ICC as a proposal for an additional official tag.

11.5 PROFILE USE

This section provides some examples of the application and development of ICC profiles

11.5.1 Color management on the web and ICC profiles

One thorn in the side of color management is images on the World Wide Web. There is no control over the environment for creating and viewing many of these. Several image formats exist for web viewing and, of course, monitors run the gamut (pun intended) of color characteristics and viewing conditions. One could attach an ICC profile to every image transported over the web and hope that the recipient has the means to use it – but this is impractical. The ICC profile can be very large, and utilizing a CMM to read the profile and use it for all images over the web is not always realistically practical today. A concerted effort is being made to educate image producers and image format software creators about using a known, defined, color space to help alleviate this problem. sRGB is a standard RGB-based color space for web images as well as other images that may be transported with no color space information attached. Some systems, like Sun's Java and Microsoft's ICM, will assume the sRGB color space for images if no other color information exists. One may argue that assuming a color space is worse (or no better) than doing nothing (as is normal today), but it is a step in the direction of getting people to use a defined, default color space. sRGB ICC color space profiles exist from several sources, including Sun's Java and Solaris development products.

11.5.2 Sample ICC profile interfaces

Included here is a sampling of the operating system vendors and other companies which support ICC profiles. Many companies use them and the color.org website includes links to most of those companies. Search for 'ICC' at their home sites.

Application programmer's interfaces (APIs) using ICC profiles in color management systems vary considerably in their designs. Generally, they provide APIs to read and write the tag data of a user-specified profile, provide some amount of data validation, allow creation and updating of profiles, process an image through profiles, and allow two or more profiles to be specified for use on an image.

A simple but effective example of an API is Sun's Kodak Color Management System (KCMS), bundled with the Solaris operating system. The interface to Kodak's color engine (CMM) was encapsulated into a few API calls, with arguments that can allow access to other CMMs by parsing the arguments into CMM-specific interfaces. This enables alternate CMMs to be plugged into the color management system. The CMM developer could overwrite the various CMM classes in the Kodak CMM with their own class. The application developer would not need to change the code to take advantage of the alternate color processing. Profile creators can use the interface to create or update profiles by providing characterization or calibration data as well as for any other supported ICC tag.

For complete documentation on KCMS, please see http://docs.sun.com and search for 'KCMS'. See also Table 11.2.

The Java 2D API includes ICC profile handling via the ColorSpace, ICC'Profile, and ICC'ColorSpace classes. Although functionality to access the individual tags of the profile for profile creation and for updating profiles is currently very limited, these classes allow the application developer to specify ICC profiles to use in processing images.

ICC_Profile.getInstance:	supply an ICC profile name
ICC_ColorSpace:	build a colorspace from the profile

Table 11.2 Summary of KCMS API calls

API call	Function
KcsLoadProfile	Provide a file ID or profile name and indicate whether to load the entire profile or just the header (load hints)
KcsGetAttribute	Provide profile ID and name of the tag to read
KcsSetAttribute	Provide the profile ID, the tag name and the data to be written in the tag
KcsConnectProfile	Provide the number of profiles, the list of profile IDs and loading hints
KcsCreateProfile	Provide a descriptive structure for creation of an 'empty' profile. (Use KcsSetAttribute and KcsUpdateProfile to complete the profile)
KcsUpdateProfile	Provide the profile ID and the characterization or calibration data to build the matrices and LUTs in the profile
KcsModifyLoadHints	Provide profile ID and new loading hints to load tags that were not previously read
KcsSaveProfile	Provide profile ID and new profile name
KcsFreeProfile	Provide the profile ID of the profile to unload, freeing up memory
KcsEvaluate	Provide the profile ID of the device link profile created by KcsConnectProfile and the source and destination of the image to be processed through it
KcsOptimizeProfile	Provide the profile ID of a device link profile and hints for optimizing its use for highest processing speed, resulting accuracy, or device profile size
KcsSetCallback	Provide a function name whose processing state would be saved in the event of processing interruption – allows restart with minimum processing loss

Java 2D bundles the Kodak color engine as its CMM, the same color engine which is included with Solaris's KCMS. For more details on the Java color classes for ICC profiles, please see http://java.sun.com/j2se for the latest release. Additional documentation and examples of ICC profile use in Java 2D are provided in Ref. 2.

Perhaps one of the best company sites for color management and ICC use in products is the Apple website [3, 4].

Microsoft's ICM 1.0 (in Windows 95) was designed to address the color management needs of applications that work with RGB color.

Microsoft's ICM2 (with Win98, Win98SE, WinME, and Win2000) supports the Lino-Color CMM from Linotype-Hell AG (bundled with the operating system), which allows support of all color spaces including HiFi color processing. ICM2 also allows developers to port alternate CMMs to work with applications using color management.

The ICClib Open Source code for reading and writing profiles, and for performing color conversions, can be obtained from http://web.access.net.au/argyll/

The lcms open source CMM is available at http://www.littlecms.com/

The Apple web site [3, 4] has full documentation on Color Sync and Profile use.

Microsoft provides structures for applications writers to use to add color management to their software [5]. These are in addition to those in the ICC header file example, provided at the ICC website, and serve to provide additional functionality.

11.5.3 ICC profile validation software

Although many CMMs perform some validation on the data read from a profile prior to processing, the need for a formal profile validation program was recognized. The purpose was not only to aid profile interoperability, but also to allow one to brand a profile as ICC compliant in the future. Although a free validation suite does not currently exist, QualityLogic has developed profile validation software [6].

Validation does not guarantee the quality of the data – only experimentation can do that. It does, however, verify that CMMs can read and make use of the data.

11.5.4 What to look for in the future at the ICC website

- Updated ICC Profile Format Specifications. Version 4 was published in 2001. Future specifications may include clarifications and additions that will enhance the usefulness of the profile.
- Source code for baseline profile generation and CMM.
- Standard profiles for some color conversions.
- Additional contributed papers and extended information on color management.

11.6 CONCLUSION

Providing a standard ICC Profile Format to contain color data is an important and significant part of providing standard color management across computer systems, color producing devices, and image manipulation software. The expanding uses of color data, particularly on the Internet, will continually present new problems, demanding innovative solutions. The ICC hopes to keep abreast of these problems and aid in their solutions.

REFERENCES

1. Wallner, D. (2000) *Building ICC Profiles – The Mechanics and Engineering*. Available as a free download at www.color.org
2. Hardy, V. J. (1999) *Java 2D API Graphics*, Sun Microsystems Press.
3. http://developer.apple.com/macros/color.html.
4. http://developer.apple.com/techpubs/macos8/MultimediaGraphics/ColorSyncManager/ManagingColorWithColorSync/frameset.html.
5. http://www.msdn.microsoft.com/library/psdk/icm/icm_6rw3.htm.
6. http://www.qualitylogic.com/genoa_test_tools/printer/icc.html.
7. http://www.color.org

12

Colour gamut determination

Marc Mahy

12.1 INTRODUCTION

The gamut of a printing device is the set of all colours that can be reproduced by the given printer. In practical applications, the gamut is mainly used in three ways. Either the gamut determines if a given set of colours can be reproduced by the given printer, parameters of the printing process are optimised to obtain a given range of colours, or the shape of the gamut is used in gamut mapping algorithms to reproduce out-of-gamut colours properly.

Due to the increasing demands on the flexibility and accuracy of colour processing workflows, an extensive gamut description is required. However, the gamut is often represented as a two-dimensional projection of the range of printable colours in the xy-chromaticity diagram or the a^*b^*-chroma plane of CIELAB [49]. As the real gamut is a volume in a three-dimensional space, such a projection only gives a first impression of the gamut but nothing can be said about the lightness distribution of the reproducible colours. To fulfil the needs for different applications, an analytical technique is presented here to obtain a three-dimensional gamut descriptor based on a general printer model.

Colour Engineering, edited by P.J. Green and L.W. MacDonald.
© 2002 John Wiley & Sons Ltd.

12.2 OVERVIEW OF GAMUT CALCULATIONS

There exist two classes of methods to obtain a gamut description [19]. According to a first approach, a large set of samples spanning the printable range is measured and empirical methods are used to estimate the gamut boundaries. In a second, more analytical approach, the gamut is based on a model describing the behaviour of the printer.

The first publications on gamut characterisation discussed the gamut of a subtractive colour process, i.e. photography [33–40]. To be able to construct a gamut that is sufficiently large, the best set of cyan, magenta and yellow pigments are calculated, for both reflection and transmission photography. These gamut calculations were mainly based on the Lambert–Beer law, which has where necessary extended to take into account optical effects in reflection photography. With this model, a number of colours along several hue values were calculated for a given luminance value. The outer contour was considered as the gamut of the printer.

For pure additive systems, on the other hand, such as CRT displays, the gamut can be easily calculated based on a linear relation between the phosphor characteristics and the XYZ space [5, 42].

Later on, gamuts were calculated for several printing systems; however, in most cases a simplified printer model was used [8–11, 33–41, 46]. Other researchers represented the gamut of the offset system in the xy-chromaticity diagram as a hexagon by connecting the primary and secondary colours of the CMY inks [48]. To obtain a three-dimensional gamut representation, either a printer model was constructed or a large number of colour patches were printed and measured. In the case of a printer model, triangulation was often used to visualise the gamut [2]. In the case of measurements, the gamut boundaries were deduced by taking the convex hull [1] or surfaces were fitted to colours at the gamut boundary [12, 14].

A more general approach to describe the gamut boundary was obtained by making use of contours. Inui presented a gamut description based on contours for a 3-ink process assuming that the gamut boundary is described by the faces of the colorant cube [16]. In the case of the Neugebauer equations, the gamut intersections with constant lightness or hue can be described analytically for well-behaved ink processes [21–23].

The gamut of a colour reproduction system has been used to deduce several printer characteristics. First of all, the gamut defines the range of all possible colours that can be printed. Hence, to compare the gamut of one system with the gamut of another reproduction device, both gamuts have to be visualised against each other. The size of the gamut such as the volume in CIELAB space does not give sufficient information [47]. In the case of a two-dimensional gamut representation, e.g. in the xy-chromaticity diagram, more information is obtained, but, because the lightness value of the darkest colour significantly determines the size of the gamut, also in this case the gamut is not well represented. A complete three-dimensional gamut visualisation is the only solution to compare gamuts properly. For this purpose, several techniques and applications to visualise gamuts have been developed in the past [3, 4, 12, 14, 17, 19, 27–29, 43, 44].

A second wide field of applications in which the gamut plays an important role is the optimisation of device parameters to obtain a large gamut or a stable colour reproduction.

Typically the effect of different sets of inks is studied in combination with paper characteristics [8–11, 33–41, 46]. Most studies are based on parameterised printing models and hence the gamut is predicted by the printer model. Typically, for a given luminance level and hue value, the most saturated printable colour is calculated. This is a time-consuming process, which results in a number of points on the gamut boundary. These sets of gamut boundary points are compared between different processes and are represented either in constant luminance cross-sections or by constant hue cross-sections.

A last important application that is in need of a gamut descriptor is gamut mapping. In this case the gamut description of both the input and output devices is required to map out-of-gamut colours from one device to another [30, 31, 50]. However, the required accuracy of the gamut description for gamut mapping is significantly higher than for the previous discussed applications. By preference, the gamut should be known exactly for several lightness levels and hue values. In the literature, several gamut descriptors have been published, such as the mountain range method that specifies the maximum chroma per lightness and hue value [3, 12], the Segment Maxima Gamut Boundary Descriptor [30, 31] and a contour description by intersecting the gamut with a number of constant lightness or hue planes [16, 21–23, 26].

12.3 TERMINOLOGY

The different independent values with which a printer can be addressed to generate a colour are called *inks* or *colorants*. Due to practical restrictions on the amount of colorant that can be laid down on paper, the maximum value for each colorant is limited. We assume that the value for each colorant ranges from 0 to 100%. If there are n colorants, the colorant values can be represented in an n-dimensional space, the *colorant space*. The physically realisable colorant combinations are contained by the *n-dimensional colorant cube* in colorant space for which the values of all colorants range from 0 to 100%. A printer with n colorants will be called an *n-ink process*.

A *printer model* is a continuous mathematical transformation from colorant space to colour space that relates colour values as functions of colorant values. The domain of the printer model is called the *colorant domain*. If there are no colorant limitations, the colorant domain coincides with the n-dimensional colorant cube. For most printers, however, there is a maximum amount of ink that should not be exceeded. Such a limitation will be specified with a linear restriction in colorant space, also referred to as a *linear ink limitation*. The region obtained by transforming the colorant domain to colour space by the printer model is called the *colour gamut* or simply the *gamut* of the printer.

A typical example of a printer is a CMYK offset system [15]. This is a 4-ink process with colorants cyan (C), magenta (M), yellow (Y) and black (K). A basic characterisation of the offset process is given by the Neugebauer equations [15, 32, 51], which transform CMYK colorant values to the XYZ space. The domain of the CMYK offset system is a four-dimensional colorant cube. However, it is common in offset to use a maximum amount of ink, typically about 360%. Hence, an ink limitation given by $C + M + Y + K \leqslant 360$ should be taken into account.

In the literature, the gamut is often calculated by transforming the boundaries of the colorant domain to colour space. These boundaries are in most cases 2-ink processes, also referred to as the *physical colorant boundaries* of the colorant domain, which are obtained by setting all colorants in the printer model except two to their minimum (0%) or maximum (100%) value. The gamut is found by transforming all the physical colorant boundaries to colour space. The outer boundaries in colour space constitute the *physical colour gamut*. For a 3-ink process, the physical colorant boundaries are the boundaries of the colorant cube. For an ink process with more than three colorants, however, the physical colorant boundaries lie on the boundaries of the colorant cube but do not determine them completely. For a CMYK process, for example, the colorant combination (0, 10, 20, 30) lies at the boundary of the colorant cube but does not belong to a physical colorant boundary. Later on we will indicate that in some cases the physical colour gamut is smaller than the real gamut and that other types of gamut boundaries should be taken into account. In most cases, however, the physical colour gamut coincides with the gamut of the printer. Such a printer will be considered as a *well-behaved printing process*; all other printers are called *less well-behaved processes*.

If for a given printer model one or more colorants are kept constant, a new printer model is obtained from the given model by setting the corresponding colorants to their constant values. The new model is called an *extracted ink process* from the given printer model. Also the colorant domain and the linear ink limitations are *inherited* by setting the corresponding colorants to their constant values. For a 3-ink process, for example, if the first colorant is kept constant at 25%, an extracted 2-ink process is obtained by setting the value of the first colorant in the 3-ink process to 25%.

The domain boundaries of a 3-ink process can be considered as 2-ink processes of which the printer model is obtained by setting the constant colorant to its minimum or maximum value in the 3-ink printer model. If there are no colorant limitations, the domain of the 2-ink process is given by all colorant combinations between 0 and 100% for the remaining two varying colorants. Such a 2-ink process is called a *2-ink boundary process of the 3-ink process*. In this case, there are six domain boundaries and hence six corresponding 2-ink boundary processes. For a 4-ink process, there are eight 3-ink boundary processes and 24 2-ink boundary processes. The 2-ink boundary processes can also be deduced from the 3-ink boundary processes. In general, *m*-ink boundary processes are extracted processes from a given *n*-ink process by setting $(n - m)$ colorants to their minimum or maximum value.

12.4 A GENERAL PRINTER MODEL

12.4.1 The Neugebauer model

Printing with three inks and three halftone screens results in eight possible combinations of ink overlap. The Neugebauer expressions [15, 32, 51] predict the resulting colour as a linear function of these ink combinations. The Neugebauer equation for the X tristimulus values, making use of the Demichel equations [6] for a 3-ink process, is given by:

$$X(c_1, c_2, c_3) = (1 - c_1)(1 - c_2)(1 - c_3)X_w + c_1(1 - c_2)(1 - c_3)X_1$$

$$+ (1 - c_1)c_2(1 - c_3)X_2 + (1 - c_1)(1 - c_2)c_3 X_3$$

$$+ (1 - c_1)c_2 c_3 X_{23} + c_1(1 - c_2)c_3 X_{13}$$

$$+ c_1 c_2(1 - c_3)X_{12} + c_1 c_2 c_3 X_{123} \tag{12.1}$$

Here, X_w is the tristimulus value of the paper, X_i the tristimulus value of ink i, X_{ij} the tristimulus value of the overlap of inks i and j, and X_{123} the overlap of all three inks. The values c_1, c_2 and c_3 are the amounts of the different colorants. The Neugebauer model predicts colour as a function of colorant values [7, 13], hence it is a transformation from colorant space to colour space. If there are no ink limitations, the domain is the colorant cube and the range is the gamut (see Figure 12.1). The Neugebauer equations can also be seen as the trilinear interpolation between the eight *corner points* of the colorant cube (see Figure 12.1). The Neugebauer expressions for the Y and Z tristimulus values are obtained in a similar way. The extension of the Neugebauer equations for n inks is straightforward.

Another way to represent the Neugebauer equations for a 3-ink process is given by

$$X = k_0 + k_1 c_1 + k_2 c_2 + k_3 c_3 + k_{12} c_1 c_2 + k_{13} c_1 c_3 + k_{23} c_2 c_3 + k_{123} c_1 c_2 c_3$$

$$Y = l_0 + l_1 c_1 + l_2 c_2 + l_3 c_3 + l_{12} c_1 c_2 + l_{13} c_1 c_3 + l_{23} c_2 c_3 + l_{123} c_1 c_2 c_3 \tag{12.2}$$

$$Z = m_0 + m_1 c_1 + m_2 c_2 + m_3 c_3 + m_{12} c_1 c_2 + m_{13} c_1 c_3 + m_{23} c_2 c_3 + m_{123} c_1 c_2 c_3$$

The coefficients k, l and m are a linear transformation of, respectively, the X, Y and Z values of the corner points of the colorant cube.

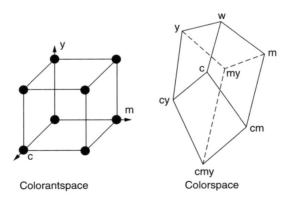

Figure 12.1 Transformation of the colorant domain to colour space for a CMY Neugebauer model. The black dots represent the corner points of the 3-ink process.

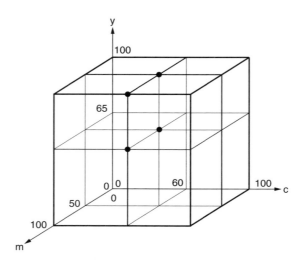

Figure 12.2   Subdivision of the CMY colorant cube in a number of cells. The four black dots are
the corner points of the common boundary of two neighbouring cells.

12.4.2 A localised Neugebauer model

For a localised printing model, the colorant space is divided into a number of cells.
For each cell the printer is modelled with an *elementary model*. Preferably the cells are
non-overlapping to obtain unique colour values for each set of colorant values. To avoid
discontinuities at the *common boundary* of two neighbouring cells, the corresponding
elementary models have to predict the same colour values at the boundary. In general,
the cells are obtained by taking a number of *sampling points* along the colorant axes and
hence a regular division of the colorant domain in cells is obtained (see Figure 12.2).

For a localised Neugebauer process [25, 45] the elementary models are Neugebauer
processes. These processes are completely determined by the corner points of the corresponding cell. In the case of a 3-ink process there are eight corner points; for a 4-ink
process 16 corner points have to be taken into account.

If the dimension of the colorant space is n, the dimension of the common boundary
of neighbouring cells is $n - 1$. The common boundary is characterised by the fact that
one of the colorants is constant. The model that describes the colour behaviour at the
common boundary can be obtained by setting one colorant in the elementary Neugebauer model to the constant value. In this way a Neugebauer process is obtained of
which the dimension of the colorant space is $n - 1$. The extracted Neugebauer process is
completely determined by the corner points of the common boundary of the two neighbouring cells. In case of a 3-ink process, the boundary is determined by four corner points
(see Figure 12.2); for a 4-ink process eight corner points have to be taken into account.
Because the extracted Neugebauer model that describes the printer at the common boundary is completely characterised by its corner points, it is not important from which cell the
elementary Neugebauer process is taken to determine the extracted Neugebauer model.
Both cells result in the same extracted Neugebauer process.

If the localised Neugebauer model is compared to other printing models, it can be made as accurate as desired by adding sampling points. If the colour target consists of colour patches that are regularly spread in colorant space, the model will predict the exact colour values for these colours. For colorant combinations between the grid points, the colour is predicted by multilinear interpolation based on the corner points of the cell. Hence, the localised Neugebauer model can be seen as a general printer model.

12.5 GAMUT CALCULATIONS

12.5.1 Introduction

Because the relation between colour and colorant values is a continuous function and the domain of the printer model is connected [18] (characteristics of the printer model), the resulting gamut in colour space is also connected.

The transformation of colorant combinations of an ink process with more than two colorants results in a volume in colour space. Such a gamut is completely determined if its boundaries are known in colour space. In a three-dimensional colour space these *gamut boundaries* are two-dimensional surfaces. Due to the continuous relation between colour and colorants, surfaces in colorant space have to be searched for that map on to the gamut boundaries in colour space. These surfaces in colorant space are called *colorant boundaries*.

A way to represent a gamut is obtained by intersecting it with a number of surfaces. In this case, the gamut description consists of contours because the intersection of two surfaces generally results in a one-dimensional curve. If the gamut is intersected with constant luminance planes, the resulting contours are called *equiluminance contours*.

12.5.2 Boundaries of a 3-ink process

Physical boundaries

Intuitively, one might think that the physical colorant boundaries transform to gamut boundaries in colour space. For a 3-ink process, the colorant domain is a cube. Each face of the colorant cube corresponds to a 2-ink process in which the third ink is kept constant at either the minimum or maximum value. Therefore, the term *physical boundary* is introduced to indicate that these boundaries correspond to the physical limitations on the ink amounts that can be printed. In colorant space these boundaries are called *physical colorant boundaries* that are transferred to the *physical colour boundaries* in colour space by the printer model.

It is possible to indicate that the physical colour boundaries may correspond to the gamut boundaries in colour space. This is demonstrated in Figures 12.3 and 12.4. For a point A inside the colorant domain, there exists a corresponding point A′ in colour space. If the printer model is linearised locally for infinitesimal changes of the colorants, a small change dc_1 of colorant c_1 corresponds to a change in colour space that can be characterised by a vector V_1. Similarly a change of the other two colorants c_2 and c_3 results in colour changes V_2 and V_3. If, however, the point A lies on a physical colorant

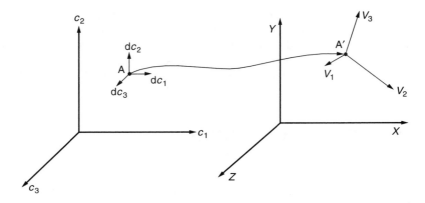

Figure 12.3 Transformation of a point A in colorant space to the point A' in colour space. The infinitesimal colorant changes dc_1, dc_2 and dc_3 induce colour changes V_1, V_2 and V_3 respectively.

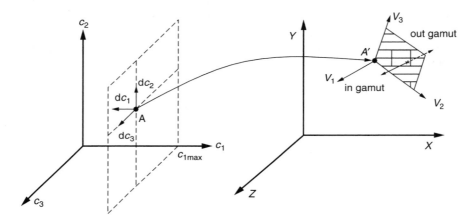

Figure 12.4 Physical boundary of a 3-ink process where $c_1 = c_{1\,max}$. The colour A' can only migrate according to the directions V_2 and V_3 or according to the direction of V_1 that corresponds to negative changes of c_1.

boundary, as in Figure 12.4 where A lies in the colorant plane with $c_1 = c_{1\,max}$, then the corresponding colour A' can only be changed in the plane defined by the vectors V_2 and V_3, and for positive changes of V_1. (in Figure 12.4 V_1 corresponds to a negative change of c_1). As a result, only migrations to half of the colour space can be achieved and as a consequence A' is a point at the gamut boundary.

If the Neugebauer equations are used as the printing model, it is shown in the following paragraphs that the physical colour boundaries have the shape of quadric surfaces. The intersection of such surfaces with any other plane yields a conic section and this is the base for an analytical method to calculate gamut descriptors consisting of, for example, slices with constant Y values (see Appendix 1).

Natural boundaries

For well-behaved 3-ink processes, such as CMY processes, the boundaries of the gamut are completely defined by the six colorant boundaries. It was found, however, that for certain printing processes colorant combinations can be found inside the colorant domain that are transformed to colours outside the physical colour gamut. Such a situation typically occurs in a 3-ink process if one of the inks can also be made with a combination of the other two inks. In this case the gamut has to be larger than the physical colour gamut, and hence additional colorant boundaries have to be looked for that define the gamut. As these boundaries relate more to the way colour is created in the printing process than to limitations on the range of realisable colorants, they are called *natural boundaries*. In colorant space, these boundaries are called *natural colorant boundaries* that are transformed to the *natural colour boundaries* in colour space by the printer model.

The derivation of the natural boundaries is as follows. If the colour A' in Figure 12.3 lies inside the gamut, an incremental colour change can be achieved in any direction by means of an appropriate selection of three colorant changes (dc_1, dc_2, dc_3). Conversely, if the colour A' lies on the boundary of the gamut, migrations are not possible in all directions. A first example of such a situation is the physical boundaries. A second possible situation is shown in Figure 12.5. Because the three vectors V_1, V_2 and V_3 lie in a plane, only infinitesimal colour migrations within this plane are possible, and hence the point A' is part of a natural boundary. The surface itself is found by working out the condition that the three vectors V_1, V_2 and V_3 lie in a plane, or in other words that they are linear dependent.

The vectors V_1, V_2 and V_3 expressed by:

$$V_1 = \left(\frac{\partial X}{\partial c_1}, \frac{\partial Y}{\partial c_1}, \frac{\partial Z}{\partial c_1} \right) \quad V_2 = \left(\frac{\partial X}{\partial c_2}, \frac{\partial Y}{\partial c_2}, \frac{\partial Z}{\partial c_2} \right) \quad V_3 = \left(\frac{\partial X}{\partial c_3}, \frac{\partial Y}{\partial c_3}, \frac{\partial Z}{\partial c_3} \right) \quad (12.3)$$

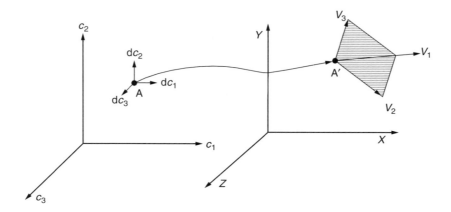

Figure 12.5 Natural boundary of a 3-ink process. Any infinitesimal change of the colorant values causes the colour A to migrate in the plane defined by any two of the three linear dependent vectors V_1, V_2 and V_3.

and the condition that the three vectors V_1, V_2 and V_3 are linear dependent is given by:

$$\det \begin{pmatrix} \dfrac{\partial X}{\partial c_1} & \dfrac{\partial Y}{\partial c_1} & \dfrac{\partial Z}{\partial c_1} \\[2ex] \dfrac{\partial X}{\partial c_2} & \dfrac{\partial Y}{\partial c_2} & \dfrac{\partial Z}{\partial c_2} \\[2ex] \dfrac{\partial X}{\partial c_3} & \dfrac{\partial Y}{\partial c_3} & \dfrac{\partial Z}{\partial c_3} \end{pmatrix} = 0 \tag{12.4}$$

If the Neugebauer equations are used and c_1 and c_2 are considered as parameters, the above-mentioned expression leads to a quadratic function in c_3:

$$c_3^2(r_0 + r_1 c_1 + r_2 c_2 + r_{12} c_1 c_2)$$

$$+ c_3(s_0 + s_1 c_1 + s_{11} c_1^2 + s_{112} c_1^2 c_2 + s_2 c_2 + s_{22} c_2^2 + s_{122} c_1 c_2^2 + s_{12} c_1 c_2)$$

$$+ (t_0 + t_1 c_1 + t_{11} c_1^2 + t_{112} c_1^2 c_2 + t_2 c_2 + t_{22} c_2^2 + t_{122} c_1 c_2^2 + t_{12} c_1 c_2) = 0 \tag{12.5}$$

This indicates that for every couple (c_1, c_2) a set of two values c_3 can be found that meet the equation. Hence, the sets of two solutions c_3 that correspond to couples (c_1, c_2) form two natural boundary surfaces.

An example of a 3-ink process in which the natural colorant boundaries intersect with the colorant domain was found in a printing process using yellow (Y), green (G) and cyan (C) colorants. Here some amounts of the green ink can also be obtained with combinations of the cyan and yellow ink. The XYZ and corresponding CIELAB values of the Neugebauer primaries of this process are given in Table 12.1. The first column corresponds to the eight possible combinations of ink overlap. Figure 12.6 shows the two natural colorant boundaries of this process, i.e. surfaces A and B. As can be seen from Figure 12.6, only surface A intersects with the colorant domain and hence only this surface should be taken into account to determine the gamut boundaries.

Table 12.1 XYZ and CIELAB values of the Neugebauer primaries of a YGC 3-ink process

	X	Y	Z	L^*	a^*	b^*
White	100.0	100.0	100.0	100.0	0.0	0.0
Y	64.8	66.2	10.5	85.1	−3.1	80.1
G	19.7	29.8	7.0	61.5	−43.1	51.2
C	25.7	33.6	69.2	64.7	−29.7	−37.8
YG	14.1	19.6	1.7	51.4	−30.0	64.6
YC	11.1	19.8	6.9	51.6	−51.0	34.4
GC	4.7	11.2	4.6	39.9	−60.9	24.6
YGC	3.0	6.8	1.1	31.4	−49.3	37.8

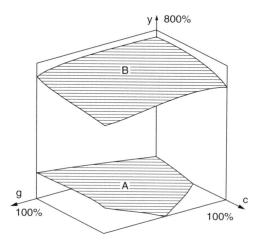

Figure 12.6 Two natural boundaries in the YGC colorant space of the process of which the Neugebauer primaries are given in Table 12.1. Only the surface A intersects with the colorant domain. The 800% value for Y lies outside the colorant cube; it is just a mathematical result if the Neugebauer model is extended to the whole colorant space.

An evaluation reveals that for this process approximately half of the realisable colorant combinations yield colours that lie outside the physical colour gamut. Figure 12.7 shows an intersection of the gamut in CIELAB at a constant L^* value. The distance between two grid lines is 10 CIELAB units. The thick solid line corresponds to the intersection of the constant L^* plane with the natural colour boundary of which the corresponding colorant combinations lie inside the colorant domain. The shaded area represents colours that lie inside the physical colour gamut. The gamut intersection, i.e. the intersection of the gamut taking into account both the natural and physical gamut boundaries, is found by taking the outer contour around the thick line (natural boundary) and the shaded area (physical boundaries).

It may look contradictory at first that the actual gamut of a printing process can be larger than that predicted by the physical colour boundaries, and hence, that a colour on a physical colour boundary sometimes lies inside the gamut. This can be explained only by the presence of at least one other colorant combination that renders that same colour and does not lie on a physical colour boundary nor on a natural colour boundary (ignoring some pathological cases). For that colour, colorant changes can be found that result in colour changes in any direction, including in a direction that goes beyond the limits set by the first colorant combination. This indicates that a close relationship exists between the presence of multiple colorant solutions and the occurrence of natural boundaries [24].

In order to demonstrate this relation, Table 12.2 shows a number of examples based on the separation of colours for the printing process presented in Table 12.1. The inversion technique given in ref. [20] is used to transform colour values into colorant values. In ref. [20] it is shown that there may be up to six colorant combinations to obtain a given colour, but not all these solutions lie inside the colorant domain. In Table 12.2, only colorant

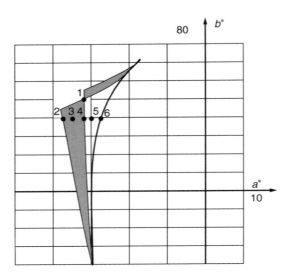

Figure 12.7 Intersection of the gamut in CIELAB of the YGC process in Table 12.1 at $L^* = 65$. The thick solid line corresponds to the intersection with the natural colour boundary. The shaded area represents the intersection with the physical colour gamut. The black dots indicate the positions of the colours in Table 12.2.

Table 12.2 Colorant values of the ink process in Table 12.1 that result in the corresponding colours

	X	Y	Z	L^*	a^*	b^*	Y	G	C
1	24.7	33.6	8.6	65.0	−32.0	50.0	32	82	0
							97	0	61
2	23.3	33.6	11.9	65.0	−38.4	40.0	0	84	11
3	24.0	33.6	11.9	65.0	−35.0	40.0	15	78	14
4	24.6	33.6	11.9	65.0	−32.5	40.0	27	71	18
							86	0	65
5	25.2	33.6	11.9	65.0	−30.0	40.0	41	62	24
							80	14	57
6	25.7	33.6	11.9	65.0	−27.8	40.0	64	40	40
							64	40	40

combinations inside the colorant domain are given. The first seven columns contain an index, followed by the XYZ and CIELAB values of the colours to be separated. The last three columns represent the colorant combination(s) that correspond with these colours. The different colours lie in the lightness plane $L^* = 65$ and are indicated with black dots and their corresponding index in Figure 12.7. Colour '1' is a special case because it can be rendered with two colorant combinations that each lie on a physical colorant boundary. This colour is the point of intersection of the two physical colour boundaries in Figure 12.7. Colour '2' also lies on a physical colour boundary, but can be rendered with only one colorant combination. Colour '3' is the most normal case since it lies inside the physical colour gamut and can be separated into just one set of colorants. Colour '4' provides an example of a colour that lies on a physical colour boundary, yet this physical colour boundary lies inside the gamut. As expected this colour can be separated into two sets of colorants: one, of course, lies on the physical colorant boundary, while the other does not. Colour '5' lies outside the physical colour gamut but inside the gamut of the process and can be separated into two sets of colorants. Finally, the last colour lies on the natural colour boundary of this printing process. If the corresponding colorant values for colours approximating the natural colour boundary are compared with each other, the two colorant combinations, as found in the case of colour '5', converge to each other if this colour is moved to the natural colour boundary, and coincide finally for colours lying on the natural colour boundary. Therefore, there are two coinciding colorant combinations that result in the same colour '6'. Figure 12.7 indicates that there is a surface in colorant space that after transformation to colour space is folded around the natural colour boundary (see also the colorant combinations in Table 12.2). As a result there will be some colours that can be obtained with two colorant combinations (white area in Figure 12.7 between the natural colour boundary and the physical colour boundaries).

12.5.3 Boundaries of a 4-ink process

Natural boundaries

For a 4-ink process, an incremental change of the colorant c_i results in a colour change represented by the vector V_i in colour space. A first kind of boundary originates if the four vectors V_i corresponding to the four colorants lie in one plane. This leads to the extension of the definition of a *natural boundary* as introduced for a 3-ink process.

Mathematically, the four vectors V_i lie in a plane if the rank of the matrix

$$\begin{pmatrix} \dfrac{\partial X}{\partial c_1} & \dfrac{\partial X}{\partial c_2} & \dfrac{\partial X}{\partial c_3} & \dfrac{\partial X}{\partial c_4} \\[2mm] \dfrac{\partial Y}{\partial c_1} & \dfrac{\partial Y}{\partial c_2} & \dfrac{\partial Y}{\partial c_3} & \dfrac{\partial Y}{\partial c_4} \\[2mm] \dfrac{\partial Z}{\partial c_1} & \dfrac{\partial Z}{\partial c_2} & \dfrac{\partial Z}{\partial c_3} & \dfrac{\partial Z}{\partial c_4} \end{pmatrix} \tag{12.6}$$

is lower than 3. This is equivalent to the conditions

$$\det \begin{pmatrix} \dfrac{\partial X}{\partial c_1} & \dfrac{\partial X}{\partial c_2} & \dfrac{\partial X}{\partial c_3} \\[2mm] \dfrac{\partial Y}{\partial c_1} & \dfrac{\partial Y}{\partial c_2} & \dfrac{\partial Y}{\partial c_3} \\[2mm] \dfrac{\partial Z}{\partial c_1} & \dfrac{\partial Z}{\partial c_2} & \dfrac{\partial Z}{\partial c_3} \end{pmatrix} = 0$$

$$
\det \begin{pmatrix} \dfrac{\partial X}{\partial c_1} & \dfrac{\partial X}{\partial c_2} & \dfrac{\partial X}{\partial c_4} \\[2mm] \dfrac{\partial Y}{\partial c_1} & \dfrac{\partial Y}{\partial c_2} & \dfrac{\partial Y}{\partial c_4} \\[2mm] \dfrac{\partial Z}{\partial c_1} & \dfrac{\partial Z}{\partial c_2} & \dfrac{\partial Z}{\partial c_4} \end{pmatrix} = 0
$$

$$
\det \begin{pmatrix} \dfrac{\partial X}{\partial c_1} & \dfrac{\partial X}{\partial c_3} & \dfrac{\partial X}{\partial c_4} \\[2mm] \dfrac{\partial Y}{\partial c_1} & \dfrac{\partial Y}{\partial c_3} & \dfrac{\partial Y}{\partial c_4} \\[2mm] \dfrac{\partial Z}{\partial c_1} & \dfrac{\partial Z}{\partial c_3} & \dfrac{\partial Z}{\partial c_4} \end{pmatrix} = 0 \qquad (12.7)
$$

Evaluation of a variety of 4-ink processes has indicated that these conditions do not occur in practical situations. This is not totally unexpected, as the coincidence of four vectors in a plane is even less likely to occur than the coincidence of three vectors in a plane in the 3-ink case. It is therefore safe to state that for the practical calculation of the gamut of 4-ink processes the natural boundaries need not be taken into account.

Hybrid boundaries

A point inside the colorant domain transforms to a gamut boundary only if the four vectors V_i lie in one plane; i.e. the point belongs to a natural colorant boundary. This immediately limits the search for other kinds of gamut boundaries that lie on the boundaries of the four-dimensional colorant domain (at least one colorant value equals 0% or 100%). Boundaries in a four-dimensional space are three-dimensional and the boundaries of the domain of a 4-ink process are completely described by eight 3-ink boundary processes.

Because any colour within the gamut of one of these 3-ink boundary processes also lies in the gamut of the 4-ink process, the remaining types of gamut boundaries of the 4-ink process coincide with the gamut boundaries of the eight 3-ink boundary processes. Therefore, it suffices to take into account the natural and physical boundaries of the 3-ink boundary processes.

In the case of a natural boundary of one of the eight 3-ink boundary processes, the gamut limitation is also affected by the physical limitation on the fourth ink and hence we prefer to call this a *hybrid boundary*. In colour space these surfaces are called *hybrid colorant boundaries* that are transformed to *hybrid colour boundaries* in colour space by the printer model.

Physical boundaries

In the case of a physical boundary of one of the eight 3-ink boundary processes, the gamut limitation is due only to physical restrictions on colorants. The physical boundaries of a

3-ink process are 2-ink boundary processes, so the corresponding *physical boundaries* of the 4-ink process are given by its 2-ink boundary processes.

For well-behaved processes, such as the CMYK process, the gamut is not affected by natural or hybrid boundaries. In that case, the gamut is only defined by the physical boundaries, i.e. 24 different 2-ink processes. If no natural boundaries are involved, the gamut is obtained by the union of the gamuts of the eight 3-ink boundary processes. For less well-behaved processes the gamut is obtained by the outer contour of the three types of gamut boundaries.

12.5.4 Analytical gamut calculation: well-behaved Neugebauer model without ink limitations

In this section, it is shown how to obtain an analytical gamut description for a well-behaved 3-ink process based on the Neugebauer model. The gamut description of a well-behaved 4-ink process is not described, but the technique presented for the 3-ink process can be easily extended to well-behaved n-ink processes. It is assumed that there are no colorant limitations; however, a more elaborate technique to determine the gamut in the case of a linear colorant limitation is given in Appendix 2.

For a well-behaved Neugebauer process, the gamut is obtained by transforming the physical boundaries to colour space. For a 3-ink process, there are six physical boundaries. These boundaries have 12 1-ink processes in common, corresponding to the 12 edges of the colorant cube. For a process modelled with the Neugebauer equations, the gamut boundaries have the following characteristics [23] (see Appendix 1):

- The 12 line segments that connect the physical colour boundaries are straight lines, as they correspond to the mapping of 1-ink processes to colour space.
- The six physical colour boundaries correspond to the transformation of the physical colorant boundaries to colour space. The physical colour boundaries all have the shape of quadric surfaces.
- Colorant combinations of the physical colorant boundaries resulting in the same Y value lie on hyperbolas in the physical colorant boundaries.
- The cross-section of the physical colour gamut boundary with planes (such as a plane with a constant Y value) results in a conic section.

One way to describe the gamut is obtained by the intersection of constant luminance planes. For each luminance intersection one or multiple contours can be found. As for well-behaved ink processes the gamut is determined by transforming the 2-ink boundary processes to colour space; these contours are found by looking for the colorant combinations in the 2-ink boundary processes that result in the given luminance value. Because colorant combinations in a 2-ink boundary process resulting in the same luminance value all lie on a hyperbola in the 2-ink boundary process, these curves will be looked for and transformed to colour space. Hence for every 2-ink boundary process, the segment of the hyperbola that lies within the domain of the boundary process is determined. As the hyperbola always intersects with the boundaries of the 2-ink boundary process (these are the 1-ink boundary processes of the 2-ink boundary process), first of all these intersection points are determined. In general there can be zero, two or four intersection points. In the case of two intersection points, they are the end point of a segment of the hyperbola belonging to the 2-ink boundary process. In case of four intersection points, the points have to be classified into pairs belonging to the same hyperbola, as in this case the 2-ink boundary process is intersected by two hyperbolas. If the end points are known, the points in between can be found based on the mathematical description of the hyperbola. In a next step, all these segments are connected to obtain one or more closed contours. These contours lie on the boundary of the colorant cube and are transformed to a contour in colour space lying in the given luminance plane (see Figure 12.8).

Hence the contours can be obtained as follows:

- From the 3-ink process, six 2-ink boundary processes are calculated. These are 2-ink processes obtained by setting one of the three inks of the 3-ink process to its minimum or maximum value.
- For each 2-ink boundary process, four extracted 1-ink boundary processes are determined by setting one ink to its minimum or maximum value.
- For every 1-ink process the colorant value that results in the required Y value is searched for (solution of a linear equation), and the solutions that lie in the colorant gamut of the 1-ink process are retained. There may be zero, two or four intersection points per 2-ink boundary process. In Figure 12.9, the three possible intersections of hyperbolas with a 2-ink boundary process are represented. The hyperbolas correspond to colorant combinations of the 2-ink boundary process that map to colours with the given luminance value.
- Because the asymptotes of the hyperbola are a horizontal and vertical line, the intersection points are classified by ordering them from low to high according to one ink value. Two succeeding points belong to the same segment of a hyperbola.
- For each two colorant pairs, the corresponding hyperbola is sampled in colorant space and transformed to colour space.

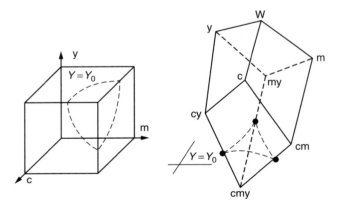

Figure 12.8 Segments of hyperbola, all resulting in colours with luminance value Y_0, lying on the faces of the colorant cube are searched for and connected with each other to get one closed contour. The segments are transformed to colour space to obtain the gamut intersection with $Y = Y_0$.

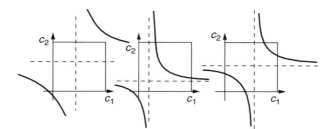

Figure 12.9 Representation of the three possible intersections of hyperbolas with the physical colorant boundaries. The hyperbola contains the colorant combinations that transform to colours with the same luminance. The dotted lines are the asymptotes.

- The different segments in the six 2-ink boundary processes are connected by evaluating their end points to find one or more closed contours. In this way all the contours are obtained that result from intersecting the gamut with the constant Y plane.

12.5.5 Gamut of a well-behaved localised Neugebauer model

For a localised printer model, the previously defined concepts of extracting processes and inheriting colorant limitations can be easily extended. An extracted ink process of a given localised process is again a localised ink process. If, for example, one ink is kept constant in a localised Neugebauer model in an n-dimensional colorant space, the model is reduced to a localised Neugebauer model in an $(n-1)$-dimensional colorant space. Hence, for a localised 3-ink process the physical boundaries are extracted 2-ink processes with the third ink set at its minimum or maximum value, i.e. localised 2-ink processes. Colorant limitations are inherited by setting the constant ink to its fixed value in the colorant limitation.

For a localised 3-ink Neugebauer process, the physical boundaries are localised 2-ink Neugebauer processes. If the gamut is represented by equiluminance contours, it is obtained by looking for hyperbolas for each cell in the physical boundaries. For each physical boundary, the hyperbolas of the different cells are concatenated and transformed to colour space. In general, there are two possibilities. In most cases a contour segment is found; however, on some occasions a closed contour is obtained (see Figure 12.10). This last situation never occurs for a global Neugebauer process. In colour space, the different contour segments are concatenated till a number of closed contours are found.

If colorant limitations also have to be taken into account, all cells have to be evaluated to see if they are cut by the ink limitation. In this case, the additional 2-ink process is a localised model that is not a localised Neugebauer model (see Appendix 2). Contour segments are searched for in the extracted 2-ink processes and the additional ink process by evaluating all the cells of the corresponding localised models. For each localised process contours are concatenated and transformed to colour space. Here the contour segments are concatenated to obtain a number of closed contours.

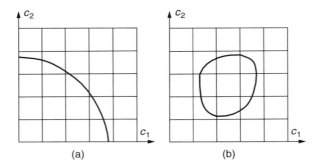

Figure 12.10 Colorant combinations in the physical colorant boundary of a localised printer model resulting in the same luminance value. The vertical and horizontal lines represent the division of the 2-ink process in cells. In image (a), a contour segment is found. In image (b), a rather exceptional case is represented in which a closed contour is detected.

12.5.6 Gamut of a less well-behaved model

For less well-behaved ink processes, an analytical gamut description is quite complex. Therefore, similar techniques as those applied for the calculation of the gamut of a well-behaved process can be used if the printer model is transformed or expanded to a localised Neugebauer model with a proper number of sampling points. In this case, the gamut of a less well-behaved printing process can be approximated by the union of the gamuts of the physical gamuts (i.e. considering the cells as well-behaved processes) of all the cells of the localised process.

12.6 GAMUT REPRESENTATIONS

In general the gamut is represented in the xy-chromaticity diagram. For a CMY or CMYK printer, the convex hull of the chromaticity coordinates of the primary and secondary colours is taken. This results in a hexagon obtained by drawing lines between the primary and secondary colours red, yellow, green, cyan, blue and magenta (see Figure 12.11(a)). However, if other colorants are printed or the CMYK process is extended with additional colorants, there is no easy way to represent the gamut in the xy-chromaticity diagram. A possible solution is obtained if all the colorant combinations are transformed to the XYZ space by the printer model, and these colours projected in the xy-chromaticity diagram. This is shown in Figure 12.11(b) for a 6-ink process (cyan, magenta, yellow, black, orange, green).

The representation of a gamut in an xy-chromaticity diagram only gives a broad idea about the colour range, because colours are specified with three values and hence the gamut is a volume in a three-dimensional colour space such as XYZ or CIELAB.

A more detailed representation of a gamut is obtained by intersecting the gamut in a psycho-visual space such as CIELAB at different lightness and hue angles and counting the number of colours per cross-section. A good gamut representation is obtained with two

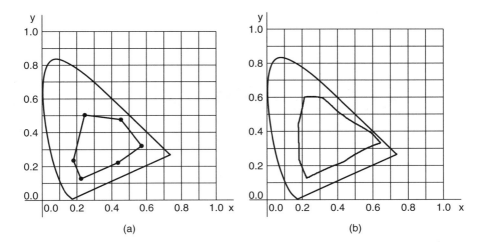

Figure 12.11   (a) A gamut representation of a CMYK process in the xy-chromaticity diagram based on the hexagon defined by the primary and secondary colours. (b) The projection of the gamut of the CMYKOG process in the xy-chromaticity diagram.

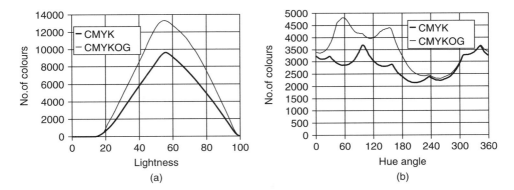

Figure 12.12   The number of colours as a function of (a) lightness and (b) hue, based on gamut cross-sections in CIELAB space.

graphs, one that specifies the number of colours per lightness cross-section and another that relates the number of colours per hue intersection. This is shown in Figure 12.12. Colours are counted in CIELAB space on a regular grid with a one CIELAB unit as grid unit.

If more detailed information is needed, a three-dimensional gamut descriptor is required. This can be done by rendering the shape of the gamut with 3-D visualisation tools or by representing cross-sections of the gamut with constant lightness or constant hue planes. In Figure 12.13, the cross-section of a CMYK gamut in CIELAB is represented for a constant lightness value of 50. In Figure 12.13(a) no ink limitation is applied; in Figure 12.13(b)

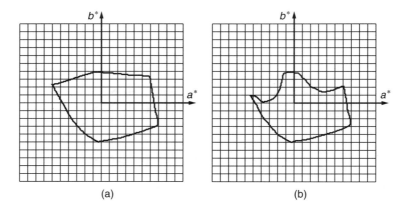

Figure 12.13 Gamut cross-section in CIELAB space of a CMYK process with lightness value 50. (a) No ink limitation; (b) maximum amount of ink 200%. The grid spacing is 10 CIELAB units.

the maximum amount of ink is 200%. As can be clearly seen in the latter, the gamut is not convex at all.

12.7 CONCLUSIONS

In the literature, several techniques are represented to calculate the gamut of a colour reproduction device. These techniques can be divided into two categories. First of all, empirical methods are based on collecting a large number of measured colour patches of a given printer and describing the outer boundaries of this set of colours. In most cases the convex hull is used, even if in general the shape of a gamut is not convex at all (see Figure 12.13). A second class of techniques can be considered as analytical methods because they are based on a description of the behaviour of the printer. However, most analytical techniques depend on a typical printer model and cannot be generalised for any printer description. In this chapter, a general printer model, i.e. a model that can be used to describe any printer, is presented based on the Neugebauer process.

It is shown that several types of gamut boundaries have to be taken into account, depending on the printer model and the number of colorants. For well-behaved printers, the gamut boundaries can be described analytically, whereas for less well-behaved printers a good approximation of the gamut boundary can be obtained.

Gamut calculations are used for several purposes. The way a gamut is represented or in other words the amount of information that is needed varies from application to application. Hence, several gamut representations and visualisations are required. A simple gamut representation of CMYK processes is given in the xy-chromaticity diagram. The largest shortcoming of this method of presentation is that no information is provided about the luminance range of the colours. The most detailed information is given by a complete description of the gamut boundaries. In this chapter, a complete gamut description is obtained by intersecting the gamut with equiluminance planes. However, for a number of applications this provides more information than is needed. Hence, a practical gamut representation, somewhat in between the xy-chromaticity diagram and a complete gamut boundary description, is given by two graphs representing the number of colours per lightness and hue intersection of the gamut in the CIELAB space.

REFERENCES

1. Balasubramanian, R. and Dalal, E. (1997) A method for quantifying the color gamut of an output device. *SPIE* Vol. 3018, 110–116.
2. Bell, I. E. and Cowan, W. (1993) Characterizing Printer Gamuts Using Tetrahedral Interpolation. *Proceedings of the IS&T/SID Color Imaging Conference*, 108–113.
3. Braun, G. J. and Fairchild, M. D. (1997) Techniques for gamut surface definition and visualization. *Proceedings of the IS&T/SID Color Imaging Conference*, 147–152.
4. Cholewo, T. J. and Love, S. (1999) Gamut boundary determination using alpha-shapes. *Proceeding of the IS&T/SID Color Imaging Conference*, 200–204.
5. Clapper, F. R., Gendron, R. G. and Brownstein, S. A. (1973), Color gamuts of additive and subtractive color reproduction systems. *J. Opt, Soc. Am*, **63**(5), 625–629.
6. Demichel, E. (1924) *Procédé*, **26**, 17–21, 26–27.
7. Giorganni, E. J. and Madden, T. E. (1998) *Digital Color Management*, Addison-Wesley, Reading, Massachusetts.
8. Gustavson, S. (1997) Color Gamut of Halftone Printing. *Journal of Imaging Science and Technology*, **41**(3).
9. Engeldrum, P. G. (1986) Four color reproduction theory for dot formed imaging systems. *Journal of Imaging Technology*, **12**(2).
10. Engeldrum, P. G. (1996) The color gamut limits of halftone printing with and without the paper spread function. *Journal of Imaging Science and Technology*, **40**(3), 239–244.
11. Gustavson, S. (1996) The color gamut of halftone reproduction. *Proceedings of the IS&T/SID Color Imaging Conference*, 80–85.
12. Herzog, P. G. and Hill, B. (1995) A new approach to the representation of color gamuts. *Proceedings of the IS&T/SID Color Imaging Conference*, 78–81.
13. Herzog, P. G. and Roger, T. (1998) Comparing different methods of printer characterization. *ICPS 98*, Vol. 2, September 7–11, Antwerp, pp. 23–30.
14. Herzog, P. G. (1999) Specifying and visualizing colour gamut boundaries. *Colour Imaging: Vision and Technology*, John Wiley & Sons.
15. Hunt, R. W. G. (1995) *The reproduction of colour*, 5th ed. Fountain Press, London.
16. Inui, M. (1993) Fast algorithm for computing color gamuts. *Color Res. Appl.*, **18**(5), 341–348.
17. Kalra, D. (1994) GamOpt: A tool for visualization and optimization of Gamuts. *SPIE* **2171**, 297–304.
18. Kinsey, L. C. (1993) *Topology of surfaces*, Springer – Verlag, New York.
19. Kress, W. and Stevens, M. (1994) Derivation of 3-dimensional gamut descriptors for graphic arts output devices. *Taga proceedings*, 199–214.
20. Mahy, M. and Delabastita, P. (1996) Inversion of the Neugebauer equations. *Col. Res. Appl.*, **21**(6).
21. Mahy, M. (1996) Gamut Calculation of Color Reproduction Devices. *Proceedings of the IS&T/SID Color Imaging Conference*, 145–150, Nov. 19–22.
22. Mahy, M. and De Baer, D. (1997) HIFI Color Printing within a Color Management System. *Proceedings of the IS&T/SID Color Imaging Conference*, 277–283, Nov. 17–20.
23. Mahy, M. (1997) Calculation of Color Gamuts Based on the Neugebauer Model. *Col. Res. Appl.* **22**(6), 365–374.
24. Mahy, M. (1998) Insight into the solutions of the Neugebauer equations. *Electronic Imaging '98, Proceedings of SPIE*, Vol. 3300, 76–85, Jan. 28–30.

25. Mahy, M. (1998), *Output characterization Based on the Localized Neugebauer Model*. ICPS 98, 30–34, Sept. 7–11.
26. Mahy, M. and De Baer, D. (1999) Measurement inversion and gamut mapping encoding, *International Colour Management Forum*, 24–25, March, University of Derby.
27. Marcu, G. and Abe, S. (1995) Ink Jet Printer Gamut Visualization. *Proceeding of 11th IS&T International Congress on Advanced Non-Impact Printing Technology*, 459–462.
28. Marcu, G. and Abe, S. (1995) Three-dimensional histogram visualization in different color spaces and applications. *Journal of Electronic Imaging*, **4**(4), 330–346.
29. Meyer, G. W., Peting, L. S. and Rakoczi, F. (1993) A color gamut Visualization tool. *Proceedings of the IS&T/SID Color Imaging Conference*, 197–201.
30. Morovic, J. (1998) *To Develop a Universal Gamut Mapping Algorithm*, Ph.D. Thesis, University of Derby.
31. Morovic, J. and Luo, M. R. (1998) Developing Algorithms for Universal Gamut Mapping. *Colour Engineering: Vision and Technology*, L. W. MacDonald (ed.), John Wiley & Sons.
32. Neugebauer, H. E. J. (1937) Die theoretischen Grundlagen des Mehrfarbenbuchdrucks. *Zeitschrift für wissenschaftliche Photographie, Photophysik und Photochemie*, Band, **36**, Heft 4, 73–89.
33. Ohta, N. (1972) Reflection density of multilayer color prints, II. Diffuse reflection density. *J. Opt. Soc. Am.*, **62**(2), 185–191.
34. Ohta, N. (1971) Structure of the color solid obtainable by three subtractive color dyes. *Die Farbe 20*, 1/3, 115–134.
35. Ohta, N. (1971) The color gamut obtainable by the combination of subtractive color dyes I. Actual dyes in color film. (1) Optimum peak wavelengths and breadths of cyan, magenta and yellow. *Photographic Science and Engineering*, **15**(5), 399–415.
36. Ohta, N. (1971) The color gamut obtainable by the combination of subtractive color dyes I. Actual dyes in color film. (2) Influence of Unwanted Secondary Absorption Peak. *Photographic Science and Engineering*, **15**(5), 416–422.
37. Ohta., N. (1971) Reflection density of multilayer color prints I. Specular reflection density. *Photographic Science and Engineering*, **15**(6), 487–494.
38. Ohta, N. (1972) The color gamut obtainable by the combination of subtractive color dyes I. Actual dyes in color film. (3) Stability of gray balance. *Photographic Science and Engineering*, **16**(3), 203–207.
39. Ohta, N. (1982) The color gamut obtainable by the combination of subtractive color dyes IV. Influence of Some Practical Constraints. *Photographic Science and Engineering*, **26**(5), 228–231.
40. Ohta, N. (1986) The color gamut obtainable by the combination of subtractive color dyes V. Optimum absorption bands as defined by nonlinear optimization technique. *Journal of Imaging Science*, **30**(1), 9–12.
41. Ottinen, P., Autio, H. and Saarelma, H. (1992) Color gamut in halftone printing. *Journal of Imaging Science and Technology*, **36**(5).
42. Pointer, M. R. (1980) The gamut of real surface colors. *Col. Res. Appl.*, **5**(3), 145–155.
43. Reel, R. L. and Penrod, M. A. (1999) Gamut visualization tools and metrics. *Proceeding of the IS&T/SID Color Imaging Conference*, 247–251.
44. Robertson, P. K. (1988) Visualizing Color Gamuts: A User Interface for the effective use of perceptual color spaces in data displays. *IEEE Computer Graphics & Applications*, 50–64.
45. Rolleston, R. and Balasubramanian, R. (1993) Accuracy of various Types of Neugebauer Model, *Proceedings of the IS&T/ SID's Color Imaging Conference*.
46. Saito, R. and Kotera, H. (1999) Dot allocations in dither matrix with wide color gamut. *Journal of Imaging Science and Technology*, **43**(4), 345–352.
47. Saito, R. and Kotera, H. (2000) Extraction of image gamut surface and calculation of its volume. *Proceeding of the IS&T/SID Color Imaging Conference*, 330–334.
48. Schläpfer, K. and Widmer, E. (1993) Which color gamut can be achieved in multicolor printing and in television. *Taga proceedings*, 41–49.

49. Wyszecki, G. and Stiles, W. S. (1982) *Color Science: Concepts and Methods, Quantitative Data and Formulae*, second edition, John Wiley & Sons.
50. See http://www.colour.org/tc8-03/survey/survey_ref.html
51. Yule, J. A. C. (1967) *Principles of Color Reproduction*, John Wiley & sons, Inc., New York.

APPENDIX 1. NEUGEBAUER EQUATIONS

Neugebauer equations for a 1-ink process

The Neugebauer equations for a process with one ink c_1 are:

$$X = k_0 + k_1 c_1$$

$$Y = l_0 + l_1 c_1 \tag{12.8}$$

$$Z = m_0 + m_1 c_1$$

with k_i and l_i constants.

These equations immediately reveal that a 1-ink process transforms onto a straight line in colour space.

Neugebauer equations for a 2-ink process

The Neugebauer equations for a process with two inks c_1 and c_2 are:

$$X = k_0 + k_1 c_1 + k_2 c_2 + k_{12} c_1 c_2$$

$$Y = l_0 + l_1 c_1 + l_2 c_2 + l_{12} c_1 c_2 \tag{12.9}$$

$$Z = m_0 + m_1 c_1 + m_2 c_2 + m_{12} c_1 c_2$$

with k_i, l_i and m_i constants. Rearranging these equations leads to:

$$\left(c_1 + \frac{k_2}{k_{12}} \right) \left(c_2 + \frac{k_1}{k_{12}} \right) = \left(\frac{X - k_0}{k_{12}} + \frac{k_1 k_2}{k_{12} k_{12}} \right)$$

$$\left(c_1 + \frac{l_2}{l_{12}} \right) \left(c_2 + \frac{l_1}{l_{12}} \right) = \left(\frac{Y - l_0}{l_{12}} + \frac{l_1 l_2}{l_{12} l_{12}} \right) \tag{12.10}$$

$$\left(c_1 + \frac{m_2}{m_{12}} \right) \left(c_2 + \frac{m_1}{m_{12}} \right) = \left(\frac{Z - m_0}{m_{12}} + \frac{m_1 m_2}{m_{12} m_{12}} \right)$$

These equations reveal that for a given colour XYZ, each of the three Neugebauer equations corresponds to a hyperbola in the $c_1 c_2$ colorant plane. The asymptotes are

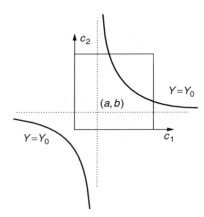

Figure 12.14 For a given colour XYZ, each of the three Neugebauer equations corresponds to a hyperbola in the c_1c_2 colorant plane. The asymptotes are horizontal and vertical lines, and the point where both asymptotes intersect is the midpoint of the conic section. The drawing shows an example for the Neugebauer expression of Y.

horizontal and vertical lines, and the point where both asymptotes intersect is the midpoint of the conic section. Figure 12.14 shows an example for the Neugebauer expression of Y.

The transformation of the domain of a 2-ink process to the XYZ space is a quadric. This is proved as follows. First the c_1c_2 terms in the first two equations in Equation (12.9) are eliminated by means of the third equation. This yields a first set of two equations from which a second set of equations can be obtained that express the c_1 and c_2 colorants as linear combinations of X, Y and Z. Substituting these two relations again in the last equation results in the formula of a quadric in the XYZ space.

If a quadric is intersected with a plane, a conic section is obtained. Hence, the hyperbola that represents a constant Y value in the c_1c_2 colorant space is mapped to a conic section in the XYZ space.

APPENDIX 2. ANALYTICAL GAMUT CALCULATION: WELL-BEHAVED NEUGEBAUER MODEL WITH LINEAR INK LIMITATION

As ink limitation, only a linear colorant limitation is evaluated. A linear colorant limitation for a 3-ink process with colorants c_1, c_2 and c_3 accepts only these colorant combinations for which

$$a_1c_1 + a_2c_2 + a_3c_3 \leqslant a_4 \tag{12.11}$$

with a_1, a_2, a_3, a_4 real values.

In the following paragraphs a gamut description will be given for a limitation on the sum of the three colorants of 260%. Nevertheless the method can be easily extended to any number of linear colorant limitations.

The linear colorant limitation for colorant amounts smaller than or equal to 260% can be written as

$$c_1 + c_2 + c_3 \leqslant 260\% \tag{12.12}$$

The domain of the printer model taking into account this colorant limitation is represented in Figure 12.15.

Suppose that the gamut descriptor consists of a set of contours in equiluminance planes. This means that hyperbolas in the six physical colorant boundaries have to be determined that result in the given luminance values. These 2-ink processes are extracted ink processes. If there is a colorant limitation, this limitation is inherited by the extracted ink process.

Take for example the physical colorant boundary with $c_3 = 0$. The linear condition is reduced to

$$c_1 + c_2 \leqslant 260\% \tag{12.13}$$

and hence all colorant combinations are allowed. The domain of this 2-ink process is represented in Figure 12.16(a). In this case, the four 1-ink boundary processes are determined

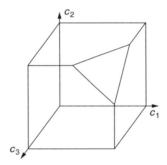

Figure 12.15 Colorant domain for a 3-ink process with a colorant limitation of 260%.

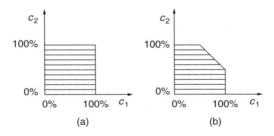

(a)   (b)

Figure 12.16 The colorant domain is represented for extracted 2-ink processes with the third ink set at its (a) minimum and (b) maximum value, taking into account a colorant limitation of 260%.

from the 2-ink process, and for every 1-ink process the colorant values are calculated that result in the given luminance value. In general zero, two or four colorant values are found that fall within the colorant gamut. For these solutions the corresponding points in the 2-ink boundary process are determined and hence a number of colorant pairs are obtained. If these colorant pairs are ordered from low to high according to one of the colorants, two succeeding colorant pairs describe a segment of a hyperbola in colorant space, which transforms to a conic section in the equiluminance plane Y.

For the 2-ink boundary process c_1c_2 with $c_3 = 100\%$, on the other hand, the linear condition becomes $c_1 + c_2 \leqslant 160\%$. The domain of this 2-ink process is represented in Figure 12.16(b). In this case, not only the solutions of the 1-ink boundary processes have to be searched for, but also possible solutions on the line $c_1 + c_2 = 160\%$. Substituting this equation in

$$Y = l_0 + l_1c_1 + l_2c_2 + l_{12}c_1c_2 \tag{12.14}$$

results in a quadratic equation in one variable. This gives at most two colorant pairs in the 2-ink boundary process. In general, there may be zero, two or four solutions in the colorant domain. If these colorant pairs are ordered from low to high according to one of the colorants, two succeeding colorant pairs will describe a segment of a hyperbola in colorant space, that transforms to a conic section in the equiluminance plane Y.

Apart from the six 2-ink boundary processes, an *additional colorant boundary* has to be analysed. This is the colorant boundary defined by the relation

$$c_1 + c_2 + c_3 = 260\% \tag{12.15}$$

Due to this linear relation, there are only two independent colorant values, for example c_1 and c_2. The colorant combinations for the c_1c_2 colorant pairs are limited by the conditions:

$$0\% \leqslant c_1 \leqslant 100\%$$
$$0\% \leqslant c_2 \leqslant 100\% \tag{12.16}$$

and due to restrictions on c_3 also by

$$S - 100\% \leqslant c_1 + c_2 \leqslant S \tag{12.17}$$

with S being the maximum sum of the colorants. Here $S = 260\%$, and hence only one condition remains, i.e.

$$160\% \leqslant c_1 + c_2 \tag{12.18}$$

The colorant domain of the additional 2-ink process is represented in Figure 12.17.

In general, there are up to six possible line segments that may limit the colorant combinations of c_1c_2. In this example the colorant domain is bounded by three segments (see Figure 12.17). For each segment one colorant is constant, whereas the other two colorants vary. Colorant combinations on these line segments resulting in colours with

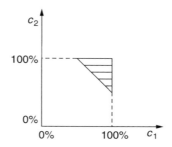

Figure 12.17   Colorant domain for the additional 2-ink process for a colorant limitation of 260%.

the given luminance value can be found as explained in the previous paragraph (solution of a quadratic equation in one colorant). Hence, a number of colorant pairs are obtained that can be easily positioned in the cross-section $c_1 + c_2 + c_3 = 260\%$. A number of colorant values are found between which equiluminance curves have to be calculated.

Because the equal sign is used at the additional colorant boundary, one colorant in the Neugebauer equations for the 3-ink process can be eliminated. Hence a 2-ink model is obtained that is no longer a Neugebauer model. This model is called the *additional 2-ink process*. If the third colorant in the Y-equation of the 3-ink Neugebauer process is eliminated, the following general formula is found:

$$Y = c_2^2(k_0 + k_1 c_1) + c_2(l_0 + l_1 c_1 + l_2 c_1^2) + (m_0 + m_1 c_1 + m_2 c_1^2) \qquad (12.19)$$

with $k_0, k_1, l_0, l_1, l_2, m_0, m_1, m_2$ being real values.

The Y-expression can be seen as a quadratic equation in c_2 with c_1 as parameter. The solutions for c_2 are given by

$$c_2 = \frac{-B \pm \sqrt{D}}{2A} \qquad (12.20)$$

with

$$A = k_0 + k_1 c_1$$

$$B = l_0 + l_1 c_1 + l_2 c_1^2$$

$$C = m_0 + m_1 c_1 + m_2 c_1^2 \qquad (12.21)$$

$$D = B^2 - 4AC$$

This means that there are two solutions c_2 for every value of c_1. Both solutions form a curve, one for the positive sign and one for the negative sign, lying at equal distances along vertical lines (i.e. constant c_1 value) from the curve $c_2 = -B/2A$. The solutions will be called the *solution-curves*, or more specifically the *+solution-curve* and

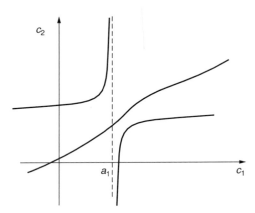

Figure 12.18 Curves in colorant space that are mapped to colours with the same luminance by the additional 2-ink process (no real roots for D). The value a_1 is the root value for $A = 0$.

the *−solution-curve*. For $A = 0$, one of the solution-curves goes to infinity, while the other has a finite value.

The discriminant D is a fourth-degree polynomial in c_1. The value of the discriminant D at infinity is always positive and hence for very large positive and negative values of c_1 there is always a real solution for c_2. The discriminant D has zero, two or four real roots. In the case of 0 real roots, there are always two solutions for c_2 for every value of c_1, resulting in two solution-curves that never cross (see Figure 12.18). For the root of A, one of the solution-curves will be finite, whereas the other solution-curve goes to infinity, i.e. at one side of the root of A it goes to +infinity, at the other side to −infinity. In the case of two real roots, there will be no real c_2 values for c_1 values between these roots. For c_1 equal to one of the roots of the discriminant D, both solutions coincide (see Figure 12.19). In the case of four roots, there will be two intervals along the c_1-axis for which no real solutions for c_2 are available (see Figure 12.20).

Because the roots of the discriminant D introduce intervals where no curves can pass, the behaviour of the solution-curves can be divided into five different classes. In class one, there is no real root for the discriminant D. In class two, an interval is represented from +infinity (resp. −infinity) to the largest (resp. smallest) real root of the discriminant D and the root of A falls outside this region. Class three is the case in which the root of A falls inside the interval of class two. In class four, there are four real roots for the discriminant D. In this case the behaviour of the two solution-curves is represented if the root of A falls outside the interval between the second and third largest roots of D. Class five is the situation of class four with the root of A within the considered interval.

The colorant values that were found before will now be classified into pairs that are the end points of equiluminance curves in the plane $c_1 + c_2 + c_3 = 260\%$. This is done as follows:

- First of all, the points are divided into intervals determined by the roots of the discriminant D because points belonging to different intervals will never lie on the same curve.

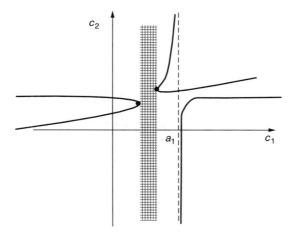

Figure 12.19 Curves in colorant space that are mapped to colours with the same luminance by the additional 2-ink process (two real roots for D). In the hatched region, no real solutions exist for c_2. The black dots indicate where both solution-curves coincide.

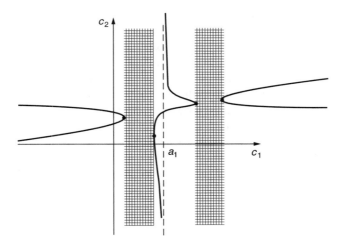

Figure 12.20 Curves in colorant space that are mapped to colours with the same luminance by the additional 2-ink process (four real roots for D). In the hatched regions, no real solutions exist for c_2. The black dots indicate where both solution-curves coincide.

- Per interval the points that belong to the solution-curve having a finite value for the root of A are collected in set one. This solution-curve is found as follows. If for the root of A, the polynomial B is positive (negative), the +solution (−solution) curve is taken. In Figure 12.21 the points e and b lie on the solution-curve for which the root of A is finite.

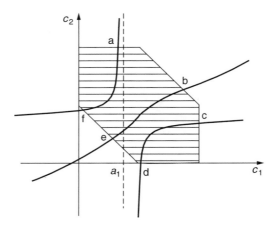

Figure 12.21 Equiluminance curves according to the additional 2-ink process (case one). The point a_1 is the root of A. The points a, b, c, d, e and f are intersections of the equiluminance curves and the boundary of the colorant domain of the additional 2-ink process.

- Then for each interval the following procedure is applied depending on the class:
  - Class 1: The remaining colorant values having a c_1 value smaller (larger) than the root of A are taken together in set two (set three). For the three sets, the colorant values are ordered from low to high according to one colorant (see Figure 12.21).
  - Class 2: The remaining points are taken together in set two. For every root of the discriminant D, there is a point in the colorant space where both curves connect. If one of these points falls within the colorant domain of the additional 2-ink process and the interval considered in this class, this point is added to set one and set two (see left example in Figure 12.22). If the interval goes from $-$infinity ($+$infinity) one set is ordered from low (high) to high (low), and the other set is ordered in the opposite direction. The first point of the last ordered set is put after the last point of the first ordered set (see Figure 12.22 for different examples).

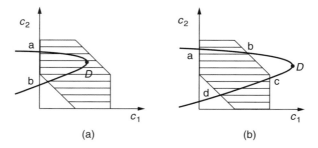

Figure 12.22 Equiluminance curves according to the additional 2-ink process (case two).

— Class 3: As in class one, the remaining points are divided into two sets. For every root of the discriminant D, there is a point in the colorant space where both curves connect. If one of these points falls within the colorant domain of the additional 2-ink process and the interval considered in this class, this point is added to the two sets to which it belongs (left image in Figure 12.23). This is set one and set two (set three) if the interval goes to +infinity (−infinity). Then the points of set one are ordered from low (high) to high (low) if the interval goes to −infinity (+infinity) and the set with points larger (smaller) than the root of A are ordered in the other direction. The first point of the last ordered set is put after the last point of set one. The points in the remaining set are ordered from low to high (see Figure 12.23 for different examples).

— Class 4: The remaining points are taken together in set two. If the second or third largest root of the discriminant D falls within the colorant domain of the additional 2-ink process, these points are added to both set one and set two. Then for each set the points are ordered in opposite directions. Finally, the sets are concatenated. If the end point of one set is equal to the starting point of the other set, the concatenation is done in such a way that these points are neighbours (see Figure 12.24).

— Class 5: The remaining points are divided into two sets as in class one. For every root of the discriminant D, there is a point in the colorant space where both curves connect. If the second or third largest root of the discriminant D falls within the colorant domain of the additional ink process, these points are added to the two sets to which it belongs. This is set one and set two (set three) if the root is smaller (larger) than the root of A. The points of set two are ordered from high to low, the points of set one are ordered from low to high, and finally the points of set three are ordered from high to low. Finally the last point of set one is put after the last

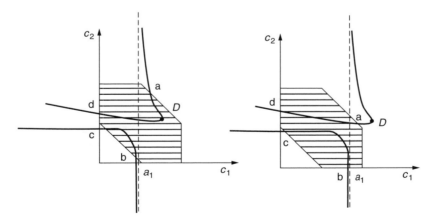

Figure 12.23 Equiluminance curves according to the additional 2-ink process (case three). The point a_1 is the root of A. The points a, b, c and d are intersections of the equiluminance curves and the boundary of the colorant domain of the additional 2-ink process.

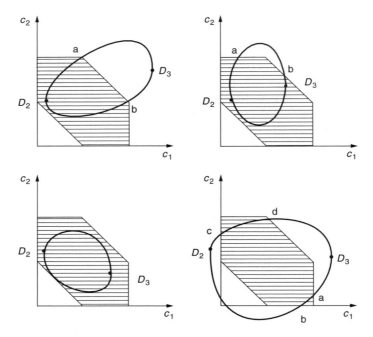

Figure 12.24 Equiluminance curves according to the additional 2-ink process (case one). The points a, b, c and d are intersections of the equiluminance curves and the boundary of the colorant domain of the additional 2-ink process. D_2 and D_3 are the in-between roots of D.

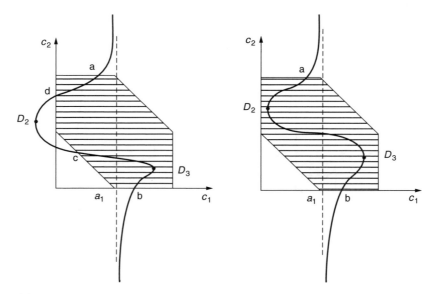

Figure 12.25 Equiluminance curves according to the additional 2-ink process (case five). The point a_1 is the root of A; D_2 and D_3 are the in-between roots of D.

point of set two and the first point of set three is put after the last point of set one (see Figure 12.25).

For each class two succeeding points describe a segment of the solution-curve to which they belong as found during the classification. Because the solution-curves are analytically available, the segments can be sampled and transformed to colour space. By comparing their end points, they can be connected properly to each other until a number of closed contours are found. This is the contour description for the corresponding luminance plane with value Y.

Colour gamut mapping

Ján Morovic

13.1 INTRODUCTION

Although not a novel task, the reproduction of colour has in recent history presented itself to us in a new form, primarily due to a rapid development of our understanding of the phenomenon of colour on the one hand and the availability of tools for using this new understanding on the other. An improved understanding of how the physical properties of objects relate to the colour perception resulting from seeing them has led to the possibility of predicting both what a given object would look like if we knew its colour-related properties and what physical properties an object ought to have so as to appear to have given colour attributes. In other words, if I would like an object to have a certain hue, colourfulness and brightness, I can determine what properties it needs to have (e.g. what combination of colorants needs to be applied to it). Furthermore, this task of relating physical properties to the resulting colour and vice versa can be divided into two components: the first relates physical properties to the stimulation that they result in for the human visual system, and the second relates the nature of the stimulation to the resulting colour percepts, as this relationship is not unique but context dependent (e.g. seeing an object against a dark background will make it look brighter than when seen against a light background, etc.). The second of these components is referred to as *colour appearance modelling* and, in the context of digital media, the first is referred to as *device characterisation*.

Colour Engineering, edited by P.J. Green and L.W. MacDonald.

Given device characterisation and colour appearance modelling, colours can be reproduced between different media, given that the original colour which is to be reproduced is reproducible on the reproduction medium. In other words, a colour having a certain hue, colourfulness and brightness can be reproduced on a reproduction medium only if that combination of hue, colourfulness and brightness is achievable on that medium. For example a very bright and colourful blue seen on a computer display might not be reproducible in newsprint. The fact that different media have different ranges of colours (referred to as *colour gamuts*) follows directly from their differing physical characteristics and implies that it is impossible to reproduce certain colours on certain media. Given this impossibility of exact reproduction of all colours between all media, a solution needs to be provided for dealing with colours that are to be reproduced on media which cannot reproduce them. Due to this impossibility, such a solution needs to assign colours from the gamut of the reproduction to all original colours that the reproduction medium cannot reproduce. As there are many possible ways of making such assignments, it is first necessary to formulate criteria for them and then to develop methods that best fulfil those criteria. These criteria include what are referred to as *rendering intents* and the methods are referred to as *gamut mapping algorithms*.

Given this framework for cross-media colour reproduction that consists of device characterisation, colour appearance modelling and gamut mapping and that is set up to achieve a given rendering intent, it is of importance to have a more detailed understanding of what it is that a system like this reproduces. While the obvious answer is *colour*, further complexity is revealed upon closer inspection. This is partly due to *colour information* having two properties – type and context. Colour information types can be physical colour-related properties (e.g. spectral reflectance or transmission characteristics), the extent to which an observer's eye is excited (e.g. cone responses, XYZ tristimulus values) or colour appearance (i.e. in terms of some perceptual attributes). In terms of context, colours can be considered on their own, against a simple background or as part of an image. Above all, however, it is important to bear in mind that the purpose of reproducing these kinds of information is beyond themselves – it is in conveying an idea, a message, a certain meaning, some feelings; to allure, please, inform, shock, warn or even disgust [49].

In this chapter the focus will be on gamut mapping for the reproduction of colour appearance in the context of complex images and all the following will be presented to this end. A review of the present state of gamut mapping research will be followed by a discussion of the connection between gamut mapping algorithms (GMAs) and rendering intents and by suggesting ways of applying these findings in current commercial colour management systems. As the vast majority of these are based on the International Colour Consortium's (ICC's) colour management framework, it will be the relationship between research findings and this framework that will be considered here. As colour gamut mapping is a fundamental element of any cross-media colour reproduction system, its understanding is of interest both when studying colour reproduction and when implementing commercial colour reproduction systems. Hence the relationship between research on the one hand and commercial application on the other is an issue of importance to academia and industry alike. Note also that this chapter is based on Morovic [50] and Morovic and Luo [56].

13.1.1 Gamut mapping aims

Before going into detail it is useful to have an idea of how various researchers have understood what gamut mapping is meant to do.

An aim for gamut mapping identified in an early paper by Stone *et al.* [68] says that as 'the relationship between [...] colors [... is] more important than their precise value', it is this relationship that gamut mapping algorithms ought to preserve. This suggests a move of the what-you-see-is-what-you-get (WYSIWYG) concept onto another level – i.e. it is applied to images rather than individual colours and could therefore be called MetaWYSIWYG [47]. In spite of playing down the importance of a match in the traditional sense, it is still crucial for a colour reproduction system to be able to reproduce individual colours exactly – even though some original colours cannot be reproduced exactly, their modifications need to be reproduced as such.

Alternatively a number of heuristically determined objectives arrived at in the past were also identified by MacDonald [39] to be common to the majority of gamut mapping studies carried out before 1993. These include the preservation of the grey axis of the original image and aiming for maximum luminance contrast, reduction of the number of out-of-gamut colours, minimisation of hue shifts and preference of increase in saturation. Note that this list represents assumptions made by some gamut mapping studies whereby the reasons for making these assumptions are in most cases based on experience from traditional colour reproduction. Even though experience from traditional colour reproduction is of great value, its maxims need to be looked at carefully when used in an environment which enables far more control over colour attributes than was previously possible. This change of framework from traditional to digital does indeed put a question mark on some results of previous research and care is therefore advised when considering research from the pre-digital era (e.g. for a difference in the understanding of lightness compression, see Morovic [48], pp. 99–102).

Another way of looking at gamut mapping is to say that its aim 'is to ensure a good *correspondence of overall colour appearance* between the original and the reproduction by compensating for the mismatch in the size, shape and location between the original and reproduction gamuts' [48]. One of the difficulties with implementing this aim is that there is as yet no good model for quantifying the appearance of complex images and neither is there one for quantifying the difference between them (though S-CIELAB [74] and the work of Nakauchi *et al.* [63] are steps in this direction).

Given these different views of what gamut mapping is or ought to be, we can now take a more detailed look at terminology, what factors affect gamut mapping, what parameters determine its performance, how gamut mapping relates to rendering intents and how gamut mapping research can be applied in a commercial context.

13.2 TERMINOLOGY

As with any subject, there is a range of possible interpretations of the basic terms used in gamut mapping as well. Therefore, to avoid misunderstandings, the definitions used by the CIE TC 8-03 on Gamut Mapping (http://www.colour.org/tc8-03/) will be given next,

whereby the definitions of an image were given by Braun *et al.* [8], those of accuracy
and pleasantness by Morovic and Luo [54] and the others by Morovic and Luo [55]:

- **Image**: two-dimensional stimulus containing pictorial or graphical information whereby
  the original image is the image to which its reproductions are compared in terms of
  some characteristic (e.g. accuracy).
- **Colour reproduction medium**: a medium for displaying or capturing colour information,
  e.g. a CRT monitor, a digital camera or a scanner. Note that in the case of printing, the
  colour reproduction medium is not the printer but the combination of printer, colorants
  and substrate.
- **Colour gamut**: a range of colours achievable on a given colour reproduction medium
  (or present in an image on that medium) under a given set of viewing conditions – it
  is a volume in colour space (Figure 13.2 later in this chapter).
- **Colour gamut boundary**: a surface determined by a colour gamut's extremes.
- **Gamut boundary descriptor (GBD)**: an overall way of approximately describing a
  gamut boundary.
- **Line gamut boundary (LGB)**: the points of intersections between a gamut boundary
  (as characterised by a GBD) and a given line along which mapping is to be carried out.
- **Colour gamut mapping**: a method for assigning colours from the reproduction medium
  to colours from the original medium or image (i.e. a mapping in colour space).
- **Colour reproduction intent**: the desired relationship between colour information in
  original and reproduction media. As a number of solutions to cross-media reproduction
  problems are possible, various colour reproduction intents can be pursued by gamut
  mapping. The most generic of these are accuracy and pleasantness but it is also possible
  to define others for specific applications (e.g. to provide an accurate reproduction of cor-
  porate identity colours while giving pleasant results for others). Note that reproduction
  intents are also referred to as rendering intents.
- **Accurate reproduction intent**: aims to maximise the degree of similarity between the
  original image and a reproduction of it as far as possible, given the constraints of the
  colour reproduction media involved. Note that the characteristic of accurate reproduction is intrinsically relative (i.e. reproduction *versus* original)
- **Pleasant reproduction intent**: aims to maximise the reproduction's correspondence
  with preconceived ideas of how a given image should look according to an individual
  whereby this criterion encompasses contrast, lack of artefacts, sharpness, etc. Note
  that unlike accuracy, pleasantness is absolute – at least as far as a given observer
  understands it at a given moment.

13.3 FACTORS AFFECTING GAMUT MAPPING

Among the factors that affect the performance of a given gamut mapping algorithm, the
most prominent ones are the characteristics of the colour reproduction system within
which the gamut mapping occurs, the colour space in which the mapping is carried out,
the differences between the original and reproduction media and the characteristics of the

image that is gamut-mapped. What follows next is a discussion of how each of these factors affects gamut mapping.

13.3.1 Colour reproduction system

As gamut mapping only ever occurs as part of a colour reproduction system, the properties of this system will influence its performance. Given the components of device characterisation (which, given our terminology here, ought to be referred to as medium characterisation), colour appearance modelling and gamut mapping, a colour reproduction system will be along the lines of the five-stage transform [39] shown in Figure 13.1.

From this transform it can be seen that the data being processed by a GMA has already passed through device characterisation and appearance modelling stages and might therefore already have errors introduced by these stages. If, for example, the characterisation model of one's original medium has any errors, then these will affect the inputs to the GMA which will therefore not be applied to the original image as such but to a modified version of that image which already includes some error. Analogously the output from the GMA will have the errors of the reproduction device characterisation and the inverse appearance transform added to it before it can be visualised. Hence, when any reproduction images are evaluated with the aim of evaluating the GMA that was used for making them, this evaluation will not be an evaluation only of the GMA but also of all the other components of the system. For example, if a reproduction shows discontinuity artefacts, it is not trivial to tell whether they are caused by one of the device characterisation stages, the colour appearance modelling or the gamut mapping or a combination of these, and determining this would involve a detailed evaluation of each of the reproduction system's components. Great care should be taken in such an evaluation before declaring one of these components to be the cause of a certain error. As with any task like this, one should always check one's raw data first (e.g. the data on which the characterisation is based as well as any parameters that determine the behaviour of the appearance model and the gamut mapping algorithm) before going into an evaluation of the algorithms involved [3].

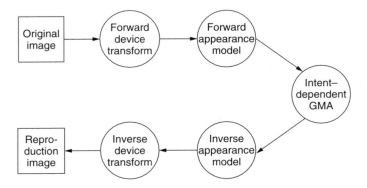

Figure 13.1 Five-stage colour reproduction transform.

13.3.2 Gamut boundary calculation

Once a colour reproduction system is in place, the GMA needs to have some information about the gamuts between which it is to map colours. Therefore it is first necessary to know the gamut boundaries of at least the reproduction and in many cases also the original gamuts, and this can be divided into two separate problems.

First, it is necessary to compute a *gamut boundary descriptor (GBD)* – i.e. some overall way of approximately describing a gamut. For media gamuts this can be done either directly from specific characterisation models – e.g. Kubelka–Munk equations [17] or Neugebauer equations [41] – or using methods which can be applied to any characterisation model [30]. Further, there are also some methods which can be used for computing the gamuts of images as well as media [35, 14].

Second, it is also important to be able to find the intersections between the gamut boundary (as computed using the above methods) and a given line along which mapping is to be carried out – the *line gamut boundary (LGB)*. The *gamulyt* method published by Herzog [22, 23], the *mountain range* method by Braun and Fairchild [5] and the *segment maxima* method by Morovic and Luo [51, 56, 58] all provide a way for doing this as well as obtaining the initial gamut boundary descriptor. The gamulyt method is intended primarily for modelling the gamuts of colour reproduction media and does so by distorting a cube to fit them. The mountain range method uses gridding and interpolation to arrive at a data structure consisting of a uniform grid in terms of lightness and hue and storing chroma values for each of the grid points.

Finally, the segment maxima method divides colour space evenly in terms of spherical angles and stores the point with the largest radius for each segment (Figure 13.2). Note that the first two of these algorithms provide LGBs for lines of constant hue and lightness and the last describes a method for finding the gamut boundary along lines of constant spherical co-ordinates and along any line of constant hue angle respectively. However, all these methods can be used to calculate the gamut boundary along any line when iterative techniques are employed.

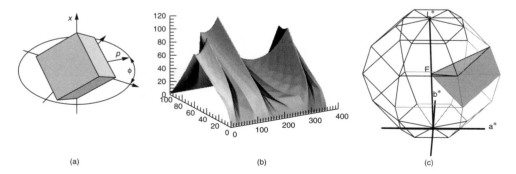

(a)   (b)   (c)

Figure 13.2 Gamut boundary calculation methods: (a) gamulyt cube, (b) mountain range grid, (c) segment maxima segmentation.

The choice of gamut calculation method should depend on its accuracy for calculating LGBs for a given gamut mapping algorithm, as errors in determining the gamut boundary will propagate throughout the rest of the colour reproduction system and may result in artefacts in reproduced images.

13.3.3 Colour space

As can be seen from the definitions of a colour gamut and of gamut mapping given above, they are both closely associated with colour spaces. Most gamut mapping algorithms intend to work with *perceptual attributes*, i.e. colourfulness, chroma, saturation, brightness, lightness, hue or colour names (e.g. red, dark green, orange, etc., whereby each of these would represent a subset of colours represented by a volume in colour space); to make this possible, they are implemented in colour spaces which predict them. More specifically, they usually intend to maintain some of a colour's perceptual attributes while changing others. If under these circumstances the *predictors* are imperfect, changes in the predictor of one attribute can also result in changes of another perceptual attribute (e.g. in some cases, if the L^* – the predictor of lightness in CIELAB – of a colour is changed, its perceived hue or chroma might also change).

In the colour spaces which are most often used for gamut mapping – CIELAB and CIELUV [12] and in some cases LLAB [37] or RLAB [18] – there are problems especially with the predictors of hue. In particular there are deficiencies in the uniformity of hue angles in the blue region of CIELAB (e.g. hue angles of around 290°) which can result in changes of perceived hue when only the L^* or C^* of a colour is changed. The performance of the hue predictor in CIECAM97s [36] is somewhat better for this region. More detail on the performance of hue predictors of various colour spaces can be found in previous papers [26, 16] as well as alternative colour spaces which were developed for the sake of better hue uniformity [4, 16, 43, 44].

When implementing or evaluating GMAs, it is important to understand the deficiencies of the colour space used for the mapping and not to confuse the colour space's predictor with the predicted perceptual attribute (e.g. in CIELAB h_{ab} is not hue and L^* is not lightness – they are only their predictors).

13.3.4 Media

The characteristics of the colour reproduction media between which a GMA is used for reproducing colour images play a major role in determining how a given GMA will perform. Where the differences between the original and reproduction media are small, the way for overcoming those small differences is relatively less important than in cases where the differences are large and a range of different GMAs can give similar results. Furthermore it seems, based on existing literature, that when medium differences are small, clipping algorithms perform better than compression algorithms [46, 59] and that the reverse is true when gamut differences are large [52]. The use of gamut clipping to smaller gamuts can introduce blocking or contouring artefacts.

In addition to medium-related gamut differences having an impact on how well a given algorithm will perform, another important way in which media affect the way a GMA is perceived to perform is the way that medium's gamut is determined. A common difficulty that arises in conjunction with using cross-medium reproduction systems involving gamut mapping is that the gamuts of the original and reproduction media are calculated under different conditions both from each other and from the actual conditions under which the system is then used [60]. For example, a CRT display's gamut could be calculated on the basis of colorimetric measurements taken in a darkroom and a print's gamut could be calculated using spectrophotometric measurements relating to the print being viewed under a standard illuminant (e.g. D50). If the system is then used to match images between the CRT and a print under office lighting conditions, any mismatch would to a great extent be caused by the difference between the conditions under which the media were characterised and the conditions under which they are used. To illustrate this point, Figure 13.3 shows the gamuts of a CRT and a print when viewed simultaneously under a range of levels of illumination [60].

13.3.5 Images

In addition to the nature of the context in which images are reproduced (i.e. the media, colour spaces, etc.), the characteristics of the images that are being reproduced themselves influence how well a gamut mapping algorithm will perform in reproducing them. While this is a phenomenon that has been long known and noted in numerous studies [40, 70, 52, 6], a systematic study of what it is about the images that causes these differences has only been started recently [59]. In previous studies it has been conjectured that it is image type (in the sense in which an image can be a business graphic, a portrait or an outdoor scene) which needs to be analysed to decide how to reproduce an image, but this claim is made without substantial evidence. While the systematic study mentioned above is still underway and has not yet found what image characteristic determines how an image needs to be reproduced, it has already been shown that it is not an image's gamut that is of importance [59]. Even though there is at present no firm understanding of what image characteristics are of importance in a cross-media reproduction context, it is a factor that needs to be considered whenever gamut mapping is carried out.

The characteristics that could be of importance can be grouped into statistical, spatial and cognitive ones. Statistical image characteristics are those which treat images simply as sets of coloured pixels and include single-valued image summaries (e.g. the mean colour), image gamut and image histogram. Spatial characteristics on the other hand also take into account the spatial arrangement of an image's pixels and include image spatial frequency, local colour contrast, sharpness and image layout. Finally, cognitive image characteristics are those obtained by treating images as entities identified by higher faculties of the mind and include image colour naturalness, image type and image content.

What is most important to bear in mind in relation to the role of images in gamut mapping is that a given algorithm is very likely to perform differently for different images. Therefore whether a GMA will work well for reproducing a particular image depends not only on that GMA, or the media between which the reproduction is to be made, but also on the image itself.

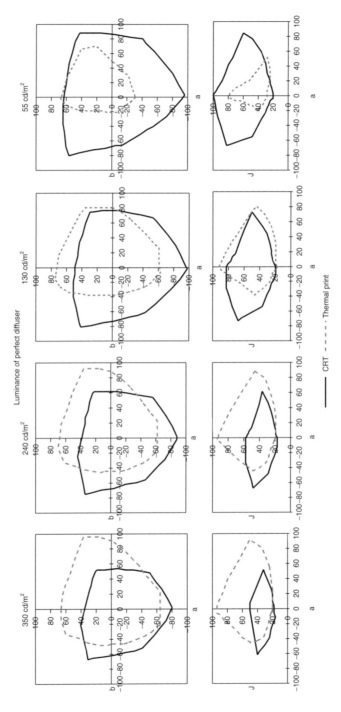

Figure 13.3 Colour gamuts of CRT and print under different levels of illumination.

13.4 PARAMETERS OF GAMUT MAPPING

Now that the principal factors affecting the performance of gamut mapping algorithms have been introduced, the parameters of gamut mapping algorithms themselves can be looked at.

13.4.1 Type of mapping

Once the necessary gamuts are known in the chosen colour space, it is possible to implement a GMA, whereby these can be broadly divided into two types – gamut clipping and gamut compression. Gamut clipping algorithms only change colours which are outside the reproduction gamut either from the very beginning or after lightness compression. For colours outside the reproduction gamut, these algorithms specify a mapping criterion, which is used for finding the point on the reproduction gamut to which a given original colour is mapped. Gamut compression algorithms make changes to all colours from the original gamut so as to distribute the differences caused by gamut mismatch across the entire range. Compression is needed when larger differences are to be overcome, as gamut clipping could result in unacceptable loss of variation (Plate 6(b)) in out-of-gamut regions under such circumstances. However, compression can also result in excessive changes to an image (Plate 6(c)) and the best results are achieved when compression is done in a more adaptive way (Plate 6(d)).

13.4.2 Choice of original gamut

When gamut compression is used, there arises a question as to which gamuts to map between. This is the case as the original gamut can be seen as either the gamut of the original medium or the gamut of the original image (i.e. a subset of the original medium's gamut – Plate 5). For current gamut clipping methods this is not an issue as it is sufficient to know the reproduction gamut to which any non-reproducible colours from the original are mapped. To make as few changes to the original image as possible, it is more advisable to use the original image's gamut as the original gamut since this means that colours are modified only when necessary. Plate 5 shows two images and their image gamuts (solid) in CIELAB when seen on a given CRT (mesh).

If, on the other hand, media gamuts are used, a given image could be modified to allow for colours which are not present in it (e.g. when the medium gamut is used, an image's colours are changed even when all of them are in the reproduction gamut to begin with). However, there is a practical advantage to mapping between media gamuts, as look-up tables (LUTs) can be calculated from them and then used for transforming an image without knowing its individual gamut.

Also bear in mind the fact that for some images (e.g. the Musicians image shown in Plate 5) the difference between image and original medium gamut is larger than for other images (e.g. the Ski image in Plate 5) whereby it makes more difference to the first type than to the second which gamut is used.

13.5 GAMUT MAPPING AND COLOUR RENDERING INTENTS

13.5.1 Colour rendering intents

Due to the differences between original and reproduction gamuts (Plate 6(e)), it is possible to generate a whole set of images which are inside the reproduction gamut and each of which will have a different relationship to the original image's appearance. This results in the need for a criterion for choosing a single member of this set, which then becomes the aim of a colour reproduction system. Indeed, a number of different criteria (referred to as reproduction requirements or rendering intents) have been defined. One of the most well-known sets is the following one given by Hunt [28]:

- *Spectral reproduction* – spectral power distributions of original and reproduction are identical.
- *Exact reproduction* – relative luminances, chromaticities and absolute luminances are identical.
- *Colorimetric reproduction* – chromaticities and relative luminances match.
- *Equivalent reproduction* – chromaticities, relative and absolute luminances of original appear as being the same in the reproduction.
- *Corresponding reproduction* – chromaticities and relative luminances in the reproduction appear to be the same as in the original when both have the same luminance levels.
- *Preferred reproduction* – equality of appearance is sacrificed in order to achieve a more pleasing result.

When reproducing images between media with different gamuts it is unlikely that any of these, except for the last one, are possible. In the case of individual colours, which are from the intersection of original and reproduction gamuts, it is also possible that equivalent and corresponding reproductions can be achieved in addition to preferred reproduction.

There is also another set of rendering intents which were specifically defined for the purposes of cross-media reproduction using colour management systems, which was compiled by the ICC and consists of the following four rendering intents [29]:

- *Perceptual intent* – 'a rendering intent that specifies the full gamut of the image is compressed or expanded to fill the gamut of the destination device. Gray balance is preserved but colorimetric accuracy might not be preserved.'
- *Saturation intent* – 'a rendering intent that specifies the saturation of the pixels in the image is preserved perhaps at the expense of accuracy in hue and lightness.'
- *Absolute colorimetric intent.*
- *Relative colorimetric intent.*

As the last two of these are not explicitly defined (especially in terms of what happens with out-of-gamut colours), they could be formulated in terms similar to those used by Hunt:

- *Absolute colorimetric reproduction* – chromaticities, relative and absolute luminances of the original appear as being the same for the reproduction of colours from the intersection of the original and reproduction gamuts, and out-of-gamut colours are clipped onto the reproduction gamut surface.

- *Relative colorimetric reproduction* – chromaticities and relative luminances of the original appear as being the same for the reproduction of colours from the intersection of the original and reproduction gamuts, and out-of-gamut colours are clipped onto the reproduction gamut surface.

It is clearly possible to define any number of other reproduction intents, also including the following two used by Morovic [48]:

- *Accurate rendering intent* – the appearance of the reproduced image is as close to the original image as is possible with respect to gamut differences.
- *Pleasant rendering intent* – the reproduced image is considered pleasant in isolation.

Unlike Hunt's reproduction requirements, these intents and the four intents defined by the ICC are explicitly aimed at *complex images* rather than individual colours. Note also that the accurate and pleasant rendering intents are targets rather than labels which could apply to a given reproduction (at least in the context of the present understanding of image appearance).

13.5.2 Relationship between gamut mapping and rendering intents

As has been pointed out in the previous section, reproduction intents in general can be seen as aims for colour reproduction systems and the question of how they are to be implemented arises. In the context of colour reproduction systems consisting of five stages [39] the most obvious place for pursuing rendering intents is at the gamut mapping stage (Figure 13.1). This has led to a very close association of rendering intents and gamut mapping algorithms, to the point where it has been suggested that they are identical. Aside from this resulting in problems of semantics (i.e. since intents are aims and gamut mapping algorithms can be tools for achieving them – it is like equating size with scaling), it also results in a need for as many gamut mapping solutions as there are rendering intents.

It might be more useful in this context to separate gamut mapping which results in the perceptually smallest differences between original and reproduced images – i.e. *accurate gamut mapping* – from the need for allowing changes in appearance to take place during the reproduction process. This separation is analogous to separating the individual stages of the colour reproduction in the first place rather than having a single transformation which is aimed at overcoming all the differences between original and reproduction media and viewing conditions. The result of such a separation (Figure 13.4) would be a six-stage transform [48]. Indeed there could be an even larger number of stages and the "image enhancement" stage could be split up into, for example, sharpness, contrast and saturation enhancement stages.

An example of using the six-stage transform would be the case when natural-looking reproductions of natural scenes are the aim and the colour reproduction system therefore might need to change the appearance of original images of natural scenes, if these do not look natural (e.g. if the skin tones in an original do not look natural then it might be desirable for reproductions of such images to have this problem corrected). This could

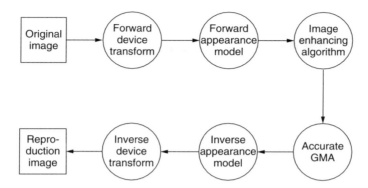

Figure 13.4 Six-stage colour reproduction transform.

be done by using an accurate GMA preceded by an algorithm for optimising the *colour quality of natural images*, e.g., the one proposed by Yendrikhovskij [73]. That particular optimisation algorithm is based on the computation of naturalness and colourfulness indices on the basis of image statistics. Naturalness is calculated by identifying the locations of prototypical memory colours for 'skin', 'grass' and 'sky' in colour space (which represent a wider range of naturally occurring colours) and then analysing image colours in relation to these. Colourfulness on the other hand is calculated from the average and standard deviation of image chroma. Finally, colour quality is expressed as a weighted combination of naturalness and colourfulness and images can then be optimised in terms of this metric. Similar algorithms could also be developed for optimising other types of images (e.g. the GMAs proposed by Braun *et al.* [9] intend to do so for business graphics).

13.6 OVERVIEW OF GAMUT MAPPING ALGORITHMS

A wide variety of gamut mapping strategies have been proposed and in some cases also psychophysically evaluated in the past. What will be presented here is an identification of the more prevalent approaches and those which seem to be particularly promising or inductive of future work.

Firstly, one of the most noticeable trends in the reviewed gamut mapping work is the agreement among different studies that the use of *image gamuts* is preferred over the use of medium gamuts and this is in some sense supported by a number of studies [19, 64, 25, 40, 46, 71]. Note that while this has been extensively demonstrated to be the case, it has also been shown recently that it is not an image's gamut that determines how the image is to be reproduced [59]. At first sight these two findings might seem contradictory. However, this is not so as the use of image gamuts in gamut mapping merely results in any algorithm making potentially smaller changes than if the original medium's gamut had been used. On the other hand, the study by Morovic and Sun [59] merely says that it is not an image's gamut that determines how well a given algorithm will perform. In other words, different GMAs perform differently for different images and this is not due to those images' image gamuts, while most GMAs will perform better for a given image if they use its gamut as the original gamut. The former point is about the performance of GMAs relative to each other, and the latter is about the performance of a single GMA depending on what it uses as the original gamut. Furthermore, remember that the size of the difference between the use of an image gamut versus the use of the medium gamut depends on the difference of these two gamuts.

Secondly, there is a significant number of studies where *clipping* is given preference over compression [65, 45, 69, 19, 64, 2, 33, 42, 15, 24, 46] whereby this is occasionally done implicitly. In some of these papers minimum ΔE clipping is used by default and in others clipping algorithms are proposed without reference to compression. In addition, there is also a good number of papers among the above which have arrived at the preference of clipping by means of well-designed psychophysical experiments [19, 64, 15, 24, 46]. Only the paper by Morovic and Luo [51], where clipping performed significantly worse than compression, is an exception to this trend. A possible explanation of this is that in all the former cases the relationship between original and reproduction gamuts was either artificial, or relatively small (when compared with gamut differences between the media [51]), or there was no lightness difference between them and therefore it was sufficient to use clipping. For overcoming larger gamut differences, therefore, it seems to be advantageous to use compression.

Thirdly, the vast majority of algorithms *start with uniform, overall, linear lightness compression* (chroma = 0 case in Figure 13.5(a)). However, there are significant exceptions to this trend in Ito and Katoh [31] where there is no initial lightness compression (chroma = 125 case in Figure 13.5(a)), in some GMAs proposed by Morovic and Luo [51, 52] where there is either no initial lightness compression or where this lightness compression is chroma-dependent (Figure 13.5(a)), and in the GMA type proposed by

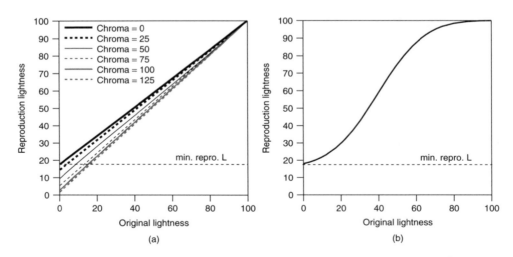

Figure 13.5 Different types of lightness compression.

Braun and Fairchild [6, 7] where sigmoidal lightness compression is used (Figure 13.5(b)). Which of these gives the best results seems to depend on gamut difference whereby uniform lightness compression has worked well in cases where only small differences needed to be overcome whereas the other solutions perform well under a wider range of conditions.

Fourthly, the *preservation of hue* (or hue angle) is also a point which occurs in most papers with few exceptions [11, 66, 52, 71, 10] and the papers where minimum ΔE clipping is used.

Fifthly, there are a good number of papers which suggest the use of *different mapping methods for different parts of colour space* [11, 21, 66, 72, 20, 31, 67, 24, 52, 71, 34, 53].

Finally, there are some papers describing interesting approaches, which have either not been tested in comparison to other methods (e.g. one which aims at preserving colour names in the process of gamut mapping [61, 62] or which defines gamut mapping as an image difference minimisation problem [63] and thus relies on a more satisfactory solution of other problems before it can become effective). More recently a gamut clipping solution that aims to preserve spatial luminance or lightness variation has also been proposed [1] and it too seems to be a promising approach well worth further evaluation.

13.7 GAMUT MAPPING ALGORITHMS AND COLOUR MANAGEMENT FRAMEWORKS

While it is still difficult at present to have image enhancement algorithms and accurate GMAs (as suggested in the previous section), there still remains the question of which GMA to use for which intent. Hence, the aim of this section is to recommend algorithms for achieving the rendering intents defined by the ICC.

First of all, the ICC's *saturation intent* seems to suggest a mapping of an original colour to the closest reproduction colour with the same saturation. Whether this is an aim which will result in acceptable reproductions is not clear, as it might well be the case that mapping to a colour with the same saturation might not be possible in some cases or that such a mapping would change a colour's appearance too significantly. Further, such a mapping could result in strongly non-monotonic relationships between original and reproduced colours, which could manifest themselves as artefacts.

The element missing from the definitions of *absolute* and *relative colorimetric intents* is the kind of clipping that should be used and a method for doing this will be recommended here.

The *perceptual intent* is perhaps the most vaguely defined ICC intent; however, as its key element seems to be the use of compression rather than clipping, a gamut compression algorithm will also be recommended.

13.7.1 Recommended gamut clipping algorithm

In terms of gamut clipping, the method proposed by Ito and Katoh [31] seems to be a good solution, not least because of its simplicity and good correlation with the results of the experimental study of Ebner and Fairchild [15]. In addition the relative importance of $L > h \geqslant C$ used in this model is also supported by Wei *et al* [71].

The suggested clipping was arrived at by first defining a functional model whose parameters were then chosen on the basis of experimental data. The model suggested that out-of-gamut colours should be clipped to colours on the reproduction gamut boundary, which have the smallest ΔE value calculated in the CIE94 colour difference metric (see Chapter 4). Based on a psychophysical experiment, the authors found that the most accurate reproductions were obtained when the $(Kl:Kc:Kh)$ coefficients in CIE94 were set to $(1:2:1)$ or $(1:2:2)$, which indicates that larger changes are acceptable in chroma than in hue and that the smallest change is tolerated in lightness. Note that the authors have since published the results of an extensive study investigating the suitability of different colour difference formulae [32] for use in minimum ΔE clipping and have found that ΔE_{94} [13] in CIELUV and ΔE_{BFD} [38] in LAB give the best results.

13.7.2 Recommended gamut compression algorithm

This gamut compression algorithm recommended here was first put together by the CIE's technical committee on gamut mapping (CIE TC8-03) and is a combination of two GMAs: GCUSP [48] and sigmoidal lightness mapping and cusp knee scaling [6]. Note that while it came about in the context of the above-mentioned technical committee, it is not endorsed by the CIE.

With this algorithm, perceived hue is kept constant, a generic (image-independent) sigmoidal lightness scaling is used and applied in a chroma-dependent way, and a 90% knee function chroma scaling is performed towards the cusp. The algorithm consists of the following steps (note that although CIELAB co-ordinates are used, the algorithm can be implemented in another colour space as well):

- Keep hue constant.
- Map lightness using the following formula:

$$L_R^* = (1 - p_c)L_O^* + p_c L_S^* \tag{13.1}$$

where L_O^* is the original lightness, L_R^* is the reproduction lightness,

$$p_c = 1 - [(C^{*3})/(C^{*3} + 5 \times 10^5)]^{1/2} \tag{13.2}$$

is a chroma-dependent weighting factor [51] which depends on the original colour's chroma C^*, and L_S^* is the result of the original lightness being mapped using a sigmoidal function (Figure 13.5(b)).

To calculate L_S^* [6], a one-dimensional look-up table (LUT) between original and reproduction lightness values is first set up on the basis of a discrete cumulative normal function (S):

$$S_i = \sum_{n=0}^{n=i} \frac{1}{\sqrt{2\pi\Sigma}} \exp\left[-\frac{\left(\frac{100n}{m} - x_0\right)^2}{2\Sigma^2}\right] \tag{13.3}$$

where x_0 and Σ are the mean and standard deviation of the normal distribution respectively, $i = 0, 1, 2, \ldots, m$, and m is the number of points used in the LUT. Hence, S_i is the value of the cumulative normal function for i/m per cent. The parameters can be determined either in an image-dependent way, or in an image-independent way where they depend on the lightness of the reproduction gamut's black point, and they can be interpolated from Table 13.1 (for details of calculating these parameters see ref. 6, p. 391).

In order to use S as a lightness mapping LUT (S_{LUT}) it must first be normalised into the lightness range of [0 100]. These normalised data are then scaled into the dynamic range of the destination device, as given in equation (13.4), where $L_{\min Out}^*$ and $L_{\max Out}^*$ are the black point and white point lightnesses of the reproduction medium respectively.

$$S_{\mathrm{LUT}} = \frac{S_i - \min(S)}{\max(S) - \min(S)}(L_{\max Out}^* - L_{\min Out}^*) + L_{\min Out}^* \tag{13.4}$$

At this point the L_S^* values can be obtained from the S_{LUT} by interpolating between the m points of corresponding L_O^* and L_S^* values it contains and using

$$100(L_O^* - L_{\min In}^*)/(L_{\max In}^* - L_{\min In}^*) \tag{13.5}$$

as the input, where $L_{\min In}^*$ and $L_{\max In}^*$ are the black point and white point lightnesses of the original medium respectively.

- Compress lightness and chroma along lines (l) towards the point (E_α) on the lightness axis having the same lightness as the cusp, and perform the compression as follows:

$$d_r = \begin{cases} d_o; & d_o \leqslant 0.9d_{gr} \\ 0.9d_{gr} + (d_o - 0.9d_{gr})0.1d_{gr}/(d_{go} - 0.9d_{gr}); & d_o > 0.9d_{gr} \end{cases}$$

where d represents distance from E_α on l, g represents the gamut boundary, r represents the reproduction and o the original (Figure 13.6).

Table 13.1 Image-independent parameter calculation

L_{minOut}^*	5.0	10.0	15.0	20.0
x_0	53.7	56.8	58.2	60.6
Σ	43.0	40.0	35.0	34.5

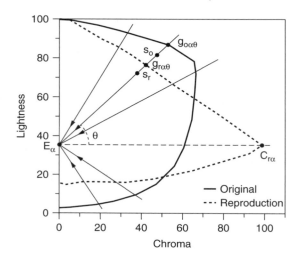

Figure 13.6 Compression towards the lightness of the cusp (g = gamut, o = original, r = reproduction, s = sample, α = hue angle, θ = second spherical angle, C = cusp, E_α = centre of compression).

13.8 SUMMARY

The aim of this chapter was to give a general overview of gamut mapping, both in terms of the factors that affect it and the parameters that determine its behaviour, and in terms of how it fits into the wider picture of colour reproduction. An overview of gamut mapping research carried out to date was also given, and it was shown how its findings can be used for the purpose of pursuing various colour reproduction intents. Furthermore, two gamut mapping algorithms were recommended for use in the context of the ICC colour management framework whereby these recommendations were made partly on the basis of the above-mentioned overview and the work of the CIE's technical committee on gamut mapping.

13.9 ACKNOWLEDGEMENTS

I would like to thank Prof. Ronnier Luo and Pei-Li Sun for their kind advice, and my family and Karen O'Toole for their invaluable support.

REFERENCES

1. Balasubramanian, R., de Queiroz, R., Eschbach, R. and Wu, W. (2000) Gamut Mapping To Preserve Spatial Luminance Variations. *Proceedings of 8th IS&T/SID Color Imaging Conference*, pp. 122–128, Scottsdale Az.

2. Berns, R. S. and Choh, H. K. (1995) Cathode–Ray–Tube to Reflection–Print Matching Under Mixed Chromatic Adaptation Using RLAB. *Journal of Electronic Imaging*, **4**/4, 347–359.
3. Botterill, G. (2000) *Personal email communication*, 28 September 2000.
4. Braun, G. J., Ebner, F. and Fairchild, M. D. (1998) Color Gamut Mapping in a Hue–Linearized CIELAB Color Space. *Proceedings of 6th IS&T/SID Color Imaging Conference*, pp. 163–168, Scottsdale Az, November 17–20.
5. Braun, G. J. and Fairchild, M. D. (1997) Techniques for Gamut Surface Definition and Visualization. *Proceedings of 5th IS&T/SID Color Imaging Conference*, pp. 147–152.
6. Braun, G. J. and Fairchild, M. D. (1999) Image Lightness Rescaling Using Sigmoidal Contrast Enhancement Functions. *Proceedings of SPIE Electronic Imaging Conference (EI'99)*.
7. Braun, G. J. and Fairchild, M. D. (1999) General–Purpose Gamut Mapping Algorithms: Evaluation of Contrast–Preserving Rescaling Functions for Color Gamut Mapping. *Proceedings of 7th IS&T/SID Color Imaging Conference*, pp. 167–172.
8. Braun, K. M., Fairchild, M. D. and Alessi, P. J. (1996) Viewing Techniques for Cross–Media Image Comparisons. *Color Research and Application*, **21**, 6–17.
9. Braun, K. M., Balasubramanian, R. and Harrington, S. J. (1999) Gamut Mapping Techniques for Business Graphics. *Proceedings of 7th IS&T/SID Color Imaging Conference*, pp. 149–154.
10. Braun, K. M., Balasubramanian, R. and Eschbach, R. (1999) Development and Evaluation of Six Gamut Mapping Algorithms for Pictorial Images. *Proceedings of 7th IS&T/SID Color Imaging Conference*, pp. 144–148.
11. CARISMA (1992) Colour Appearance Research for Interactive System Management and Application – CARISMA, Work Package 2 – Device Characterisation, *Report WP2–19 Colour Gamut Compression.*
12. CIE (1986) CIE Publication 15.2. *Colorimetry*, Second Edition.
13. CIE (1995) *Industrial Colour–Difference Evaluation*, CIE 116–1995.
14. Cholewo, T. J. and Love, S. (1999) Gamut Boundary Determination Using Alpha–Shapes. *Proceedings of 7th IS&T/SID Color Imaging Conference*, pp. 200–204.
15. Ebner, F. and Fairchild, M. D. (1997) Gamut Mapping from Below: Finding the Minimum Perceptual Distances for Colors Outside the Gamut Volume. *Color Research and Application*, **22**, 402–413.
16. Ebner, F. and Fairchild, M. D. (1998) Finding Constant Hue Surfaces in Color Space. *SPIE Proceedings*, **3300**, 107–117.
17. Engeldrum, P. G. (1986) Computing Color Gamuts of Ink–jet Printing Systems. *SID Proceedings*, **27**, 25–30.
18. Fairchild, M. D. (1998) *Color Appearance Models*, Addison–Wesley.
19. Gentile, R. S., Walowitt, E. and Allebach, J. P. (1990) A comparison of techniques for color gamut mismatch compensation, *Journal of Imaging Technology*, **16**, 176–181.
20. Granger, E. M. (1995) Gamut Mapping for Hard Copy Using the ATD Color Space. *SPIE Proceedings* **2414**, 27–35.
21. Haneishi, H., Miyata, K., Yaguchi, H. and Miyake, Y. (1993) A New Method for Color Correction in Hardcopy from CRT Images. *Journal of Imaging Technology*, **37**/1, 30–36.
22. Herzog, P. G. (1996) Analytical Color Gamut Representations. *Journal of the IS&T*, **40**, 516–521.
23. Herzog, P. G. (1998) Further Development of the Analytical Color Gamut Representations. *SPIE Proceedings*, **3300**, 118–128.
24. Herzog, P. G. and Müller, M. (1997) Gamut Mapping Using an Analytical Color Gamut Representation. *SPIE Proceedings*, **3018**, 117–128.
25. Hoshino, T. and Berns, R. S. (1993) Color Gamut Mapping Techniques for Color Hard Copy Images. *SPIE Proceedings*, **1909**, 152–164.
26. Hung, P. C. (1995) Gamut Mapping Using Lightness Adaptation. *Electronic Imaging, SPIE*, July 1995.
27. Hung, P. C. and Berns, R. S. (1995) Determination of Constant Hue Loci for a CRT Gamut and Their Predictions Using Color Appearance Spaces. *Color Res. Appl.*, **20**, 285–295.

28. Hunt, R. W. G. (1987) *The Reproduction of Colour in Photography, Printing & Television*, Fourth Edition, Fountain Press, England.

29. ICC (1998) *International Color Consortium Specification ICC.1: 1998–09 File Format for Color Profiles*. See http://www.color.org/

30. Inui, M. (1993) Fast Algorithm for Computing Color Gamuts. *Color Research and Application*, **18**, 341–348.

31. Ito, M. and Katoh, N. (1995) Gamut Compression for Computer Generated Images. *Extended Abstracts of SPSTJ 70th Anniversary Symposium on Fine Imaging*, 85–88.

32. Ito, M. and Katoh, N. (1999) Three–dimensional Gamut Mapping Using Various Color Difference Formulæ and Color Spaces. *Proceedings of SPIE and IS&T Electronic Imaging '99 Conference*.

33. Katoh, N. and Ito, M. (1996) Gamut Mapping for Computer Generated Images (II). *Proceedings of 4th IS&T/SID Color Imaging Conference*, pp. 126–129.

34. Kim, S. D., Lee, C. H., Kim, K. M., Lee, C. S. and Ha, Y. H. (1998) Image Dependent Gamut Mapping Using a Variable Anchor Point. *SPIE Proceedings*, **3300**, 129–137.

35. Kress, W. and Stevens, M. (1994) Derivation of 3–Dimensional Gamut Descriptors for Graphic Arts Output Devices. *TAGA Proceedings*, 199–214.

36. Luo, M. R. and Hunt, R. W. G. (1998) The Structure of the CIE 1997 Colour Appearance Model (CIECAM97s). *Color Research and Application*, **23**, 138–146.

37. Luo, M. R. and Morovic, J. (1996) Two Unsolved Issues in Colour Management – Colour Appearance and Gamut Mapping. *Proc. 5th International Conference on High Technology: Imaging Science and Technology – Evolution & Promise*, Chiba, Japan.

38. Luo, M. R. and Rigg, B. (1987) BFD(l:c) Colour Difference Formula, Part 1 and Part 2. *Journal of the Society of Dyers and Colourists*, 126–132.

39. MacDonald, L. W. (1993) Gamut Mapping in Perceptual Colour Space. *Proceedings of 1st IS&T/SID Color Imaging Conference*, pp. 193–196.

40. MacDonald, L. W. and Morovic, J. (1995) Assessing the Effects of Gamut Compression in the Reproduction of Fine Art Paintings. *Proceedings of 3rd IS&T/SID Color Imaging Conference*.

41. Mahy, M. (1997) Calculation of Color Gamuts Based on the Neugebauer Model. *Color Research and Application*, **22**, 365–374.

42. Marcu, G. and Abe, S. (1996) Gamut Mapping for Color Simulation on CRT Devices. *Electronic Imaging '96 Color Hard Copy and Graphic Arts*, San Jose.

43. Marcu, G. (1998) Gamut Mapping in Munsell Constant Hue Sections. *Proceedings of the 6th IS&T/SID Color Imaging Conference*, pp. 159–162.

44. McCann, J. J. (1999) Color Gamut Measurements and Mapping: The Role of Color Spaces. *Proceedings of SPIE Electronic Imaging Conference*, **3648**, 68–82.

45. Meyer, J. and Barth, B. (1989) Color Gamut Matching for Hard Copy. *SID 89 Digest*, 86–89.

46. Montag, E. D. and Fairchild, M. D. (1997) Psychophysical Evaluation of Gamut Mapping Techniques Using Simple Rendered Images and Artificial Gamut Boundaries. *IEEE Trans. Image Proc.*, **6**, 977–989.

47. Morovic, J. (1996) To Achieve WYSIWYG Colour via Adobe Photoshop. *Course Notes of Easter School '96: Colour Management for Information Systems*, 1st March, Derby.

48. Morovic, J. (1998) *To Develop a Universal Gamut Mapping Algorithm*, Ph.D. Thesis, University of Derby (page references are to the *Condensed format edition*).

49. Morovic, J. (1999) Colour Reproduction – Past, Present and Future. *Libro de Actas – V Congreso Nacional de Color (Proceedings of the 5th National Congress on Colour)*, 9–15, 9th–11th June Terrassa, Spain.

50. Morovic, J. (1999) Gamut Mapping and ICC Rendering Intents. *Proceedings of International Colour Management Forum*, Derby.

51. Morovic, J. and Luo, M. R. (1997) Cross–media Psychophysical Evaluation of Gamut Mapping Algorithms. *Proc. AIC Color 97 Kyoto*, **2**, 594–597.

52. Morovic, J. and Luo, M. R. (1997) Gamut Mapping Algorithms Based on Psychophysical Experiment. *Proceedings of the 5th IS&T/SID Color Imaging Conference*, pp. 44–49.

53. Morovic, J. and Luo, M. R. (1998) A Universal Algorithm for Colour Gamut Mapping, *Proceedings of CIM '98 Conference*.
54. Morovic, J. and Luo, M. R. (1998) The Pleasantness and Accuracy of Gamut Mapping Algorithms. *ICPS Conference Proceedings*, **2**, 39–43.
55. Morovic, J. and Luo, M. R. (1999) Developing Algorithms for Universal Colour Gamut Mapping. *Colour Imaging: Vision and Technology*, L. W. MacDonald (ed.), John Wiley & Sons.
56. Morovic, J. and Luo, M. R. (2000) The Fundamentals of Gamut Mapping: A Survey. *Journal of Imaging Science*, accepted for publication.
57. Morovic, J. and Luo, M. R. (2000) Calculating Medium and Image Gamut Boundaries for Gamut Mapping. *Color Research and Application*, accepted for publication.
58. Morovic, J. and Sun, P. L. (1999) Methods for Investigating the Influence of Image Characteristics on Gamut Mapping. *IS&T/SID 7th Color Imaging Conference*, 138–143.
59. Morovic, J. and Sun, P. L. (2000) The Influence of Image Gamuts on Cross–Media Colour Image Reproduction. *IS&T/SID 8th Color Imaging Conference*, accepted for publication.
60. Morovic, J. and Sun, P. L. (2000) How different are Colour Gamuts in Cross–Media Colour Reproduction?. *Colour Image Science Conference*, pp. 169–182, Derby.
61. Motomura, H. (1999) Categorical Color Mapping for Gamut Mapping II – Using Block Average Image. *Proceedings of SPIE and IS&T Electronic Imaging '99 Conference*.
62. Motomura, H., Yamada, O. and Fumoto, T. (1997) Categorical Color Mapping for Gamut Mapping. *Proceedings of the 5th IS&T/SID Color Imaging Conference*, pp. 50–55.
63. Nakauchi, S., Imamura, M. and Usui, S. (1996) Color Gamut Mapping by Optimizing Perceptual Image Quality. *Proceedings of the 4th IS&T/SID Color Imaging Conference*, pp. 63–67.
64. Pariser, E. G. (1991) An Investigation of Color Gamut Reduction Techniques. *IS&T's 2nd Symposium on Electronic Publishing*, 105–107.
65. Sara, J. J. (1984) *The Automated Reproduction of Pictures with Nonreproducible Colors*, Ph.D. Thesis, Massachusetts Institute of Technology (MIT).
66. Schläpfer, K. (1994) *Color Gamut Compression – Correlations Between Calculated and Measured Values*, IFRA Project, EMPA, 8 August 1994.
67. Spaulding, K. E., Ellson, R. N. and Sullivan, J. R. (1995) UltraColor: A New Gamut Mapping Strategy. *SPIE Proceedings*, **2414**, 61–68.
68. Stone, M. C., Cowan, W. B. and Beatty, J. C. (1988) Color Gamut Mapping and the Printing of Digital Color Images. *ACM Transactions on Graphics*, **7**, 249–292.
69. Taylor, J. M., Murch, G. M. and McManus, P. A. (1989) Tektronix HVC: A Uniform Perceptual Color System, *SID Digest of Technical Papers*.
70. Viggiano, J. A. S. and Moroney, N. M. (1995) Color Reproduction Algorithms and Intent. *Proceedings of the 3rd IS&T/SID Color Imaging Conference*, pp. 152–154.
71. Wei, R. Y. C., Shyu, M. J. and Sun, P. L. (1997) A New Gamut Mapping Approach Involving Lightness, Chroma and Hue Adjustment. *TAGA Proceedings*, 685–702.
72. Wolski, M., Allebach, J. P. and Bouman, C. A. (1994) Gamut Mapping. Squeezing the Most out of Your Color System. *Proceedings of the 2nd IS&T/SID Color Imaging Conference*, pp. 89–92.
73. Yendrikhovskij, S. N. (1998) *Color Reproduction and the Naturalness Constraint*, Ph.D. Thesis, Technische Universiteit Eindhoven, The Netherlands.
74. Zhang, X. M. and Wandell, B. A. (1996) A spatial extension to CIELAB for digital color image reproduction. *Proceedings of the SID Symposiums*.

Implementation of device-independent color at Kodak

Kevin Spaulding and Edward Giorgianni

14.1 INTRODUCTION

There are many ways to represent and communicate the color of a digital image. Historically, most digital imaging systems have been built around *device-dependent* color spaces, where the digital code values stored for each pixel of an image are those that will produce the desired result on a specific output device. For example, the color of the image can be represented in terms of the *video RGB* code values for a target CRT, or the CMYK values appropriate for a certain graphic arts printing process. However, these device-dependent color values can only be related to the corresponding color appearance by taking into account the characteristics of the intended output device, the intended viewing environment, and the human observer.

Communicating color using device-dependent color spaces can work well for *closed systems* where all of the components of the imaging chain can be designed with an understanding of the characteristics of an invariant target device. However, the need for interchanging images within and among *open systems*, where such solutions are less

Colour Engineering, edited by P.J. Green and L.W. MacDonald.

satisfactory, is becoming increasingly important. By definition, an open system is not restricted to using only certain types of inputs and outputs, but can make use of all available types of imaging devices and media. Consequently, open systems can frequently suffer from interoperability problems because different components in the system may communicate color differently.

Naïvely, one might expect that if the color of an image simply were represented in terms of a *device-independent* color space (such as CIE XYZ tristimulus values or CIELAB), the interoperability problems would be resolved. This is the basic thesis of many *color management* systems. However, standard CIE colorimetry alone is not enough to unambiguously communicate the color of an image. For example, two images having the same colorimetry can look quite different depending on the environment in which they are viewed (luminance level, ambient illumination, surround conditions, etc.). The development of color appearance models to account for these viewing environment factors is a topic of active research; but even if a perfect color appearance model were used to represent a color image, this would still be insufficient to ensure the unambiguous communication of color images. As will be discussed further in the next section, it is further necessary to know something about the state of the image that is being encoded. For example, if someone specified the CIE colorimetry of an image, together with all of the parameters needed to compute the color appearance, but did not specify whether the color values were those of an original scene or those intended to be produced on an output print, it would not be possible to determine unambiguously how the image should be printed.

Therefore, unambiguous *color encoding* requires more than specifying device-independent color values. It must include:

- Information about the state of the image (scene, print, negative, etc.)
- Description of the intended viewing environment (parameters necessary to relate color values to color appearance)
- Color metric used to represent the image (XYZ, CIELAB, Rec. 709 RGB, etc.)
- Digital encoding (how color values are represented as digital code values).

Without all of this information being defined, either explicitly or implicitly, a color encoding specification cannot be considered to be complete.

14.2 IMAGE STATES

Digital images exist in many different *image states*. The image state is one of the fundamental attributes of a digital color encoding and relates to the basic characteristics of an image. The image state will be a function of how the image was captured, as well as the processing that may have been applied to an image. While the concept of an image state can be applied to all attributes of an image, such as sharpness or noise, the aspect of image state that is of importance for the current discussion relates to the interpretation of the color values of the image. For example, the code values of an image could correspond to the sensor RGB values from a digital camera, the CIELAB values of a reflection print, or the ISO Status M density values of a photographic negative. These image examples vary in two distinct aspects. First, they each use a different *color space* to encode the image (sensor RGB, CIELAB, and ISO Status M densities). However, just as importantly, each of these image encodings corresponds to a distinctly different *image state* (a scene, a print, and a photographic negative). Even if the same color space, CIELAB for example, were used to encode all of these images, it would still not be possible (or at least not optimal) to treat the images identically. Obviously, something quite different would have to be done with the CIELAB values of a color negative relative to the CIELAB values of a print. As will be discussed in more detail later, the same also would be true for the CIELAB values of a scene relative to those of a print.

Most digital images can be broadly categorized into two types of image states: *unrendered* and *rendered*. Images in an unrendered image state are directly related to the colorimetry of a real or hypothetical original scene. Such images are sometimes called *scene-referred* images. Images in this category would include raw digital camera captures and images stored in the Kodak PhotoYCC color interchange space [1]. Images in a rendered image state are representations of the colorimetry of an output image (such as a print, a slide, or a CRT display) and are sometimes called *output-referred* images. Many common color encodings, such as sRGB [2] and SWOP CMYK [3], fall into this category. A third category of image state would be encodings of *photographic color negatives*. While raw images captured by photographic color negative scanners will be in such an image state, it is typically a temporary stop on the way to forming a rendered image, or determining a corresponding scene-referred image.

In order to enable the optimal use of digital images, it is important to distinguish images in an output-referred image state from those in a scene-referred image state. It is well known that the colorimetry of a pleasing rendered image generally does not match the colorimetry of the corresponding scene. Among other things, the tone/color reproduction process that 'renders' the colors of a scene to the desired colors of the output image must compensate for differences between the scene and rendered image viewing conditions [4, 5]. For example, rendered images generally are viewed at luminance levels much lower than those of typical outdoor scenes. Consequently, an increase in the overall contrast of the rendered image usually is required in order to compensate for perceived losses in reproduced luminance and chrominance contrast. Additional contrast increases in the shadow regions of the image are also needed to compensate for viewing flare associated with rendered-image viewing conditions.

Psychological factors such as color memory and color preference also must be considered in image rendering. For example, observers generally remember colors as being of higher purity than they really were, and they typically prefer skies and grass to be more colorful than they were in the original scene. The tone/color reproduction aims of well-designed imaging systems will account for such factors.

Finally, the tone/color reproduction process must also account for the fact that the dynamic range of a rendered image is usually substantially less than that of an original scene. It is therefore typically necessary to discard and/or compress some of the highlight and shadow information of the scene to fit within the dynamic range of the rendered image. This is shown in Plate 7, which illustrates a typical backlit scene. In this example, the approximate scene colorimetry was determined from a scan of a color negative. The upper image (a) shows a rendering of the scene appropriate for the foreground information, and the lower image (b) shows a rendering of the scene appropriate for the background information. In the first case, much of the highlight information was clipped by the rendering process. And, likewise, in the second case, much of the shadow information was lost. This is illustrated further in Figure 14.1, which shows a (smoothed) histogram of the scene luminance data for the image shown in Plate 7. A conventional reflection print of this scene can reproduce only about six stops (1.8 log luminance units) of scene information within the dynamic range of the output medium. The indicated ranges show the subsets of the scene luminance information corresponding to the two images in Plate 7. It can be seen that only a portion of the total scene information is reproduced in either of the rendered images.

Because the colorimetry of scenes and their corresponding rendered images are intentionally and necessarily different, it would be ambiguous to try to represent images in both image states using the same color encoding. For example, if someone were to send the CIELAB values for a particular image, but with no information about whether the color values were scene color values or rendered image color values, the recipient would not know what to do with the image in order to make a good print. If the CIELAB values were rendered color values appropriate for the output viewing environment, it would be necessary simply to determine the device code values needed to produce the specified colorimetry. However, if the color values corresponded to original scene color

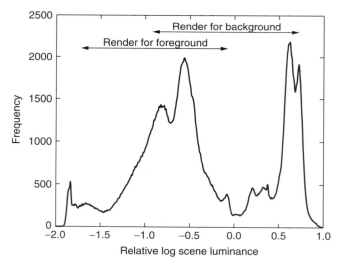

Figure 14.1 Histogram of relative log scene luminance values for the scene shown in Plate 7. A scene luminance range of about 1.8 log units can be reproduced on a typical output reflection print. Since the dynamic range of the original scene is substantially larger than this, a subset of the image date must be selected during the rendering process. Different results are obtained depending on whether the foreground or background region of the image is selected.

values, it would be necessary to modify the image colorimetry by applying an appropriate tone/color reproduction transformation before printing the image. Directly reproducing the scene colorimetry on an output print would produce results that generally would be judged inferior. For example, Plate 8(a) and (b) shows a pair of images generated from the same scene. The upper image (a) approximately matches the colorimetry of the original scene; whereas an appropriate tone/color reproduction transformation has been used to modify the colorimetry of the lower image (b) to produce an image that generally would be judged to have improved color reproduction.

One of the advantages of encoding images in a scene-referred image state is the capability of retaining the maximum amount of image information. As illustrated in Plate 7, once an image is committed to a rendered output-referred image state appropriate for printing or display, any extended dynamic range information is permanently lost. Retaining the scene-referred image data preserves the maximum flexibility for the use of an image. This allows for the correction of image capture exposure errors, and it also enables customers to make decisions about the preferred way to render any particular image. For example, a photographer could decide at the time an image is printed whether to optimally render the foreground or the background information in a backlit scene. It is valuable to preserve this option because many times there will not be a single 'best' choice that can be made when the image is captured. For the image shown in Plate 7, the final decision would probably depend on whether the photographer was most interested in the boys in the foreground, or the scenic Alps in the background. Retaining the extended dynamic-range scene information also enables other options, such as employing advanced image processing to perform a digital 'dodge-and-burn' operation to produce a print where both the foreground and the background are properly rendered, as shown in Plate 8(c). Comparable results could not be attained starting from one of the conventionally rendered images shown in Plate 7.

14.3 STANDARD IMAGE STATE COLOR ENCODINGS

The fact that images exist in many different image states and color spaces significantly complicates the development of software applications that use and manipulate images. For example, an image-processing algorithm that works in one color space might not have the expected behavior when used in another color space. To reduce complexity of imaging system design, it is desirable to define standard color encodings for each of the main classes of image states. This will enable:

- Unambiguous communication of color information
- Development of standard image-manipulation algorithms
- Development of standard color-processing paths.

In the past, proposals for the standardization of color encodings typically have involved the specification of a particular output-device-dependent color space that is central to the workflow for a certain market segment. Examples of such color spaces include sRGB and SWOP CMYK. While these solutions can work well within the limited scope of a particular application, significant compromises would be necessary to use them in other applications. For example, hardcopy media and CRT displays typically have very different color gamuts. Therefore, using sRGB (which is based on a particular CRT model) as a standard color encoding would necessarily involve clipping many colors that could have been produced on a given hardcopy medium. This would be unacceptable in many hardcopy-based market segments, such as consumer photofinishing and graphic arts.

The International Color Consortium (ICC) has defined a Profile Connection Space (PCS) [6] that comprises a device-independent color encoding specification that can be used to explicitly specify the color of an output-referred image with respect to a reference viewing environment. It could be argued that the PCS could serve as one of these standard color encodings. However, it was never intended that the PCS be used to directly store or manipulate images. Rather, it was simply intended to be a color space where device profiles could be joined to form complete input-to-output color transforms. Neither the CIELAB nor the *XYZ* color encodings supported for the PCS are particularly well suited for many common kinds of image manipulations. Additionally, quantization errors that would be introduced by encoding images in PCS would be significantly larger than necessary because a large percentage of the code values correspond to unrealizable colors.

Given the limitations of the existing solutions, Eastman Kodak Company has developed a family of color encodings for use in the development of its digital imaging products [7, 8]. These color encodings are being offered for use by our customers and other companies, and they have also been proposed for international standardization. *Reference Input Medium Metric RGB (RIMM RGB)* is ideal for the manipulation, storage, and interchange of images from sources such as digital cameras that naturally capture *scene-referred* image data. *Reference Output Medium Metric RGB (ROMM RGB)* serves a similar purpose for images from sources such as print scanners and other devices that produce images in a rendered *output-referred* image state. A diagram illustrating how these standard color encodings can be used as the basis for a general imaging system architecture is shown in Figure 14.2.

Before images can be sent to an output device, such as a printer, it generally will be necessary to convert scene-state images to rendered-state images using a tone/color rendering operation. However, in the same way that a negative is much more versatile than a print, an image in a scene-state will be much more versatile than one in a rendered-state. Therefore, it will be desirable in many imaging systems to delay any conversion to a rendered-state until the time when an output image is generated. For example, consider the case where a color negative is used to capture an image of a backlit scene having a brightly lit background and a dimly lit foreground. The negative generally will contain information for both regions of the image, and that information can be used to make a print that is properly exposed for either the foreground or the background. In order to make a rendered-state image from this negative, it is necessary to make a choice about which part of the image is important. (In a conventional photographic system, this is done when the negative is printed optically.) However, scene-state images can retain all of the information on the negative, which allows the delay of any decision as to how the scene is to be rendered. This maintains the maximum amount of flexibility in the system.

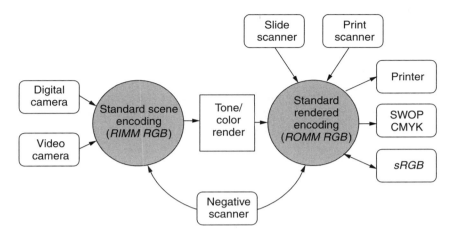

Figure 14.2   Image state diagram showing standard color encodings.

14.3.1 Criteria for selection of *RIMM/ROMM RGB* color encodings

It was desirable that the *RIMM RGB* and *ROMM RGB* color encoding specifications be defined such that they are as similar as possible to one another. Doing so simplifies the development of image-manipulation algorithms across the two color encodings. It also simplifies the rendering process in which a rendered *ROMM RGB* image is created from an original-scene image encoded in *RIMM RGB*. This is best achieved by basing the two encodings on the same color space. The criteria that were used to select this color space include the following:

- Direct relationship to the color appearance of the scene/image
- Color gamut large enough to encompass most real-world surface colors
- Efficient encoding of the color information to minimize quantization artifacts
- Simple transformation to/from ICC PCS
- Simple transformation to/from video RGB (e.g., sRGB)
- Well suited for application of common image manipulations such as tone scale modifications, color-balance adjustments, sharpening, etc.
- Compatible with established imaging workflows.

An additive RGB color space with an appropriately selected set of wide-gamut primaries is ideal for satisfying all of these criteria. When images are encoded using any such set of 'big RGB' primaries, there is a direct and simple relationship to scene/image colorimetry because the primaries are linear transformations of the CIE XYZ primaries. Big RGB color spaces have the additional advantage that simple LUT−matrix−LUT transformations can be used to convert to/from additive color spaces such as PCS XYZ, video RGB (sRGB), and digital camera RGB.

Two of the criteria that affect the selection of the particular RGB primaries are somewhat conflicting. First, the chromaticities of the primaries should define a gamut sufficiently large to encompass colors likely to be found in real scenes and images. At the same time, their use should result in efficient digital encodings that minimize quantization errors.

Increasing the gamut to encompass more colors can only be achieved by trading off against correspondingly larger quantization errors. If the chromaticities of the primaries are chosen to include the maximum possible color gamut (for example, choosing the XYZ primaries would encompass the entire spectrum locus), a significant fraction of the color space would correspond to imaginary colors and to colors that would not be commonly encountered in real images. Therefore, in any encoding using such a color space, there would be large numbers of 'wasted' code value combinations that would never be used in practice. This would lead to larger quantization errors in the usable part of the color space than would be obtained with different primaries defining a smaller chromaticity gamut. It is, therefore, desirable to choose primaries with a gamut that is 'big enough' but not 'too big'.

Figure 14.3 shows the primaries selected for *RIMM/ROMM RGB*. These primaries encompass the gamut of real-world surface colors, without devoting a lot of space to nonrealizable colors outside the spectrum locus. Also shown for comparison are the sRGB primaries. It can be seen that the area defined by the sRGB chromaticity boundaries is inadequate to cover significant portions of the real-world surface color gamut. In particular,

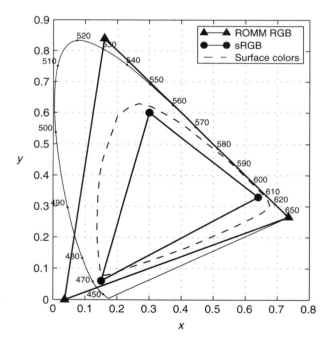

Figure 14.3   Comparison of *ROMM RGB* and *sRGB* primaries in $x - y$ chromaticity coordinates.

it excludes many important high-chroma colors near the yellow-to-red boundary of the spectrum locus.

Another important requirement for the *RIMM RGB* and *ROMM RGB* color encodings is that they be well suited for application of common image manipulations. Many types of image manipulations include the step of applying nonlinear transformations to each of the channels of an RGB image (e.g., tone scale modifications, color balance adjustments, etc.). The process of forming a rendered image from a scene is one important application of this type. One way to accomplish the rendering operation is by applying a nonlinear tonescale transformation to the individual channels of an RGB image in a scene image state. A well-designed transformation of this type will have the desirable effects of:

- Increasing the luminance and color contrast in the mid-tones
- Compressing the contrast of the highlights and shadows
- Increasing the chroma of in-gamut colors
- Gamut mapping out-of-gamut colors in a simple but visually pleasing way.

If an input scene is represented using the *RIMM RGB* color encoding, the result of applying such rendering transforms will be a rendered image in the *ROMM RGB* color encoding.

Nonlinear channel-independent transforms will, in general, modify the relative ratios of the red, green, and blue channel data. This can lead to unwanted hue shifts, particularly for high chroma colors. Hue shifts are particularly problematic in reproductions of natural chroma gradients, having constant hue and saturation. Such gradients tend to occur when rounded surfaces are illuminated by a moderately directional light source. In these situations, chroma increases with distance from the specular highlight and then decreases again as the shadows deepen.

The hue shifts induced by the application of the nonlinear channel-independent transforms can never be completely eliminated, so one objective when optimizing the location of the primaries was to eliminate or minimize objectionable hue shifts, sometimes at the expense of less noticeable or less likely hue shifts. Hue shifts for a particular color can be eliminated when the color lies on one of the straight lines passing through the primaries and the white point on a chromaticity diagram.

Hue shifts introduced by the application of nonlinear transformations were studied using a chroma series for eight color patches from the Macbeth Color Checker. These patches included red, yellow, green, cyan, blue, magenta, light skin, and dark skin. Hue shifts in skin tones and yellows, particularly in the direction of green, are considered the most objectionable. These hue shifts are most strongly affected by the location of the blue primary. Other colors that were considered particularly important during the optimization process were blues and reds.

There is a trade-off between the color gamut of the primaries, quantization artifacts, and the extent of the hue shifts that occur during rendering. If the primaries are moved out to increase the color gamut, quantization artifacts will increase, and the hue shifts introduced during the application of a nonlinear transformation generally will decrease. This results from the fact that the RGB values in real images will be distributed over a smaller range, thereby reducing the impact of nonlinear transformations. If the color gamut is decreased by moving the primaries closer together, quantization artifacts diminish; but hue shifts are generally larger and color gamut is sacrificed.

Finally, a basic requirement for any commercially useful color encoding is that it be compatible with typical commercial imaging workflows. In many cases, Adobe Photoshop software is an important component in such imaging chains. Conveniently, version 5.0 of the Adobe Photoshop software has incorporated the concept of a 'working color space', which is different from the monitor preview color space. This is consistent with the notion of storing/manipulating images in a 'big RGB' color space. Adobe has placed a constraint on the definition of valid working color spaces that requires the primaries to have all positive $x-y-z$ chromaticity values. This condition is satisfied for the *ROMM RGB* primaries. Because the Adobe Photoshop software operates within a rendered-image paradigm, it is inappropriate to use *RIMM RGB* as a Photoshop software working color space. (For more information about using *ROMM RGB* as a Photoshop software working space, see the white paper posted at www.Kodak.com – search on 'ROMM'.)

During the selection of the *RIMM/ROMM RGB* primaries, an extensive optimization process was used to determine the best overall solution to satisfy all of these criteria. The hue shifts associated with the selected *RIMM/ROMM RGB* primaries are shown in Figure 14.4. This plot shows a series of line segments connecting the a^*, b^* values before and after a nonlinear tone scale transformation was applied to a chroma series in each of the eight color directions. It can be seen that only relatively small hue shifts are introduced for the highest chroma colors in the blue and cyan directions, and the hue shifts elsewhere are virtually negligible. Overall, these hue shifts are very small compared to those associated with many other sets of primaries.

14.3.2 *ROMM RGB* color encoding

Reference Output Medium Metric RGB (ROMM RGB) is designed to be an extended-gamut color encoding for representing the color appearance of an output-referred image. In addition to specifying the image state and color space, it is also necessary to specify an intended viewing environment in order to unambiguously define an encoding of color-appearance. One of the requirements for *ROMM RGB* is that it be tightly coupled to the ICC Profile Connection Space (PCS). Color values in the PCS represent the CIE colorimetry of a defined reference medium that will produce the desired color appearance when viewed in a reference viewing environment. The reference viewing environment for *ROMM RGB* was based on that defined in the latest ICC draft specification [9], and is defined to have the following characteristics:

- The luminance level for observer adaptive white is $160 \, \text{cd/m}^2$.
- The observer adaptive white has the chromaticity values of CIE Standard Illuminant D_{50}: $x = 0.3457$, $y = 0.3585$.
- The viewing surround is average. (In other words, the overall luminance level and chrominance of the surround is assumed similar to that of the image.)
- There is 0.75% viewing flare, referenced to the observer adaptive white.
- The image color values are assumed to be encoded using flareless (or flare corrected) colorimetric measurements based on the CIE 1931 Standard Colorimetric Observer.

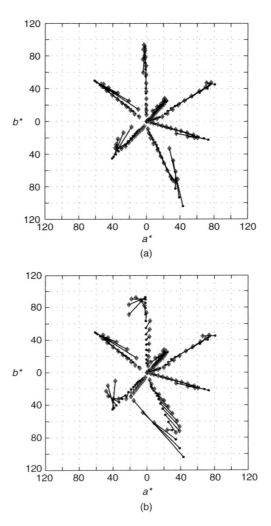

Figure 14.4 Hue shifts resulting from a typical nonlinear rendering transform for (a) the *RIMM/ROMM RGB* primaries, and (b) an alternate set of wide-gamut primaries. The hue shifts for the most important colors are visually negligible for the *RIMM/ROMM RGB* color encoding.

The *ROMM RGB* color encoding is defined in the context of a reference imaging medium associated with a hypothetical additive color device having the following characteristics:

- Reference primaries defined by the CIE chromaticities given in Table 14.1.
- Equal amounts of the reference primaries produce a neutral with the chromaticity of D_{50}: $x = 0.3457$, $y = 0.3585$.

Table 14.1 Primaries/white
point for reference imaging
medium

Color	x	y
Red	0.7347	0.2653
Green	0.1596	0.8404
Blue	0.0366	0.0001
White	0.3457	0.3585

- The capability of producing a white with a luminance factor of $F_W = 0.89$, and a black with a luminance factor of $F_K = 0.0030911$.

Images intended to be viewed in other viewing environments, or on a medium different from the reference medium, can be encoded in *ROMM RGB* by first determining the corresponding tristimulus values that would produce the intended color appearance on the reference medium when viewed in the reference viewing environment. The corresponding tristimulus values can be determined by using an appropriate color appearance transformation to account for the differences between the viewing conditions. Additionally, it may be necessary to account for differences in the media characteristics.

The conversion of the PCS XYZ tristimulus values to *ROMM RGB* values can be performed by a matrix operation, followed by a set of one-dimensional functions. This is equivalent to the operations associated with a basic monitor profile. This means that *ROMM RGB* can be used conveniently in a system employing ICC profiles using an appropriately designed display profile.

Most current implementations of the ICC PCS [10] incorporate the concept of a reference medium where the black point of the reference medium is mapped to $Y_{PCS} = 0$, and the white point of the reference medium is mapped to $Y_{PCS} = 1.0$. Therefore, to relate actual CIE image colorimetry to PCS XYZ values, an appropriate normalizing transformation is required:

$$X_{PCS} = \frac{(X - X_K)X_W}{(X_W - X_K)Y_W}$$

$$Y_{PCS} = \frac{Y - Y_K}{Y_W - Y_K} \tag{14.1}$$

$$Z_{PCS} = \frac{(Z - Z_K)Z_W}{(Z_W - Z_K)Y_W}$$

where X, Y and Z are the CIE image tristimulus values, X_{PCS}, Y_{PCS} and Z_{PCS} are the PCS tristimulus values, X_W, Y_W and Z_W are the tristimulus value of the reference medium white point ($X_W = F_W X_0 = 85.81$, $Y_W = F_W Y_0 = 89.00$ and $Z_W = F_W Z_0 = 73.42$, where $X_0 = 96.42$, $Y_0 = 100.00$ and $Z_0 = 82.49$), and X_K, Y_K and Z_K are the tristimulus value of the reference medium black point ($X_K = F_K X_0 = 0.2980$, $Y_K = F_K Y_0 = 0.3091$ and $Z_K = F_K Z_0 = 0.2550$).

14.3.3 *ROMM RGB* conversion matrix

Given the defined primaries shown in Table 14.1, the following matrix can be derived to compute the linear *ROMM RGB* values from the PCS image tristimulus values:

$$
\begin{bmatrix} R_{ROMM} \\ G_{ROMM} \\ B_{ROMM} \end{bmatrix} = \begin{bmatrix} 1.3460 & -0.2556 & -0.0511 \\ -0.5446 & 1.5082 & 0.0205 \\ 0.0000 & 0.0000 & 1.2123 \end{bmatrix} \begin{bmatrix} X_{PCS} \\ Y_{PCS} \\ Z_{PCS} \end{bmatrix} \tag{14.2}
$$

As required by the definition of the *ROMM RGB*, this matrix will map image tristimulus values with the chromaticity of D$_{50}$ to equal *ROMM RGB* values. A neutral with a Y_{PCS} value of 1.0, corresponding to the reference medium white point, will map to linear *ROMM RGB* values of 1.0. Likewise, the reference medium black point will map to linear *ROMM RGB* values of 1.0.

14.3.4 Nonlinear encoding of *ROMM RGB*

A nonlinear quantization function is used to store the *ROMM RGB* values in an integer form. A simple gamma function nonlinearity incorporating a slope limit at the dark end of the intensity scale is defined for this purpose:

$$
C'_{ROMM} = \begin{cases} 0; & C_{ROMM} < 0.0 \\ 16\,C_{ROMM}\,I_{\max}; & 0.0 \leqslant C_{ROMM} < E_t \\ (C_{ROMM})^{1/1.8}\,I_{\max}; & E_t \leqslant C_{ROMM} < 1.0 \\ I_{\max}; & C_{ROMM} \geqslant 1.0 \end{cases} \tag{14.3}
$$

where C is either R, G, or B, $I_{\max}$ is the maximum integer value used for the nonlinear encoding, and

$$
E_t = 16^{1.8/(1-1.8)} = 0.001953 \tag{14.4}
$$

For the baseline 8-bit configuration, $I_{\max}$ is equal to 255. The linear segment of the nonlinearity is used to impose a slope limit to minimize reversibility problems because of the infinite slope of the gamma function at the zero point. Twelve- and 16-bit versions of *ROMM RGB* are also defined. The only difference is that the value of $I_{\max}$ is set to 4095 or 65 535, respectively. In cases where it is necessary to identify a specific precision level, the notation *ROMM8 RGB, ROMM12 RGB*, and *ROMM16 RGB* is used. Table 14.2 shows some sample encodings for a series of neutral patches of specified Y_{PCS}.

14.3.5 *RIMM RGB* color encoding

Reference Input Medium Metric RGB (RIMM RGB) is a companion color encoding specification to *ROMM RGB* that can be used to encode the colorimetry of an *unrendered scene*. Both encodings utilize the same 'big RGB' color space defined by the primaries and white point given in Table 14.1. The reference viewing conditions used to encode

Table 14.2 Sample neutral patch encodings

Y_{PCS}	ROMM8 RGB	ROMM12 RGB	ROMM16 RGB
0.00	0	0	0
0.001	4	66	1049
0.01	20	317	5074
0.10	71	1139	18 236
0.18	98	1579	25 278
0.35	142	2285	36 574
0.50	174	2786	44 590
0.75	217	2490	55 855
1.00	255	4095	65 535

scene color values for *RIMM RGB* are typical of outdoor environments, and are defined as follows:

- The luminance level for the observer adaptive white is $> 1600\,\text{cd/m}^2$.
- The observer adaptive white has the chromaticity values of CIE Standard Illuminant D_{50}: $x = 0.3457$, $y = 0.3585$.
- Viewing surround is average. (In other words, the overall luminance level and chrominance of the surround is assumed similar to that of the scene.)
- There is no viewing flare for the scene other than that already included in the scene colorimetric values.
- The scene color values are assumed to be encoded using flareless (or flare corrected) colorimetric measurements based on the CIE 1931 Standard Colorimetric Observer.

Scenes captured under conditions different from the reference viewing environment can be encoded in *RIMM RGB* by first determining the corresponding tristimulus values that would produce the intended color appearance in the reference viewing environment. The corresponding tristimulus values can be determined by using an appropriate color appearance transformation to account for the differences between the viewing conditions.

14.3.6 *RIMM RGB* conversion matrix

Since *ROMM RGB* and *RIMM RGB* use a common color space, the conversion from the scene tristimulus values to the corresponding linear *RIMM RGB* values can be accomplished using the same conversion matrix that was given in equation (14.2), except that the input tristimulus values are the scene XYZ values rather than the PCS XYZ values.

$$
\begin{bmatrix} R_{RIMM} \\ G_{RIMM} \\ B_{RIMM} \end{bmatrix} = \begin{bmatrix} 1.3460 & -0.2556 & -0.0511 \\ -0.5446 & 1.5082 & 0.0205 \\ 0.0000 & 0.0000 & 1.2123 \end{bmatrix} \begin{bmatrix} X_{D50} \\ Y_{D50} \\ Z_{D50} \end{bmatrix}
\tag{14.5}
$$

(Note that the scene *XYZ* values are normalized such that the luminance of a correctly exposed perfect white diffuser in the scene will have a value of $Y_{D50} = 1.0$.)

14.3.7 Nonlinear encoding of *RIMM RGB*

Because the dynamic range of unrendered scenes is generally larger than that of the medium specified for *ROMM RGB*, a different nonlinear encoding must be used. The *RIMM RGB* nonlinearity is based on that specified by Recommendation ITU-R BT.709 (Rec. 709) [11]. (This recommendation was formerly known as CCIR 709.) This is the same nonlinearity as is used in the *Kodak PhotoYCC* color interchange space encoding implemented in the *Kodak Photo CD* system [1], and is given by:

$$
C'_{RIMM} = \begin{cases}
0; & C_{RIMM} < 0.0 \\
\left(\dfrac{I_{\max}}{V_{clip}}\right) 4.5\, C_{RIMM}; & 0.0 \leqslant C_{RIMM} < 0.018 \\
\left(\dfrac{I_{\max}}{V_{clip}}\right)(1.099 C_{RIMM}^{0.45} - 0.099); & 0.018 \leqslant C_{RIMM} < E_{clip} \\
I_{\max} & C_{RIMM} \geqslant E_{clip}
\end{cases}
\tag{14.6}
$$

where C is either R, G, or B; $I_{\max}$ is the maximum integer value used for the nonlinear encoding; $E_{clip} = 2.0$ is the scene luminance level that is mapped to $I_{\max}$; and

$$
V_{clip} = 1.099 E_{clip}^{0.45} - 0.099 = 1.402
\tag{14.7}
$$

For the baseline 8-bit/channel *RIMM RGB* configuration, $I_{\max}$ is 255. In some applications, it may be desirable to use a higher bit precision version of *RIMM RGB* to minimize any quantization errors. Twelve- and 16-bit/channel versions of *RIMM RGB* are also defined. The only difference is that the value of $I_{\max}$ is set to 4095 or 65 535, respectively. In cases in which it is necessary to identify a specific precision level, the notation *RIMM8 RGB, RIMM12 RGB*, and *RIMM16 RGB* is used.

14.3.8 *ERIMM RGB* color encoding

The *RIMM RGB* color space is defined to have a luminance dynamic range that can encode information up to 200% of the luminance value associated with a normally exposed perfect (100%) diffuse white reflector in the scene. This should be adequate for many applications such as digital cameras, which themselves have a somewhat limited dynamic range. However, for some applications, most notably scanned photographic negatives, a greater luminance dynamic range is required to encode the full range of captured scene information. For example, consider the histogram of scene luminance data previously shown in Figure 14.1. The *RIMM RGB* encoding would only retain scene information up to a relative log scene luminance value of 0.3. A significant portion of the scene information would be lost with a *RIMM RGB* encoding in this case. In order to provide an encoding that can retain the full range of captured scene information, a variation of the *RIMM RGB* color space is defined, known as *Extended Reference Input Medium Metric RGB (ERIMM RGB)*.

As with *RIMM RGB, ERIMM RGB* is directly related to the colorimetry of an original scene. The nonlinear encoding function is the only encoding step that is altered. For

ERIMM RGB, it is desirable both to increase the maximum scene luminance value that can be represented, as well as to reduce the quantization interval size. The size of the quantization interval is directly related to the minimum scene luminance value that can be accurately represented. In order to satisfy both the extended luminance dynamic range and reduced quantization interval requirements simultaneously, it is necessary to use a higher minimum bit precision for *ERIMM RGB*. A minimum of 12-bits/color channel is recommended.

14.3.9 Nonlinear encoding of *ERIMM RGB*

A modified logarithmic encoding is used for *ERIMM RGB*. A linear segment is included for the very lowest luminance values to overcome the noninvertibility of the logarithmic encoding at the dark end of the scale. The encoding was defined such that the linear and logarithmic segments match in both value and derivative at the boundary. In equation form, this encoding is represented by

$$C'_{ERIMM} = \begin{cases} 0; & C_{RIMM} \leqslant 0 \\ \left(\dfrac{0.0789626}{E_t}\right) C_{RIMM} I_{max}; & 0 < C_{RIMM} \leqslant E_t \\ \left(\dfrac{\log C_{RIMM} + 3.0}{5.5}\right) I_{max}; & E_t < C_{RIMM} \leqslant E_{clip} \\ I_{max}; & C_{RIMM} > E_{clip} \end{cases} \tag{14.8}$$

where C is either R, G, or B, I_{max} is the maximum integer value used for the nonlinear encoding; $E_{clip} = 10^{2.5} = 316.23$ is the upper scene luminance limit that gets mapped to I_{max}; and

$$E_t = e/1000 = 0.00271828 \tag{14.9}$$

is the breakpoint between the linear and logarithmic segments, e being the base of the natural logarithms. For a 12-bit encoding, I_{max} is 4095, and for a 16-bit encoding, I_{max}

Table 14.3 Sample scene luminance encodings

Relative luminance	Rel. log luminance	*RIMM8 RGB*	*RIMM12 RGB*	*ERIMM12 RGB*
0.001	−3.00	1	13	119
0.01	−2.00	8	131	745
0.10	−1.00	53	849	1489
0.18	−0.75	74	1194	1679
1.00	0.00	182	2920	2234
2.00	0.30	255	4095	2458
8.00	0.90	NA	NA	2906
32.00	1.50	NA	NA	3354
316.23	2.50	NA	NA	4095

is 65 535. In cases in which it is necessary to identify a specific precision level, the respective notations *ERIMM12 RGB* and *ERIMM16 RGB* are used.

To compute *ERIMM RGB* values, equation (14.8) should be used in place of equation (14.6) in the procedure described above for determining *RIMM RGB* values. Examples of *RIMM RGB* and *ERIMM RGB* encodings for neutral patches at different scene luminance levels are shown in Table 14.3. It can be seen that the range of luminances that can be represented in *ERIMM RGB* is greatly extended relative to *RIMM RGB*.

14.4 IMAGE STATES IN A COLOR-MANAGED ARCHITECTURE

The use of color management systems, such as that developed by the ICC, is becomingly increasingly common in a variety of digital imaging systems. Color management systems are typically built around an architecture where the color response of an input device is characterized using an *input profile* that describes the relationship between the device code values and device-independent color values in some profile connection space (PCS). Similarly, the color response of an output device is characterized using an *output profile* that describes the relationship between the PCS color values and the corresponding device code values needed to produce those color values. The PCS specified in the ICC color management architecture, as well as that in virtually every other color management system, is defined (explicitly or implicitly) to be in a rendered output-referred image state. That is, the PCS is a representation of the colorimetry of an output image (typically a reflection print viewed in a typical indoor viewing environment). As a result, this complicates the use of color management architecture for implementing an imaging system based on the image states paradigm. For example, a traditional input profile cannot be used for an input device that captures scene-referred data if it is desired to convert the image data to the standard scene-referred color encoding (i.e., *(E)RIMM RGB*). This is because the output of such a profile would be PCS color values in a rendered image state. As discussed above, the process of rendering an image from a scene image state to a rendered image state will typically involve an irreversible loss of information. As a result, it would not be possible to combine a device-to-PCS profile with a PCS-to-*RIMM RGB* profile to transform the image into *RIMM RGB* without seriously compromising the quality of the resulting image due to the information loss.

However, this does not mean that traditional color management architectures must be discarded altogether in order to build an imaging system around the image state paradigm shown in Figure 14.2. Rather, it simply means that conventional input/output profiles cannot be used in the imaging chain until the point where the image is ready to be committed to a final output rendering. Fortunately, most color management systems provide for the concept of a 'device link profile' that can be used to bypass the PCS and go directly from an input color space to an output color space. (Typically, such device link profiles would be created by cascading an input profile with an output profile, but this is not a requirement.)

Figure 14.5 illustrates this approach in more detail. Device link profiles are used to transform scene-referred input images into *RIMM RGB*. This includes input devices such

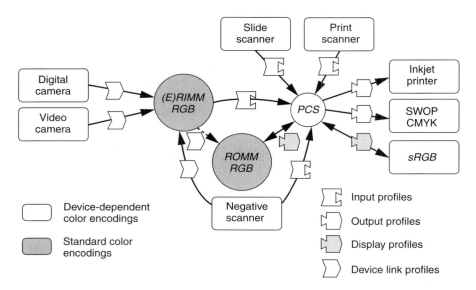

Figure 14.5   Image state architecture using color management.

as digital cameras (when unrendered sensor RGB values are available), as well as color negative film scanners. In this case, not only is *RIMM RGB* used as a stopping point where images can be stored or edited, it also becomes the output color space for the device profiles, effectively serving the role of a 'scene-referred profile connection space'. At the point when it is finally desired to produce an image on an output device such printer or CRT, a conventional input profile can be used to render the *RIMM RGB* image to the PCS with the desired system tone/color reproduction aims. A conventional output profile can then be used to transform the image to the appropriate output device code values.

Conventional input profiles can be used for input devices, such as print/slide scanners and CRTs, where the input images are already in a rendered image state. These input profiles can be combined directly with output profiles to produce an image for a particular output device. Alternatively, the input profile can be combined with a *ROMM RGB* profile to convert the image to *ROMM RGB* for the purposes of storage, interchange or editing. Because *ROMM RGB* is a simple LUT/matrix away from ICC PCS XYZ values, it falls within the class of color encodings that can be represented with a simple display profile.

It should be noted that the input profiles used for rendered images that are intended to be viewed in environments significantly different from the reference viewing environment defined for the PCS must also include appropriate viewing environment transformations. For example, photographic slides are typically intended to be viewed in a darkened room. The color reproduction characteristics of slide films are designed with a higher luminance contrast in order to produce pleasing images in that viewing environment. Therefore, an input profile for a slide scanner must not only account for the colorimetric characteristics of the scanner, but should also include an appropriate transformation that will determine the visually equivalent colorimetry for the PCS reference viewing environment. It may also be necessary for the profile to perform some amount of 're-rendering' of the image to map the extended dynamic range of the slide film into the print-like dynamic range of the PCS reference medium.

During the process of working with images that are stored in the *RIMM/ROMM RGB* color encodings, it will frequently be necessary to preview the image on a video display. In a color-managed system, this can be accomplished by combining the appropriate *RIMM RGB* or *ROMM RGB* profile with a display profile for the particular video display. Because *RIMM/ROMM RGB* are based on a simple additive color space, a simple display-type profile using only a LUT followed by a matrix generally can be used to get to PCS *XYZ*. Likewise, the output profile for the video display would comprise a matrix followed by a gamma-function nonlinearity. For cases where processing speed is a critical concern, these operations can be combined, yielding a simple LUT−matrix−LUT processing chain that could be implemented directly and optimized for speed.

An example of an imaging chain for a representative system utilizing the standard image state architecture is shown in Figure 14.6. The input device for this example is a color negative film scanner. A device link profile can be used to convert the raw film scanner image to a corresponding *ERIMM RGB* image. This profile would account for the characteristics of the scanner, as well as the characteristics of the film used to capture the image. Once the image is in *ERIMM RGB*, many different types of algorithms could be used to operate on the image. For example, a 'scene balance algorithm' (SBA) can be used to automatically color balance the image to correct for any variations in capture illumination and/or film processing, or a digital 'dodge-and-burn' operation could be used to darken the background of a backlit scene. Although *ERIMM RGB* is an appropriate color encoding for most types of image manipulations, it is particularly important that any algorithms that can benefit from the extended dynamic range scene information be applied before the image is rendered to an output-referred state.

After all of the scene-state image manipulations have been applied, the image can be rendered to produce a corresponding rendered-state image. In this example, the image is converted to a *ROMM RGB* representation where further operations will be applied. This conversion can be applied by combining an *ERIMM RGB* input profile with a *ROMM RGB* profile. The *ERIMM RGB* input profile is used to impart the system tone/color reproduction aims relating the scene color values to the corresponding rendered image color values. These aims may be application dependent. For example, consumer photographers generally prefer a higher contrast and color saturation look than that preferred by professional portrait photographers. In many cases, acceptable tone/color reproduction characteristics can be achieved by applying a simple tone reproduction curve to the *ERIMM RGB* scene exposure values. In this case, the *ERIMM RGB* to *ROMM RGB* transformation will collapse down to a simple one-dimensional LUT.

Once the image is in *ROMM RGB*, additional rendered-state image operations can be applied. For example, text annotations and a creative border could be added to the image, or the image could be composited with an image from a print scanner, etc. The final *ROMM RGB* image can then be printed by applying a *ROMM RGB* profile and an output profile for the particular output device.

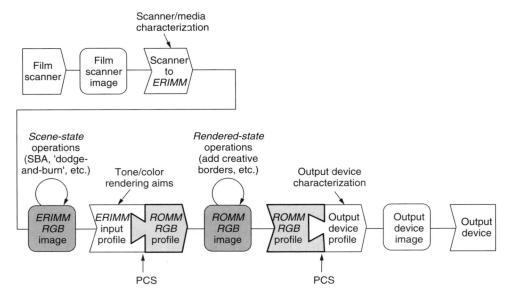

Figure 14.6 Imaging chain example using standard color encodings and color management.

14.5 IMPLEMENTATION WITH JPEG 2000

Historically, many desktop imaging applications have been built around the assumption that the digital image stored in the file is ready to display directly on a CRT. This has caused significant interoperability problems for applications that have attempted to store images with other color encodings. For example, if an application were to open a *ROMM RGB* image and send the color values directly to a video display, the image would appear to be very desaturated because it was encoded using a set of high-chroma primaries. This has made it necessary to use special software to open and/or color-manage images stored in any color space other than video RGB. As a result, such images cannot be used by a large number of applications. The overall situation has effectively made it impractical to use other color spaces for most consumer applications.

JPEG 2000 is a new file storage format that is in the final stages of development. One of the requirements that has been built into the format specification is that all JPEG 2000 compliant file readers must be able to properly decode an image stored in any color encoding that can be defined using a restricted class of ICC profiles. In particular, the supported ICC profile formats include any display-type profile that utilizes a LUT–matrix transformation to get to PCS XYZ. Both the *(E)RIMM RGB* and *ROMM RGB* color encodings can be represented using profiles that fall within this restriction. As a result, images can be stored using these color encodings without sacrificing interoperability. Applications that understand how to manipulate images in these color spaces will be able to do so, while any other application can simply use the attached ICC profile to convert the image to a video RGB color space (e.g., sRGB) or some other color space that it understands.

REFERENCES

1. Eastman Kodak (1991) *KODAK Photo CD System – A Planning Guide for Developers*, Eastman Kodak Company, Rochester, NY.
2. IEC (1999) *Multimedia Systems and Equipment – Colour Measurement and Management – Part 2-1: Colour Management – Default RGB Colour Space – sRGB*, IEC 61966-2-1.
3. ANSI (1995) *Graphic Technology – Color Characterization Data for Type 1 Printing*, ANSI/CGATS TR 001.
4. Hunt, R. W. G. (1995) *The Reproduction of Colour*, Fifth Edition, Fountain Press, Tolworth, UK.
5. Giorgianni, E. J. and Madden, T. E. (1998) *Digital Color Management: Encoding Solutions*, Addison-Wesley, Reading, MA.
6. International Color Consortium (1999) *File Format for Color Profiles*, Specification ICC.1A: 1999–04.
7. Spaulding, K. E., Woolfe, G. J. and Giorgianni, E. J. (2000) Reference Input/Output Medium Metric RGB color encodings (RIMM/ROMM RGB), In *IS&T's 2000 PICS Conference*, pp. 155–163.
8. Spaulding, K. E., Woolfe, G. J. and Giorgianni, E. J. (2000) Image States and Standard Color Encodings (RIMM/ROMM RGB), In *Eighth Color Imaging Conference: Color Science and Engineering: Systems, Technologies, Applications*, pp. 288–294.
9. Borg, L. (2000) *The Profile Connection Space*, ICC Votable Proposal Submission 19.10.
10. *Interpretation of the PCS, appendix to Kodak ICC Profile for CMYK (SWOP) Input*, ANSI CGATS/SC6 N 254, June 3, (1998).
11. *Basic parameter values for the HDTV standard for the studio and for international programme exchange*, Recommendation ITU-R BT.709 (formerly CCIR Recommendation 709).

Engineering color at Adobe

James King

15.1 INTRODUCTION

This chapter is a mix of part history, part tutorial, and part more serious color issues. In all cases it provides the view of an Adobe Systems Incorporated employee as the desktop color management systems were introduced and have evolved.

15.2 THE EARLY BLACK AND WHITE PostScript® DAYS

The first significant product produced by Adobe® Systems Incorporated was the PostScript software for the first Apple® LaserWriter® printer in January 1985. PostScript was invented and developed by Adobe and this was the first PostScript printer to enter the market. By the end of the century Adobe, in conjunction with its partner companies, had developed thousands of different PostScript products.

Apple's Macintosh computer (the Mac), the Apple LaserWriter printer based upon a Canon® laser print mechanism and using Adobe's PostScript software, together with the PageMaker® application program produced by the Aldus® Corporation, are credited

Colour Engineering, edited by P.J. Green and L.W. MacDonald.
© 2002 John Wiley & Sons Ltd.

with creating the *desktop publishing* revolution. In 1985 for about US $9000 a person could own a complete system. Of course, prices came down fairly rapidly not long thereafter, primarily due to the drops in the cost of RAM. The financial relationship between Apple Computers and Adobe Systems was that Adobe would receive a royalty payment from Apple for each LaserWriter that Apple sold. It turned out that the Apple Laser-Writer was a very popular product and Adobe shared in its financial success, which helped Adobe become a healthy start-up without repeated trips to the venture capitalists. This royalty model intrigued the financial community and has continued to be the one favored by Adobe when delivering its PostScript software into products sold by other companies.

Of course, the first Apple LaserWriter was a black-and-white laser printer. Adobe's early venture into products that support color came on two related fronts with an unusual twist. In 1985 Adobe signed an agreement with the Linotype® Corporation to produce the Linotronic® 300 phototypesetting machine, the first typesetter to include PostScript software. This machine accepted PostScript output from Mac applications like Aldus PageMaker and produced high-resolution film output that was used to make plates for printing presses for black-and-white runs and, interestingly enough, also for color runs.

In 1987 Adobe introduced Adobe Illustrator®, an application program for graphic artists that also ran on the Macintosh computer. It was the first professional quality application that allowed graphic artists to create detailed, precise and elaborate drawings on the computer. They loved the ability to make instant changes in their drawings without the limitations of paper and drawing boards. With Illustrator, the artist easily could provide multiple choices from a base drawing to present to their customers and this was something not reasonable to do with traditional methods.

The unusual twist, noted above, comes from the fact that the output on the Linotronic was black-and-white film but when used to create plates for the cyan, magenta, yellow and black (CMYK) inking stations of a printing press would ultimately produce color output. So PostScript software, very similar to that used for the black-and-white Apple LaserWriter, was used in the Linotronic to ultimately produce color output on a press. The early adopters found it exciting and valuable to produce draft output on their Apple LaserWriter and then make film and plates for a press run on the Linotronic, all from the same PostScript job and producing identical results.

Adobe Illustrator had a companion product called Adobe Separator that would take the single composite color PostScript output page that included operators for color, and produce a four-page PostScript file that would produce the black-and-white film for the C, M, Y and K plates. In fact, if the press was to have another special ink station with a 'spot ink' it could also be identified in Adobe Illustrator and film and plates made for it as well. We are going to discuss this in more detail in the next section because it provides some insight into later developments. This separation technique was pioneered by Aldus in Aldus PageMaker and remains in widespread use today.

The first high-quality PostScript composite color output device was a 35 mm film recorder from Agfa®, which came out in 1989. At about the same time the first PostScript composite color desktop product, the QMS® ColorScript® 100, a thermal-transfer desktop printer using a multicolor banded film ribbon, was also released.

15.3 A LITTLE BIT ABOUT POSTSCRIPT

We want to introduce enough about PostScript so that the context of our comments on how it handles color can be understood. PostScript is a programming language specially designed to create page images from simple primitive operations for text, graphics and images. It is imagined that a PostScript program is creating a drawing a step at a time on a large imaginary surface (like a sheet of paper). An important aspect of PostScript is its *device-independent* design. The normal PostScript program does not use the details of the device, e.g., the number and kind of pixels. The imaginary surface also has a coordinate system used to locate positions on the surface and to denote the size of things, again independent of the ultimate output mechanism's properties. With PostScript, text is represented as strings of characters referencing 'outline' fonts, graphics are represented as lines and curves creating 'paths' which are either 'stroked' or 'filled', and pictures are supplied as sampled image data that can rotated, scaled, and halftoned by the PostScript processor. A page image is not supplied to a PostScript output device but rather a programmatic description of each object and how to draw it on the page is sent to the device as a PostScript program. The first Apple LaserWriter included all of these advanced features. PostScript is and always has been well documented in a series of Reference Manuals authored by Adobe and published by Addison-Wesley Publishing Company [1].

PostScript programs establish various settings, collectively call the *graphics state*, and then use drawing primitives that operate under control of those settings. For example, to output the text string 'Hello' in the center of a page, one might use the PostScript program:

```
%%!PS-Adobe-3.0
/Helvetica findfont 24 scalefont setfont
260 600 moveto
0.0 setgray
(Hello)show
showpage
```

The first line is a comment (indicated by starting with %) and identifies this program as obeying the rules for PostScript, 3rd Edition [1]. The second line selects the font with the name `Helvetica` (`findfont`), then scales that font to make 24-point characters (`scalefont`) and then sets it into the current graphic's state (`setfont`). The 'Post' in the name PostScript comes from the fact that the operators occur *after* the things that they operate upon and are stacked and retrieved from a 'push-down stack' similar to a stack of plates in a cafeteria. For example the next line, `260 600 moveto`, stacks the number 260 followed by the number 600 and then executes the `moveto` operator which takes the two operands off the top of the stack and moves the current point of attention to the page coordinates (260, 600). The default coordinate system for the imaginary drawing surface has its origin in the lower left corner and has units of 1/72 of an inch. The PostScript program can change this.

The operator `setgray` operating on the operand 0.0 sets the color to 100% black. In this case `setgray` accepts any number that is between 0.0 (black) and 1.0 (white). It is a *device* color setting, which means that the device is directly controlled by this value with no intervening conversion, normalization or calibration taking place.

Next, the operator `show` is the text imaging operator. It takes the preceding string of text enclosed in '()' and images it at the current position on the page in the current font and using the current color. And finally, the operator `showpage` indicates that all imaging for this page has been completed and the page can be output.

If we wanted the word 'Hello' to appear as 100% red using whatever means available to the device we might code the example as:

```
%%!PS-Adobe-3.0
/Helvetica findfont 24 scalefont setfont
260 600 moveto
1.0 0.0 0.0 setrgbcolor
(Hello)show
showpage
```

Here with the operator `setrgbcolor` we have used the device-specific RGB color space and selected 100% of the first component (Red) and 0% of the other two (Green and Blue). And if we wanted strict control over a CMYK device we might code this example as:

```
%%!PS-Adobe-3.0
/Helvetica findfont 24 scalefont setfont
260 600 moveto
0.0 1.0 1.0 0.0 setcmykcolor
(Hello)show
showpage
```

Here using the `setcmykcolor` operator we have used the device-specific CMYK color space and selected 100% of the Magenta, 100% of the Yellow, and 0% of Cyan and Black. Overprinting of translucent Magenta and Yellow ink in a subtractive model will produce a red color.

All three of these examples are using device-dependent color spaces. There is no attempt to use any color management, and the reds produced on different devices using the last two examples will differ according to how those devices respond to signals to produce full Red or full Magenta and full Yellow overprinted.

All of the three examples above can be submitted to any PostScript device and reasonable things will happen. More specifically, if we submit the last example to a PostScript based system displaying its results on an RGB display the device CMYK directives will be converted to RGB signals to send to that display. Likewise, if we send the second example to a CMYK driven printer the device RGB will have to be converted to CMYK printer directives. Since there are no definitions for these device-dependent color spaces and they change with each device, these conversions have nothing much to work with, so very naive conversions are used. The complete conversion from RGB to CMYK is as follows, where *BG* (*k*) and *UCR* (*k*) are invocations of the user-defined black-generation and undercolor-removal functions, respectively:

$$c = 1.0 - \text{red}$$

$$m = 1.0 - \text{green}$$

$$y = 1.0 - \text{blue}$$

$$k = \min(c, m, y)$$

$$\text{cyan} = \min(1.0, \max(0.0, c - UCR(k)))$$

$$\text{magenta} = \min(1.0, \max(0.0, m - UCR(k)))$$

$$\text{yellow} = \min(1.0, \max(0.0, y - UCR(k)))$$

$$\text{black} = \min(1.0, \max(0.0, BG(k)))$$

Conversion of a color value from CMYK to RGB is a simple operation that does not involve black generation or undercolor removal:

$$\text{red} = 1.0 - \min(1.0, \text{cyan} + \text{black})$$

$$\text{green} = 1.0 - \min(1.0, \text{magenta} + \text{black})$$

$$\text{blue} = 1.0 - \min(1.0, \text{yellow} + \text{black})$$

Conversion from device RGB to device Gray uses the NTSC video formula used in American television sets:

$$\text{Gray} = 0.3 \times \text{red} + 0.59 \times \text{green} + 0.11 \times \text{blue}$$

Similar formulae are used for the rest of the conversions (see the PostScript Language Reference Manual [1] for more details).

To a color scientist these conversions are hopelessly simplistic but in this particular situation, when dealing with undefined and varying device colors, they have served quite well. As we will see later, more scientifically grounded conversions are provided for more properly defined CIE based color spaces [11, 12].

15.4 PostScript PROCEDURES

One more feature of PostScript is needed to understand fully the color processing that was done in the beginning with PostScript on black-and-white devices – the PostScript procedure or subroutine. The idea can be conveyed by a simple modification to our first example:

```
%%!PS-Adobe-3.0
/setblackink {1.0 exch sub setgray} def
/Helvetica findfont 24 scalefont setfont
260 600 moveto
0.9 setblackink
(Hello)show
showpage
```

The careful reader will have noticed in the previous examples that 100% black is indicated by zero in the device Gray color space whereas 100% red is denoted by 1.0 in the device RGB color space. If this is bothersome we could invent our own procedure, say setblackink, that takes the argument 1.0 to indicate 100% black ink. A definition for such a procedure has been inserted as the second line above and consists of three parts: the name of the procedure to be defined, /setblackink, the body of the procedure as PostScript coding enclosed in '{ }' and the word def that assigns the coding to the name. Here we define setblackink to be the PostScript operations {1.0 exch sub setgray}. Then setblackink is used later with an argument of 0.9. That 0.9 is placed on the pushdown stack and the new setblackink procedure is run which means that the segment of operators within the '{ }' in its definition are executed. So next a 1.0 is placed on the stack. The exch operator exchanges the top two elements of the stack, and the sub subtracts them. Thus we have computed the formula '$1.0 - b$' where 'b' is the value argument to the setblackink procedure (0.9). Then the setgray operator is invoked with a value of 0.1, setting the color to 90% black as desired.

Throughout the examples in the rest of this section there will be occurrences of the bind operator. This is a technical detail which allows a procedure to be immune to any further redefinitions of the operators and procedures that they themselves use. Bind says to solidify the procedure definition against subsequent change and so it is a recommended operator for procedures that are to have a global scope over a whole document. There is a lot more to PostScript procedures [1] but establishing the basic notion is all that we need.

15.5 TYPICAL WORKFLOW FOR MAKING BLACK AND WHITE FILM

Typically a user of Adobe Illustrator and Aldus PageMaker would create colored objects using a CMYK color space. No precise definition of that color space was given, but common practice among companies within a particular workflow allowed usage to implicitly define their CMYK color space. Adobe Illustrator would produce a composite color PostScript file making use of a PostScript color operator setcmykcolor specifying colors in CMYK. If that file was sent to an actual color printer, a composite color page would be produced.

However, instead of sending that file directly to a color printer, the file can be processed by the Adobe Separator program to produce four or more black-and-white page descriptions. Using the fact that PostScript supports the definition of procedures and, in fact, the redefinition of any of the built-in procedures, this could be done by prefacing different redefinitions of the setcmykcolor operator before the page each time it was sent to the film-making typesetter. For example, if we wanted to produce the cyan plate for printing we could put the following definition for setcmykcolor at the beginning of the PostScript program for that page:

```
/setcmykcolor
        {pop pop pop 1.0 exch sub setgray} def
```

The code fragment within the '{ }' first pops off the last three parameters (MYK) that were provided to the setcmykcolor operator, leaving the cyan value as the only thing on the operand pushdown stack. Then 1.0 exch sub subtracts that value from 1 (reversing for the gray colorspace) and sets that shade of gray using the setgray operator. This essentially turns all setcmykcolor operators into ones that set the gray level according to the cyan values presented to the setcmykcolor operator. For example, if the operator within the composite color page was

```
0.3   0.0   0.4   0.1 setcmykcolor
```

the operand stack would have the values (0.3, 0.0, 0.4, 0.1) on it when our new definition of setcmykcolor was executed. The three pops would remove the rightmost values from the stack, leaving the value 0.3 or the cyan value. Doing the 1.0 exch sub will compute the value 0.7 which is then given to setgray to set the proper gray level. To obtain the pages for the other three colors, the original composite color file is run three more times with three other similar redefinitions being done, one for each color. The one for magenta leaves the magenta value for the setgray, the one for yellow leaves the yellow value, etc.

Each setcmykcolor operator in the composite job will always be processed when using a redefinition like the one shown above. This means that even if no black ink is to be used for an object it will still be drawn, but in 'white' or with no ink. This has the effect of creating a 'knockout'. Suppose that we have a black square with a yellow circle imaged on top of it as shown in Figure 15.1.

There cannot be any black underneath the yellow circle or the yellow will not show, since printing inks are translucent. So the black has to have a 'knockout' of the circle as shown in Figure 15.2 where the yellow circle is lifted off so we can see beneath it.

When the black plate is created the yellow circle will still be imaged but with white or no ink. This will knock out the previously imaged black square.

It is possible that this effect is not wanted, but it is desirable for the inks to actually overprint with no knockouts. This can be done, but is slightly more complicated. Adobe Illustrator has an object property called 'overprint' which can be turned on or off for each object drawn. If overprint is on, knockouts under that object are suppressed. Instead of drawing with no ink, this is done by not drawing the object at all, by temporarily redefining

Black square

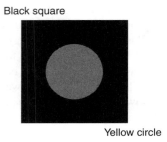

Yellow circle

Figure 15.1   Yellow circle on black square.

Black square with knock out

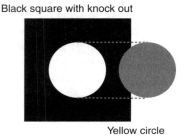

Yellow circle

Figure 15.2   Yellow circle lifted off to see beneath.

all the drawing operators like show to do nothing and thus causing the knockout to be skipped.

A rather elaborate set of conventions for redefining operators and controlling overprinting and knockout have been established even to handle the case when an encapsulated PostScript file (EPS) is included within another file. Adobe documented these conventions in a Technical Report [2]. The introduction of new, more powerful color management features in PostScript Level 2 should have displaced the usage of these color conventions, but they are still in use today.

15.6 POSTSCRIPT LEVEL 2

In early 1989 it became clear that the way in which Adobe was developing the PostScript language and software lacked discipline and was allowing Adobe's product developers to add features and to make product-specific changes to the PostScript language too easily. The original idea was to have a standard device-independent language for describing pages that could be accepted by any device and produce high quality output. This original vision was being threatened, yet there were strong demands to add features to the PostScript language, including color management, image compression, halftoning improvements, and many more. It was also becoming clear that the device-dependent nature of the color operators was in conflict with the desired device-independent goal of the PostScript language. This prompted the idea to standardize on one version of the PostScript language that included all basic features and then to require that all products use this standard version. So a list of all the non-standard features that had been added over the preceding years was added to the list of features that were being asked for, and a plan to develop PostScript Level 2 was developed. Most of this effort began in the Adobe Advanced Technology Group (ATG) and early prototype implementations were completed by that group.

Adobe had received many requests to add more sophisticated color management to the PostScript language. In particular, Kodak had sent Adobe a document that was written in the style of the PostScript Language Reference Manual [1] and it set out in detail the exact language features they thought should be added for color management. During 1989 and early 1990 Adobe designed PostScript Level 2 and added new features for what has come to be known as 'device-independent color' or 'CIE-based color'. The final design did not end up looking much like the proposal from Kodak, but their document certainly spurred Adobe's development. Many other companies also presented requirements and suggestions for enhancing the color processing in the PostScript language to Adobe during this period.

A thorough and specific description of the color management features in PostScript Level 2 can be found in the PostScript Language Reference Manual, 3rd Edition [1], and in a paper by one of the Adobe color architects [3]. The important notion of separating the specification of the desired color space and color from its device-specific rendering is key to the color management in PostScript Level 2. See Figure 15.3. As is done in the International Color Consortium (ICC) architecture [5], PostScript Level 2 defines a profile connection space (PCS) as CIE XYZ, although that term is not used in the documentation. The ICC also defines CIE L*a*b* as an alternative PCS but PostScript only provides for CIE XYZ to be used as the PCS.

Originally Adobe had decided to add only one CIE color space to the PostScript color processing model: CIE L*a*b*. All managed color was to be submitted to a PostScript device using CIE L*a*b* and a draft document of this specification was circulated to Adobe's partner companies. Many, if not most, found this approach unacceptable. The general feeling was that desktop applications processed colors in the display's RGB color space and the personal computer processing power was not then available to do any real-time conversions while generating output for the display. They also felt that converting

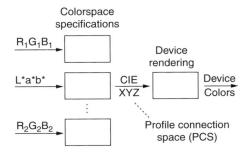

Figure 15.3 Separation of specification and rendering.

from the display's *RGB* to *L*a*b** for printing would add an unnecessary time burden. There was also a belief that the more you convert the colors from one color space to another the more problems you introduce and that delaying conversions and doing them all at once was a good strategy. Most, if not all, of the problems stemming from color conversions come from 'quantization' or truncation of color values into fixed computer units that are really too small. Often the results of a color conversion are stored in the computer as 8-bit quantities and for many color spaces this is just not enough to capture the detailed changes in color values needed. The more compute, store, compute, store cycles that are performed with inadequate storage unit size, the more the real color values get damaged.

The negative feedback on the first design sent the Adobe designers back to the drawing board. A method to support arbitrary RGB color spaces was devised and added to the draft specification. This RGB path first involves three component-wise functions to linearize each of *R*, *G* and *B*, followed by the use of a 3 × 3 matrix to map to CIE XYZ. See Figure 15.4. Thus any linearly additive RGB device can be defined in terms of CIE XYZ. This was deemed to be a superior design, since the user's display RGB would be sent to the PostScript device unchanged and converted once within the device as needed. No color processing would be required within the computer applications.

For example, when coded into the PostScript language the specification of the sRGB [4, 13] color space found in the IEC 61966 Standard would look like:

```
[/CIEBasedABC
<<   /DecodeLMN
         [{  dup 0.03928 le
              {12.92321 div}
              {0.055 add 1.055 div 2.4 exp} ifelse
           } bind dup dup]
     /MatrixLMN [      0.412457      0.212673      0.019334
                       0.357576      0.715152      0.119192
                       0.180437      0.072175      0.950301]
     /WhitePoint [     0.9505        1.0   1.0890]
>>
] setcolorspace
```

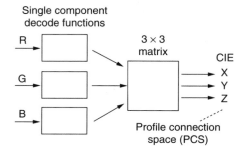

Figure 15.4 Specification of RGB color spaces.

It uses the ITU-R BT.709-2 reference primaries and the CCIR XA/11–recommended D65 white point. The notation '[]' encloses a PostScript language array which contains a list of elements. The notation '<<>>' encloses a PostScript language dictionary which contains pairs of items each having a name (e.g., /WhitePoint) and a value (e.g., the array [0.9505 1.0 1.0890]). In this example the dictionary contains three elements, for the /WhitePoint, the 3 × 3 matrix /MatrixLMN, and the three single-component decode functions. The single-component decode functions are provided in the /DecodeLMN dictionary entry as functions coded in the PostScript language. The coding is enclosed in '{ }' and involves divides (div), exponentiation (exp) and a conditional if-then-else (ifelse) and computes the gamma curve specified for sRGB. The coding dup dup makes two duplicate copies of the preceding function code so that the containing array ends up holding three identical functions, one for each of the R, G and B channels. The interpretations of the MatrixLMN and WhitePoint are more obvious, being arrays of numbers. The terminology 'CSA', which is the acronym for 'Color Space Array', is commonly used to talk about these PostScript language color space specifications. The ICC terminology for the equivalent function is 'source profile'.

So the PostScript Level 2 proposal for color specifications after this change contained this mechanism for defining RGB color spaces, the device color spaces from the earlier days, and CIE L*a*b*. After one more review the final design emerged after it was pointed out that two stages of the single-component adjustments and the 3 × 3 matrix would allow one common mechanism to handle both RGB and L*a*b*. So the explicit L*a*b* interface was dropped and this more complicated common mechanism was introduced. See Figure 15.5. The ability to support L*a*b* with this two-stage mechanism is dependent upon the fact that the user-defined PostScript functions can be used to compute simple functions in the single-component decode functions. Since this one specification schema now was to cover both RGB and L*a*b* colorspaces it was called CIEBasedABC, the ABC referring to the three input channels, be they RGB, L*a*b* or something else.

The CSA for the CIE 1976 L*a*b* colorspace using the CIEBasedABC colorspace mechanism and the CCIR XA/11-recommended D65 white point is shown below. In this definition the a* and b* components are defined to lie in the range −128 to +127 and L* from 0 to 100.

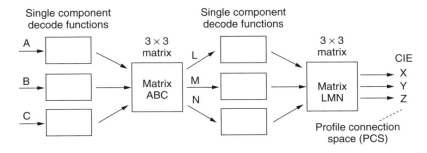

Figure 15.5 Two repeated stages for CIEBasedABC color space specifications.

```
[/CIEBasedABC
    << /RangeABC   [0 100    -128 127    -128 127
       /DecodeABC [
            {16 add 116 div} bind
            {500 div} bind
            {200 div} bind]
       /MatrixABC   [
            1 1 1
            1 0 0
            0 0 -1]
       /DecodeLMN   [
            {dup 6 29 div ge
                {dup dup mul mul}
                {4 29 div sub 108 841 div mul}
            ifelse
            0.9505 mul} bind             .
            {dup 6 29 div ge
                {dup dup mul mul}
                {4 29 div sub 108 841 div mul}
            ifelse} bind
            {dup 6 29 div ge
                {dup dup mul mul}
                {4 29 div sub 108 841 div mul}
            ifelse
            1.0890 mul} bind]
       /WhitePoint [
            0.9505 1.0 1.0890 ]
    >>
] setcolorspace
```

This PostScript CSA codes the following functions:

1. DecodeABC computes the three values:

$$a = (L^* + 16)/116$$

$$b = a^*/500$$

$$c = b^*/200$$

2. Then the MatrixABC takes those three values and computes three new values:

$$l = a + b = (L^* + 16)/116 + (a^*/500)$$

$$m = a$$

$$n = a - c = (L^* + 16)/116 - (b^*/200)$$

3. Next the `DecodeLMN` computes:

$$X = X_W \cdot g(l)$$

$$Y = Y_W \cdot g(m)$$

$$Z = Z_W \cdot g(n)$$

where the function $g(x)$ is defined as

$$g(x) = x^3 \qquad\qquad x \geqslant (6/29)$$
$$g(x) = (108/841) \cdot (x - (4/29)) \quad x < (6/29)$$

and X_W, Y_W, Z_W are the white point values (0.9505, 1.0, 1.0890).

4. This computation should be recognized as the formulae defining CIE L*a*b*.

The PostScript Language Reference Manual [1] explains this CSA specification schema thus: '*The formulation of **CIEBasedABC** color spaces models a simple zone theory of color vision, consisting of a nonlinear trichromatic first stage combined with a nonlinear opponent-color second stage. This formulation allows colors to be digitized with minimum loss of fidelity, an important consideration in sampled images.*' This model covers all linear additive RGBs including all television standards, L*a*b*, Yuv, Ycc and several other similar spaces, but not L*u*v*.

As noted earlier, the PostScript language is a programming language and a PostScript language program is executed step by step by a PostScript software interpreter. A 'graphics state' is maintained holding the current values of most important settings, such as which font to use for text operations, how wide lines should be, etc. With the introduction of the color management features of PostScript Level 2, the color space and the color chosen within that color space became values in the graphic state. To establish the color space used by subsequent imaging, the [...] `setcolorspace` operator was introduced. Its single operand is a CSA as shown in the above examples. Once a color space is chosen then a color within that color space is established by using the `setcolor` operator. This operator takes as many operands as there are components in the current color space. So, for example, if the current color space is a three-component RGB color space, then a yellow color (combination of 90% red and 80% green) would be set in the graphics state by the following operation:

```
0.9   0.8   0.0 setcolor
```

Before the color management features of PostScript Level 2 were introduced, `setgray`, `setcmykcolor` and `setrgbcolor` were the color operators available and they set the device-dependent colorspace and established a value within that colorspace simultaneously. The new, separate operators are required for setting device-independent colorspaces.

The PostScript language can express programs for what are often called 'compound' pages: pages containing a variety of text, pictures and graphics objects collected together.

Since the `setcolorspace` operator can be issued at any time, each object in the compound document can have its own source colorspace. If the philosophy of the software that creates the compound document is to delay color conversions as long as possible, it will just propagate the color space found with each object as it is included into the compound document. This means that one can find PostScript page descriptions with many, perhaps dozens, of source color spaces on a single page. As each `setcolorspace` operator is encountered the current color space is redefined.

Note that at this point (1990) there was no facility for device-independent CMYK in the specification. In fact, this feature was not introduced into PostScript until 1995. The reason that there was no device-independent CMYK originally introduced into the PostScript color management was because of Adobe's heavy experience with device CMYK support. The belief and experience was that those designers and pre-press professionals were actually controlling the printing press by very specifically manipulating the CMYK values and that any attempt to get into the path between those users and the press would be of little to no value. Some of us felt that the phrase 'device-independent CMYK' was an oxymoron. In fact, even though Adobe did introduce the CIEBasedDEFG colorspace specification in 1995 to handle device-independent CMYK and RGB colorspaces that are not linearly additive, many troublesome aspects of the device nature of CMYK are still with us today. We will discuss these issues in the next section on 'Handling black in CMYK workflows'.

The CSA for the CIEBasedDEFG class of color spaces builds from the CSA for CIEBasedABC described in detail above. An initial table lookup is added before the normal CIEBasedABC processing to allow nonlinear and four component color spaces to be reduced to a space representable as CIEBasedABC. See Figure 15.6 for the schematic. CIEBasedDEFG and the three component version, CIEBasedDEF, provide support for calibrated CMYK, nonlinear RGB, L*u*v* and many other color spaces.

The complete picture of how the PostScript language manages color today is summarized nicely in the PostScript Language Reference Manual [1] in Figures 4.5 and 4.6 on pages 212–213 and includes facilities for handling spot colors, indexed access, patterns as color, and controlling separations.

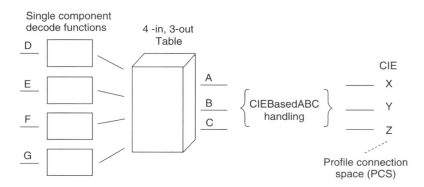

Figure 15.6 Pre-stage for CIEBasedDEFG color space specifications.

15.7 HANDLING BLACK IN CMYK WORKFLOWS

In theory a three-component color space using cyan, magenta and yellow translucent inks (CMY) in a subtractive color system can represent a wide range of colors, including black. However, due to the fact that sometimes the overprinted black has a color cast or is too translucent and/or involves too much ink on the paper (three layers) or is too expensive, a solid black ink is usually introduced to create a four-component colorspace. This redundancy and the special uses to which that black ink is put are a huge source of problems when using color management involving CMYK.

The typical color management system will convert all colors to a Profile Connection Space (PCS) which is usually CIE XYZ and/or CIE L*a*b*, both of which are three-component color spaces. Thus we have four color components reduced to three and then perhaps back to four. If the original CMYK colors exploited the redundancy inherent in a four-color system, it is lost when it is reduced to three components.

Compounding the situation is the confusion of several different problems involving black. Let's look at several of them.

1. **Rich black**. Even the black ink has a considerable translucency and if a very opaque strong black is desired, printers will often print cyan ink underneath the black. This gives the area more ink density and creates what is often called 'rich black'. An extra complication occurs when using rich black because of the press mis-registration between the black and cyan ink layers. To handle this, the cyan is usually 'choked' to a slightly smaller image area so that it all lies comfortably under the black area within the mis-registration tolerances.

   Handling more than one kind of black in most color management systems is extremely difficult, if not impossible.

2. **Knockout**. Again because of the fact that on most presses the exact position of the inking of the paper at each successive inking station can only be controlled within certain tolerances, one has to constantly compensate for the mis-registration errors that inevitably occur. If one were printing translucent yellow lettering on a black background, the yellow would hardly show because the black shows through. So one has to 'knock out' the text in the black background to allow the yellow ink to have the paper as its background. To allow for the mis-registration the yellow lettering is usually 'spread' to be slightly larger than the lettering knocked out of the black background. It looks better to have two blacks, the single black ink and the black ink with yellow over it, than it does to have the lettering reveal the white of the paper in places where there is mis-registration.

   Fine, but what if the colors are reversed and one wants to print rather small black lettering on a yellow background? Since the black on top of the yellow will be an even stronger black, there is no need to knock out the lettering on the underlying yellow layer. Doing the knockout when small letter sizes are involved can often cause more problems than it solves. So a general technique is to do no knockout if black ink is used for small letters over a light colored background.

   To a naïve colorimetric system, black over yellow or black over white paper is just black. Yet these distinctions are very important to the printer and tasteful control over these situations is desired. Any reduction to a three-component PCS loses this information and hence the control.

3. **Black generation**. For picture halftoning we also face issues with how to get the right black appearance within a colored picture given that we can create black by overprinting cyan, magenta and yellow or by using black ink or, in fact, using some combination. This is most easily discussed if we assume that the picture has been converted to CMY values initially. Practitioners have long found that substituting amounts of black ink in pixels where all three of CMY occur will often enhance the crispness of the picture and bring out interesting details. So various methods are employed for determining where to do 'black generation' and how the CMY components should be reduced or eliminated in those areas using 'under color removal'. This is far from a science since the colors are sometimes overprinted and sometimes not when halftoning. Halftoning depends upon the viewer being a sufficient distance from the surface that averaging effects in the visual system take effect. For the best known attempt to model this situation, see Neugebauer's famous paper [8].

If a person has carefully edited a digital picture and done custom black generation, then they will be very unhappy if subsequent processing by a color management system treats black as black and changes the careful balance that was established between the black ink and the overprinting of cyan, magenta and yellow.

4. $r = g = b = 0$ **means black**. When originating material on an RGB display screen, the underlying software most commonly uses an RGB colorspace model. If one creates black text on the screen it is easily represented in RGB with each component set to zero ($rgb = (0, 0, 0)$). Problems occur when we want to print on a CMYK device. This is a variation on the black generation problem. For the same reasons mentioned above, printing small-sized black text using overprinted cyan, magenta and yellow is usually not good. Using solid black ink, if it is available, gives much better results. So during the RGB to CMYK conversion, the black-generation/under-color-removal steps are required to generate solid black ink. Well, what if $r = g = b$ but they are not equal to zero? One might believe that this should be a level of gray and hence some percentage of black ink should be used. This gets trickier since we need to establish the relationship between percentages of equal amounts of RGB and black. In addition, it may easily be true that for some RGB colorspaces equal amounts of red, green and blue does not represent a gray color.

The situation is further aggravated if composite material containing text, graphics (with shading) and images is merged before being processed by color management software. The black treatment for one kind of material may be exactly the wrong treatment for another kind, and if the kinds are intimately mixed in a fancy overlaid design it is impossible.

The key to handling this situation is to allow color space switching within a screen or page description and to switch to a monochromatic gray color space when we know that we want solid black ink used to render a particular object. If that information is carried throughout the workflow then there is no guessing. Many operating system vendors have failed to discover the subtleties of this situation and only offer RGB support.

Both the CIEBasedDEFG color space support introduced into PostScript Level 2 for CMYK and the design of the ICC color architecture [5] map CMYK into either CIE XYZ or CIE L*a*b* as it is transformed into some other colorspace. This means that the important redundancy used in the CMYK color space is lost during these conversions. Imation® Corporation [6] has recently introduced a method for Apple Colorsync® that uses ICC profiles but does not lose the black information during the conversions. This goes a long way to solving the first three problems but does not address the fourth $r = g = b = 0$ problem.

15.8 DESTINATION PROFILES IN PostScript

The reader is referred back to Figure 15.3, a diagram showing the separation of input color space specification and output color rendering. Our discussion this far has concentrated on the input color space specification or the source profiles in the ICC terminology, and we have learned that these structures in PostScript are called color space arrays or CSAs. What about the rendering or destination profiles? In PostScript these are called color rendering dictionaries (CRDs), have a similar but different structure from the CSAs, and play the role of destination profiles of an ICC system. The setcolorrendering operator accepts a CRD as its operand. For PostScript output devices the CRDs are the means for color calibration. Devices usually come with default CRDs installed in their ROM or initial state, but a user can reset the CRDs in effect by use of the setcolorrendering operator. Some printer drivers offer the service of downloading CRDs before a print job. Unlike CSAs, which may be set with the setcolorspace operator many times during the processing of a single page, CRDs are associated with the output device and only change when something affecting the way that device generates color changes.

Figures 15.7 and 15.8 show the schematic for color rendering dictionaries. Figure 15.8 is an elaboration of the leftmost box shown in Figure 15.7 that represents the step of adjusting the colors component-wise to compensate for differences in white and black points between the source and destination. Note that many of the functions denoted by boxes in these diagrams may reduce to 'do nothing' steps if they are defined as identity

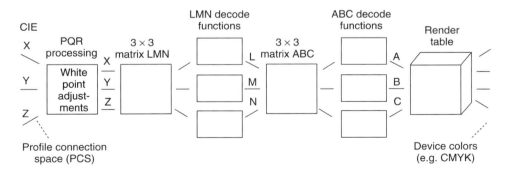

Figure 15.7 Color Rendering Dictionary (CRD).

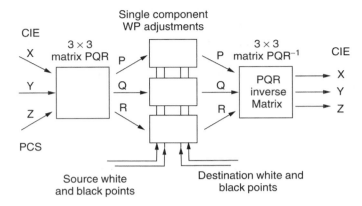

Figure 15.8 White/black point adjustments in a CRD.

functions. For example, if one is producing a CRD for an additive linear display, the table lookup (Render Table) is not needed and can be omitted.

As shown in Figure 15.7 the colors represented in the PCS first undergo adjustment for white and black point differences, then are possibly converted into another color space more suitable for indexing into a table lookup, and after that lookup are ready for presentation to the device in its native color space. The conversion to an alternate color space for table lookup is the inverse to the two-phase conversion done in a CSA: a 3 × 3 linear transformation followed by a component-wise adjustment done twice (see Figure 15.5).

The adjustments for white and black point differences shown in Figure 15.8 are noteworthy because the ICC design departed from this feature found in PostScript. The PCS CIE XYZ values first undergo a linear transformation defined by the 3 × 3 MatrixPQR. This allows for the white and black point corrections to take place in a color space other than CIE XYZ. The inverse of the MatrixPQR then returns the adjusted values to CIE XYZ to continue down the pipeline.

The actual white and black point adjustments are done independently a component at a time. The white point and black point values for the source are provide by the CSA and the white point and black point values for the destination are provided in the CRD. The CRD contains three arbitrary functions written as PostScript procedures and represented as the three central boxes in Figure 15.8. These procedures perform the white/black point adjustments as desired by the creator of that CRD. For each component of the PQR color space, five inputs are sent to the PostScript procedure defined for that component which then provides the adjusted component value as output. This can be seen more clearly below in the example coding for a CRD.

```
<</ColorRenderingType 1
  /MatrixLMN
         [   3.2410-0.9692  0.0556
```

```
                  -1.5374 1.8760-0.2040
                          -0.4986    0.0416 1.0570 ]
/EncodeLMN
   [    {dup -0.018 le
         {neg 0.45 exp -1.099 mul 0.099 add}
         {dup 0.018 lt
               {4.5 mul}
                     {0.45 exp 1.099 mul 0.099
               sub}ifelse
         } ifelse
     } bind
   dup
   dup
]
/RangeLMN
       [    -0.3751.125-0.125   1.0-0.25    1.0 ]
/MatrixABC
     [ 1.0    0.0    0.0
       0.0    1.0    0.0
       0.0    0.0    1.0 ]
/EncodeABC
       [ { } bind { } bind { } bind ]
/RangeABC
       [    -0.3751.125 -0.125    1.0-0.25    1.0 ]
/WhitePoint
     [ 0.9505    1.0    1.0890 ]
/BlackPoint
     [ 0.0    0.0    0.0 ]
/MatrixPQR
     [ 0.40024   -0.22630   0.0
       0.70760    1.16532    0.0
      -0.08081    0.04570    0.91822 ]
/RangePQR
   [ -0.125 1.375 -0.125 1.375 -0.125 1.375 ]
/TransformPQR
   [ { 4 index 0 get div 2 index 0 get mul
         4 { exch pop } repeat
       } bind
       { 4 index 1 get div 2 index 1 get mul
         4 { exch pop } repeat
       } bind
       { 4 index 2 get div 2 index 2 get mul
         4 { exch pop } repeat
       } bind ]
/RenderTable
```

```
[ 25 19 21
[ %< table values would be here >%]
  3
          { 1.0 exch sub } bind
          { 1.0 exch sub } bind
          { 1.0 exch sub } bind
]
>>
```

Understanding the contents of this CRD is keyed off Figures 15.7 and 15.8. The Range entries define what the mathematically valid range of the arguments are at that point in the flow. The TransformPQR entry contains three PostScript procedures which perform the white/black point adjustments on the P, Q, and R values respectively. In this case each procedure does the identical transformation on its value, which is a simple von Kries [9] chromatic adaptation.

The first procedure in the TransformPQR is:

```
{ 4 index 0 get div 2 index 0 get mul
      4 { exch pop } repeat
} bind
```

When this procedure executes, the following four values are placed on the PostScript operand stack by the PostScript interpreter. For example, the first procedure for adjusting P would receive [X_{Wsp}, Y_{Wsp}, Z_{Wsp}, P_{Wsp}, Q_{Wsp}, R_{Wsp}], [X_{Bsp}, Y_{Bsp}, Z_{Bsp}, P_{Bsp}, Q_{Bsp}, R_{Bsp}], [X_{Wdp}, Y_{Wdp}, Z_{Wdp}, P_{Wdp}, Q_{Wdp}, R_{Wdp}], [X_{Bdp}, Y_{Bdp}, Z_{Bdp}, P_{Bdp}, Q_{Bdp}, R_{Bdp}], and P, where P is the top of the stack and is the first color component of source color processed to this point, and the other four parameters are arrays containing all P, Q, and R components for each of the white point of the source, the black point of the source, the white point of the destination and the black point of the destination, respectively. The first, P component of each of these arrays is shown as X_{Wsp}, X_{Bsp}, X_{Wdp}, and X_{Bdp}. These arrays actually contain six values; the first three are the components for the white or black point as CIE XYZ values and the next three have those same three values but converted into the PQR colorspace for convenience.

The 4 index reaches into the stack and copies the fourth argument to the top of the stack. That fourth element is the source white point array [X_{Wsp}, Y_{Wsp}, Z_{Wsp}, P_{Wsp}, Q_{Wsp}, R_{Wsp}]. Next, 0 get replaces that array with its 0th element, which is the value X_{Wsp} itself. So the top of the stack now contains P, X_{Wsp} and the div then divides, yielding (P/X_{Wsp}). Then 2 index 0 get similarly obtains the value X_{Wdp} and the stack contains (P/X_{Wsp}), X_{Wdp} which are then multiplied, yielding ((P/X_{Wsp}) × X_{Wdp}). The coding 4 {exch pop}repeat cleans all but that computed value from the stack (trust me on this one) so the overall procedure returns with ((P/X_{Wsp}) × X_{Wdp}) on the stack. In a similar way the Q and R procedures compute corresponding values. They differ in that the one for Q uses 1 get and the one for R uses 2 get which reach into the arrays further to get the corresponding Q and R values.

The `RenderTable` includes the actual large table which has been omitted for clarity. The table size is indicated by the three numbers 25 19 21 preceding the table and they indicate the size of the table in their respective dimensions. Following the table are three 'T' single-component transform procedures which are used to adopt the table outputs to any device peculiarities. In this example they invert the values in the range 0–1. That is, 0 becomes 1 and 1 becomes 0 and similarly for the other values between 1 and 0. Even though the T functions could be absorbed into the render table itself, it is often useful to allow the interpolation computed using the render table entries to work in other than the peculiar device space. The interpolations may be more accurate.

15.9 DIFFERENCES BETWEEN ICC PROCESSING AND PostScript PROCESSING

The ICC profiles are defined to be for D50 white points. If you wish to make a profile for a device that operates at a white point other than D50, then you have to make the matrix in the profile not only convert from the device's color space to CIE XYZ but also make any white point conversions necessary at the same time. So the matrix contains two transformations combined, one for color space differences and one for white point differences. The same is true for the lookup table transformations provided for in an ICC profile; they must convert not only color spaces but white points to/from D50 as well.

The PostScript model keeps these two conversions separate and allows the white point conversions to do both source and destination conversions at the same time, as shown in the above example of PQR processing defined in the CRD. The matrices and/or table lookups convert only color spaces. The PCS in PostScript does not have a defined white or black point. The values are processed across the PCS from the source white/black points into the destination white/black points during the transformation. So the PCS conversions may have a source white point of D65 and a destination white point of 9500 K.

Another source of difference between PostScript color management and that of the ICC is in handling 'rendering intents'. Typically, display RGB colorspaces have a larger gamut than CMYK printing colorspaces. When converting for printing, a 'gamut compression' must take place to map the richer RGB gamut into the less rich CMYK one. Unfortunately there is no single gamut compression technique that is satisfactory for all purposes [10]. For example, experience has shown that the relative differences among the colors is important to maintaining the 'look' of a picture, even more so than maintaining the individual colors exactly. For this reason compression methods that shrink all colors, not just those that are out of gamut, do better on pictures than more conservative methods that alter fewer of the colors.

However, if one is processing logos or catalog colors, one would like any colors that can be matched in RGB and CMYK to be matched during such a conversion. The ICC has determined a basic four categories of gamut compression that are common: relative colorimetric, absolute colorimetric, perceptual, and saturated. The first two are for processing logos, etc.; perceptual is for processing pictures; and saturated is for business graphics or other cases where a solid ink color is preferred over a mixing of inks that may be more accurate colorimetrically.

The color conversions for each of these different intents require different matrix and table values for the output profiles. In addition, if there are many objects on a given page and each requires a different intent processing, one has to be able to switch to different output profile values per object.

In PostScript this is handled by having four different CRDs available within the output device or previously downloaded to it, one corresponding to each of the four rendering intents. In essence, the output device is characterized four times, one for each of the four gamut compression techniques. Now a problem arises when one is trying to maintain device-independent processing as has been the goal of the PostScript designers. The kind of processing is determined by the kind of object in the input, be it a picture, a logo, or a business graphic. Yet the processing to reflect this is a property of the output device. An output device with a very restricted gamut will require different output profiles than one with a gamut more close to RGB, even in addition to their color differences. The output profiles, CRDs, cannot be included with the input because then that input would only be suitable for the single output device.

A trick is employed. The input simply indicates with each object which of the four kinds of processing it should be treated with. This four-valued indicator is called the 'rendering intent'. In PostScript the operator for establishing the rendering intent is called `findcolorrendering` and it takes one parameter, the rendering intent. This operator will attempt to find the proper Color Rendering Dictionary (CRD) for this intent. Since the proper rendering also depends upon the color of the paper in use, whether the output is to paper or transparency material, the halftoning technique employed, etc., the `findcolorrendering` operator is charged with the more general task of finding a CRD to match all of the current printing conditions. Its definition can also be downloaded so that a custom algorithm can be used to find the proper CRD for any given situation, include a rendering intent. A general resource management system is available in PostScript to assist in keeping many CRDs and their properties available to choose from, within the output device.

The ICC architecture similarly uses a rendering intent and one supplied by the input, either by the input profile or more preferably by the application that is using the ICC system. However, it packages all four different processing parameters (matrices and tables) into a single output profile. Thus an ICC output profile really contains four sets of numbers and corresponds to four PostScript CRDs, not one.

15.10 'THE TAIL WAGGING THE DOG'

The Adobe group that defined and built PostScript Level 2 was confident that its release in 1992 heralded the beginning of a device-independent color management era where it would be much easier to produce the same output on widely varying devices and get acceptably similar results. The era was begun, but the speed at which the industry has moved to accept color managed workflows has been very much slower than that group had envisioned. Only in the most recent years has significant work been shifted to take advantage of this advanced function in the PostScript language.

One reason for this gradual adoption is that PostScript language devices are a little like the tail of the dog. The main part of the dog is the applications and printer drivers that generate the PostScript language output. Having all this wonderful color management power in the PostScript language wasn't as immediately profound as anticipated. Another significant factor was an Adobe marketing requirement that PostScript Level 2 devices had to be excellent PostScript Level 1 implementations. That made PostScript Level 2 products readily acceptable in the marketplace but may have considerably delayed adoption of the advanced features of Level 2 because there was no urgency to convert application software to use the features of PostScript Level 2. PostScript is now at the 3rd Edition (Level 3 implementation) but the Level 1 to Level 2 color changes were much more significant than the Level 2 to Level 3 changes.

One other factor that slowed adoption of color management was the device-dependent nature of much of the pre-press work and its strong dependency on CMYK workflows. Even today, mixing device-independent processing and CMYK workflows invites problems for the naïve. And finally, color managed workflows require calibration of both the printing and the display devices, necessitating the development of specialized tools and disciplines for calibration.

15.11 THE FORMATION OF THE INTERNATIONAL COLOR CONSORTIUM

Coming out with advanced color management features in PostScript Level 2 in 1990 positioned Adobe Systems well ahead of the desktop computer, color management adoption curve. We emphasize 'desktop computer' here because sophisticated and successful color management systems had been developed much earlier than 1990 but in the previous floor-standing proprietary system era. Much of what has been done on the desktop systems has been an adaptation of the lessons learned in those earlier systems.

In early 1991, Apple Computers started a color management development effort, which has become the very successful ColorSync® system. It had an initial code name of 'Primavera'. Apple had one-on-one discussions with many companies as it was establishing its plans to roll out the first announcement of ColorSync in January 1993. These discussions were all conducted under non-disclosure agreements, so even though Company X had suspicions that Apple had been discussing ColorSync with Company Y the non-disclosure agreements prevented those Companies X and Y from discussing the topic between them.

The birth of the International Color Consortium (ICC) took place at a Fogra Conference in Munich, Germany in spring 1993 (at least in this author's opinion). A panel session on color was part of the program and key speakers from Adobe, AGFA, Apple, EFI, Kodak, LightSource, Linotype, Logo, Pre-Press Technologies, Scitex and Sun were present. After the program ended, the Fogra hosts invited the panel to a small meeting room to have a beer and to talk seriously about cooperating. Nearly everyone in the room had had one-on-one talks with Apple over the preceding year. With Apple's consent, for the first time in the context of this meeting the companies were able to talk about ColorSync and their interests in color management in a much more open fashion. Instead of a wheel topology with Apple at the center it was more of a group discussion. Earlier, during the panel discussion, the author suggested that 'We are likely all keeping the same secrets'.

Due to the proprietary and vertical nature of the previous generation of floor-standing color equipment and due to the photographic industry's similarly proprietary history in color, it had been extremely difficult to get any information to cross company boundaries. Many companies felt that they held valuable intellectual property that should never be revealed to 'the competition'.

The more modern view, that had a strong beginning in Munich in 1993, is that all companies are in this together and none will succeed unless all succeed. The key difference with the desktop world is that the complete system that the user deploys comes from several, if not many, companies. In the past with more vertical systems most, if not all, of the parts came from one company. Instead of the system being designed behind the closed doors of Company X, it is now being designed piecemeal by the introduction of new product pieces and finally within a consortium where all parties can meet and openly discuss the evolving color architecture – the ICC.

After the meeting in Munich several more meetings were held in quick succession at a Seybold Conference, at an ACM Siggraph and at Adobe headquarters in California. Out of this came first the 'Apple ColorSync Consortium', but that didn't last long because Apple had a strong desire to encourage intercompany cooperation and felt that tying the consortium closely to Apple would inhibit it. The Fogra participants pushed for a totally independent organization, which was then named the 'International Color Consortium'. This was a very open and liberal view on the part of Apple for which we are all now very grateful. Not long after, the group made an agreement with Fogra to be the 'technical secretary' and with the NPES to provide an administrative envelope, and this arrangement has worked very well.

Adobe was involved in all of these meetings and was one of the founding members of the ICC. Partly because of the early work with color management in the PostScript Language, Adobe realized that intensive cooperation between companies was essential to building successful color workflows because so many different companies supply the pieces needed to build such a desktop system. The ICC now has over 70 member companies.

Included in the founding members of the ICC were Microsoft, Sun, Silicon Graphics, and Apple. Color management is a very fundamental function and belongs within the operating system (OS) of your computer. In the same way that the OS knows the resolution, bits per pixel and other properties of your display device, it should also know its color characteristics. These companies were the key OS developers and it was important to get them to agree to include ICC processing under the covers. They have all done that.

One of the important things for Adobe was to ensure that the ICC design and the features that had already been incorporated into the PostScript language were compatible, or at least interoperated well. The first release of the ICC Specification corresponded to ColorSync 2.0, and Adobe had considerable influence on many of the changes between ColorSync 1.0 and 2.0 to bring the ColorSync/ICC model closer to the established PostScript model. This was and is important not only for Adobe but for all those using desktop color management together with PostScript devices. There do remain significant differences between the two color management models. Most notably, the ICC standardized on D50 for all profiles and the PostScript architecture didn't (see Figure 15.8 and the associated discussion). As discussed earlier, the CSA corresponds to the ICC source profile but can be based upon any white/black points, and the CRD corresponds to the ICC destination profile and can also be based on any white/black points. While the ICC has a unified representation for both source and destination profiles, the PostScript model has two different representations, one for the source CSA and a similar but different one for the CRDs.

Adobe and Apple have both developed algorithms for converting most ICC profiles to either CSAs and CRDs and vice versa. They add white point adjustments to the matrix values to/from D50 as appropriate and reformat the other information into the proper forms. These were developed so that the PostScript printer drivers could convert ICC profiles when creating PostScript language programs.

15.12 PORTABLE DOCUMENT FORMAT (PDF)

PDF was introduced by Adobe in 1993 as the underlying file format supporting its Acrobat® product line. PDF has a large overlap with the PostScript Language and its color management closely follows what has been described earlier for PostScript. There are two differences worth mentioning. The definitions for PDF were done after a couple of years of experience with PostScript Level 2 and it was decided not to support the double-staged transformation method shown in Figure 15.5. Rather it was decided that a simpler model using only the single stage shown in Figure 15.4 together with a special L*a*b* colorspace invoked by name would be better than the combined model of PostScript that required L*a*b* to be specified by filling obscure numbers and functions into the two-stage model. In addition, at the time, PDF did not have support for arbitrary function computation that is required to define L*a*b* completely in that way. This is what the L*a*b* Color Space Resource (analogous to the PostScript CSA) looks like in PDF:

```
<</ColorSpace
   << /CS1 [ /Lab
         <<
             /Range [ -128  127  -128  127 ]
             /WhitePoint [ 0.951  1  1.089 ]
         >> ]
   >>
>>
```

This specifies the white point to be used when converting to CIE XYZ and it also specifies that the a* and b* values range between 128 and 127. The L* value always ranges between 0 and 100. The formulae needed to process L*a*b* are built-in and no further information is needed.

The second difference is that in a subsequent release of PDF (1.3) the use of ICC source profiles as alternatives to the PDF/PostScript CSAs was allowed. When the CIEBased-DEFG colorspace was introduced into PostScript to handle calibrated CMYK, instead of

adding yet another type of CSR to PDF, ICC profiles were allowed in general and especially for this purpose. Eventually Adobe expects that the use of ICC source profiles in
PDF files will predominate. Note that the differences between CSRs and source profiles
are compensated for as Acrobat processes them. Here is an example of the embedding of
an ICC profile in a PDF file:

```
<</ColorSpace
   << /CS1 [ /ICCBased
      <<
        /N 3
        /Range [ 0 1.0 0 1.0 0 1.0 ]
      >>
      stream
      %(<the actual ICC profile is included here>)%
      endstream
      ]
   >>
>>
```

Here the number of components of the ICC source colorspace is given as $N = 3$ and the
range over which each of those values may vary is in each case 0–1. The actual ICC
profile is embedded into the PDF file as a PDF 'stream' which was invented for this and
other purposes [7].

15.13 COLOR MANAGEMENT IN OTHER ADOBE PRODUCTS

Most of Adobe's products, like Adobe Photoshop®, Adobe Illustrator, Adobe
PageMaker®, Adobe InDesign®, Adobe FrameMaker® and other products, have always
required and implemented color management. If they have data in the CMYK colorspace
and are asked to display it on an RGB display, they have to do a colorspace conversion.
Until the last few years when ICC became well established, these products had their
own built-in methods for doing accurate color conversions primarily between CMYK and
RGB and often fixing on a few particular common CMYK colorspaces. The conversion
from RGB to CMYK together with the separation of each of cyan, magenta, yellow and
black into separate 'pages' is usually called 'color separation' and most of these products
supported such conversions. Since the users of these products are serious production
printing professionals, they demanded and got high quality conversions for these purposes.
The main distinction between what has been done in the past and doing these things using
ICC profiles is the uniformity across products and the ability to use a variety of tools
from many vendors to create and use ICC profiles.

Besides Acrobat/PDF and PostScript Level 2, the most significant Adobe color management event was when in early 1998 Adobe Photoshop 4.0 introduced the use of ICC

profiles as standard, and nearly required, usage for color conversions. Getting color management into Photoshop in 4.0 was a major breakthrough for the advocates of desktop computer color management. And the way in which it was done rather forced everyone using Photoshop to learn to use the features. It is interesting to note that a different approach was taken with Photoshop than with PostScript Level 2. This was not likely a conscious strategy but nonetheless interesting. As noted above, a marketing requirement for the introduction of PostScript Level 2 was strong upward compatible support for previous version of the PostScript language. The Photoshop 4.0 introduction, on the other hand, sacrificed some backward compatibility in favor of converting its user community quickly to learn to deal with ICC profiles and more managed color. It suffered short-term anxiety from customers but passed through the transition relatively quickly.

Another important development with respect to color management and Adobe was the realization that a common 'Adobe Imaging Model' was evolving around PostScript and PDF and that it needed to have a well-defined and supported color management component. Internally Adobe has worked hard to get each product group to subscribe to this common view and to share and reuse programming code to the greatest extent possible, particularly for color management. This has made a big difference in how our products interoperate with each other and, since it is closely aligned with the ICC model, Adobe products also interoperate well with other ICC based products.

Couldn't Adobe use the ICC color management that is within the various common operating systems instead of building its own? The products do offer that choice when the OS offers a suitable color management facility, but an Adobe system is also available. For too long the color management systems supported by the OSs were inconsistent among themselves and contained limitations or serious bugs. If Adobe software had to be developed to support products on even one OS, then that same software can be used on all OSs and provides a common interface and function, and can be fixed and improved without lobbying another company to cooperate.

15.14 WHEN THINGS GO WRONG

There are common things that can cause the color management workflow not to be as effective as we would wish. Here are four highlights.

1. *Quantization.* This term has its roots in converting analog information into digital form, thus making decisions about how to 'quantize' it with a fixed digital precision and range. This usually means losing some of the analog information, because the range or precision decided upon for the digital data is not large enough to capture the full accuracy of the analog data. Today when you hear this term being used with respect to color management, just think of the low order digits of your values being set to zeroes. To a computer scientist, this is pure and simple digital 'truncation' of data.

    Perhaps it was unfortunate that computer manufacturers settled upon the basic unit of computer memory and organization as the 8-bit byte. With this, one can represent a number between 0 and 255. Since the range of human discrimination is within this range of 'a couple of hundred' values, if used wisely the 8-bit byte suffices nicely.

However, if one has color values in a color space that does not have the same linearity as the human eye, the values are not well distributed within 0–255 and 'quantization' occurs. So we ride this ridge of almost adequate but in important cases not adequate. We should really move from 8-bit units to 16-bit units so that the more careless among us can survive.

2. *Gamut compression.* This is the process of reducing the range of color values to fit a restricted gamut device or application. Once this is done it is usually impossible to reconstruct the richer data that one started with. The simple rule is once compressed, always compressed. For this reason, it is best to avoid gamut compression until late in a process when it is absolutely essential to do. People who just have to work in a restricted gamut device color space will do well to learn how to work in richer spaces and how to use their tools effectively to reduce the gamut late in the workflow.

3. *White/black point compensations.* This is a very confused topic since there are many reasons to make adjustments for white and black points. In addition, there are many different white/black points that get confused easily. For example, there is the white illuminant for measuring colors on a reflective surface. There is the white of a surround when viewing a screen or surface. There is the white of the paper that is used as a color when rendering a picture or scene. Often the particular white points being discussed are not carefully defined, thus leading to confusion and incorrectly implemented and less interoperable systems. The ICC has been struggling with this issue since the beginning. As a credit to Adobe, the white points discussed above during the PostScript and PDF expositions are quite well defined in the Adobe documentation [1, 7].

4. Another source of problems is just doing bad computer arithmetic. This is particularly aggravating during table lookup interpolation. Numerical analysis is a mathematical specialty and the lessons learned and methods developed by its practitioners are well applied to color management software. One has to know what happens to the error factors when they are multiplied, divided, etc. Quantization errors caused by saving intermediate values into containers with too little precision or too little range is often confused with bad arithmetic. But both occur far too frequently.

15.15 THE FUTURE

The future looks bright for color management, both in Adobe's products and with respect to the ICC. The ICC has been revitalized in the last couple of years, making significant revisions to the ICC Profile Specification, supporting the development of a prototype second-generation system, and holding frequent, energetic, well-attended meetings.

The more widespread common support is within Adobe products, the more synergism develops. The concepts are now well known and understood by the various development teams, which means that many more of the smaller day-to-day decisions will be made in a more consistent and better way.

For both groups the current challenge is to make the overall 'system' work better and more accurately. We also need to find ways to hide some of the complexity from the casual user. The ICC is taking important steps to improve the interoperation of profiles and code modules from different vendors.

15.16 ACKNOWLEDGEMENTS

This chapter is better because of the critical review by Ed Taft and Lindsay MacDonald. My thanks to them.

REFERENCES

1. Adobe Systems Incorporated (February 1999) *PostScript Language Reference Manual*, 3rd Edition. Addison-Wesley Publishing Company, p. 912, ISBN 0-201-37922-8 (This book can be downloaded from Adobe's web site in soft copy form: http://www.adobe.com or http://partners.adobe.com/asn/developer/technotes/postscript.html.)
2. Adobe Systems Incorporated, (1996) Technical Note #5044: *Color Separation Conventions for PostScript Language Programs*, May.
3. Gentile, R. (1993) Device independent color in PostScript, February, *IS&T/SPIE Int'l Symposium on Electronic Imaging*, http://imaging.org.
4. Hewlett-Packard (1996) *sRGB Specification*, http://www.srgb.com.
5. King, J. (1998) *On the Desktop Color System Implied by the ICC Standard*, November, *IS&T/SID Color Imaging Conference*, http://www.imaging.org.
6. Imation Corporation, (1999) Color Fidelity Module (CFM) for Apple ColorSync, August, (On the Imation web site search for 'CFM' http://www.imation.com.)
7. Adobe Systems Incorporated, (July 2000), *PDF Reference Manual*, 2nd Edition. Addison-Wesley Publishing Company, p. 695, ISBN ISBN 0-201-61588-6 (This book can be downloaded from Adobe's web site in soft copy form: http://www.adobe.com or http://partners.adobe.com/asn/developer/technotes/postscript.html.)
8. Neugebauer, H. E. J. (1937) Die theoretischen Grundlagen des Mehrfarbendruckes. *Z. wiss Photogr.*, **36**, 73–89.
9. Von Kries, J. A. (1911) In *Handbuch der Physiologisches Optik*, Vol. II (W. Nagel, ed.), pp. 366–369. Leopold Voss, Hamburg.
10. Fairchild, M. (1997) *Color Appearance Models*, Addison-Wesley, Reading, MA.
11. Hunt, R. W. G. (1996) *The Reproduction of Colour*, 5th edn. Fisher Books, England.
12. Wyszecki, G. and Styles, W. (1982) *Color Science: Concepts and Methods, Quantitative Data and Formulae*, 2nd ed., John Wiley and Sons, New York.
13. International Electrotechnical Commission (IEC) (1998) *Colour measurement and management in multimedia systems and equipment, Part 2: Default RGB color space – sRGB*, 9 January. Available at http://www.adobe.com.

Colour management in digital film post-production

Wolfgang Lempp and Leonardo Noriega

16.1 INTRODUCTION

Digital film post-production is a relatively recent development in the century-old tradition of film making. In its 15 or so years of existence it has seen vertiginous growth, almost completely replacing the traditional methods of optical visual effects. Today it is rare that a feature film does not contain at least one digitally manipulated sequence. The treatments range from simple scratch removal to completely computer-generated characters and landscapes. Strongly project-driven, and largely consisting of small specialist companies, little research has been devoted to digital representation and manipulation of film colour.

As digital film post-production gains importance and moves towards the digital mastering of complete feature-length films, the requirement for a better understanding of colour science amongst its practitioners becomes apparent. This chapter looks at the development of this specialist sector within the film industry and tries to outline the challenges for development of a fully digital film process chain.

Colour Engineering, edited by P.J. Green and L.W. MacDonald.
© 2002 John Wiley & Sons Ltd.

16.2 TRADITIONAL FILM POST-PRODUCTION

Colour management in the traditional film process is relatively self-contained, as the only medium employed in the critical judgement of colour is the projected film print. This simple statement, however, conceals a process of considerable complexity, since the positive prints are produced from original camera (or intermediate) negatives which have no colorimetric meaning. Artistic intent thus has to be interpreted indirectly from film negatives. This section discusses the various stages in rendering a print from a negative.

16.2.1 The film process chain

Figure 16.1 illustrates the process of bringing the original scene onto the cinema screen. The many stages of the process meet the various artistic and technical needs of producing a movie. The film maker (or director) has essentially two opportunities to determine the colour and mood of a film in order to generate a particular 'look': first during the original photography and second at the 'answer print' stage. Colour adjustments after this point exist primarily as laboratory control procedures intended to remove any inconsistency due to variations in the film processing. Colour science has largely been confined to the realm of the film stock manufacturers.

As the overall visual appearance of a film is an artistic matter in the hands of the cinematographer, defined during the original photography and 'answer print' stages, the negative assumes a particular significance in the process. The latitude of negative film is considerably greater than that of print film. The dynamic range of recorded exposures can easily exceed 10000:1 [5] which equates to a (logarithmic) density range of 4. This enables the recording of a considerable degree of scene information at both high and low light levels, and makes the negative a particularly sensitive image capture medium.

In order to produce a positive, Red, Green, and Blue printer lights are transmitted through the negative in an optical printing process to expose print film. The only variables

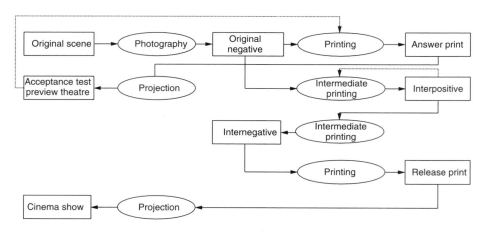

Figure 16.1   Traditional film process chain.

available to the cinematographer after the negative is exposed and developed are variations in exposure for the three primary colours during printing (hence the term 'colour timing' for adjusting the final colour balance of a feature film). There is no mechanism to undo the lack of contrast in an underexposed or overexposed negative within the conventional laboratory process.

The cinematographer specifies what is required in the print viewed by the final audience, and colour management becomes a process of extracting these requirements from the negative. The negative is self-evidently crucial as the basic raw material from which the cinematic visual experience is created. Rendering a scene in a believable or pleasant manner, within the constraints of the colour gamut and dynamic range of the film material, has always been the responsibility of the cinematographer and requires considerable skills and experience.

In some cases it has been necessary to go to extreme lengths in order to produce a pleasing effect – an example from the early days of black-and-white cinema was the use of special facial make-up, in order to counteract the unnatural spectral sensitivity of the film stock. The task has been made a lot easier over the years with ever more faithful colour reproduction, and the vastly improved dynamic range of modern film stocks, but it has shifted the process of determining the colour rendition of a scene more towards the printing stage. Also in the situation where the film is destined for television broadcast, and picture is transferred to video via a telecine suite, a culture of 'let's fix it in post' has been fostered.

The main visual impact is still achieved by controlling the lighting, colour balance and exposure during shooting, and film stocks and processing are consistently pushed to their limits in trying to produce a new visual experience. Sometimes this includes special instructions for negative or print development, which emerge as a result of the grading process. It is important to realise that this grading process is iterative and strongly dependent on the way the film scenes are cut together sequentially. The adaptation of the viewer to each scene in the absence of a reference white in the dark cinema is used to good effect. The process of producing a release print from an original negative is today at a stage that there is hardly a colour or situation that cannot be reproduced in a perfectly believable manner on the cinema screen.

16.2.2 The tools of the cinematographer

A consistent 'colour look' has traditionally been achieved by visual assessment, with professional graders and Directors of Photography (DOPs) judging the projected image. The issue of consistency is partly addressed by special instructions to the laboratory, but other techniques are also used in the film development process to ensure consistency of the chemical processing stages.

The principal means of achieving this is to correlate the positive and negative densities. Densitometry, rather than colorimetry, is the preferred means of quality control in the cinema industry for a number of reasons. The density of film (both positive and negative) is specified in a triplet of RGB values, derived from Status A (for positive) or Status M (for negative) measurements [8]. These are based on defined sets of narrowband Red, Green and Blue filters, designed to correspond with the absorption troughs in the film spectral dye sensitivities. Separate illuminants are also included in the specifications for reflecting and transmitting materials.

Unlike the *XYZ* sensitivity functions of conventional colorimetry, the Status A and M curves are physically realisable, and are used in commercially available densitometers. These devices are used in the control of the printing process, via the use of the Laboratory Aim Density (LAD) technique [15]. This technique is used to ensure the colour consistency of prints from negatives. LAD values are given on the data sheet (for example, see ref. 19) for any cine film (positive or negative), and describe the density required for a neutral mid-grey patch. A perfectly exposed negative representation of the LAD patch should produce RGB densities close to (i.e. within the specified tolerance levels of) those stated on the film data sheet. If these densities do not match, then compensating adjustments can be made to the printing exposures. The printed LAD patch should also produce a close match to the LAD densities specified on the positive print film data sheet. The use of LAD values ensures that all prints made from a given negative have a similar density.

16.3 DIGITAL VISUAL EFFECTS

The complex world of traditional film post-production has grown another dimension since the advent of digital film technology. Originally confined to occasional shots unachievable through conventional technology, digital visual effects today make a significant contribution to most feature films. They encompass all kinds of digital manipulation of film sequences, from wire removals and sky replacements to blue screen compositing, computer graphic character animation, set extension and crowd replication. In some cases digital effects can provide the source material for entire scenes, or even entire films.

The composition of computer-generated source material with photographic film images can complicate the grading process, necessitating a realistic image preview on a computer-driven display under controlled viewing conditions. In order to achieve good image fidelity, a sound understanding of the underlying colour science is necessary. In this section the issues involved in digitisation and colour management are discussed in relation to some of the problems associated with the grading of digital effects in film.

16.3.1 Digitising the negative

One aspect tends to dominate the look of the final image and is a fixed characteristic within the process: the greyscale response or film *gamma*. The film gamma and the resulting image contrast is not adjustable to a significant degree within the conventional colour film post-production process. A faithful reproduction of the film gamma at both input and output stages is the most significant requirement of a digital film system.

As discussed previously, the camera negative holds significantly more information than a print made from it, in terms of both spatial resolution and dynamic range [2, 5, 12]. The colour gamut of the negative is in this respect a somewhat ill-defined quantity. Seen in conjunction with the intended print stock on which the final image is to be viewed, it may be said that the negative holds a slightly larger range of colours than the print can reproduce. Because scanners based on CCD arrays generally have limited dynamic range and the attendant signal-to-noise problems, negative film with low gamma values gives better results when digitised than positive film with higher gamma values. It is therefore possible to devise a meaningful encoding scheme that can capture all of the information on the camera negative and also be used to generate an identical copy on a suitable film recorder. Unfortunately, given the physical properties of the negative, with its reduced contrast and orange base (due to coloured couplers), it is in itself almost meaningless in colorimetric terms. It would appear to be the worst possible choice for deriving quantitative colour information for the film post-production process and its multi-stage creative colour control.

16.3.2 Scanner encoding scheme

One standard for colour management, the Kodak *Cineon* digital film system [16], was partly adopted by SMPTE in the specification of a digital file format for the interchange of images between facilities [17]. This encoding scheme, a 10-bit logarithmic code for RGB film densities, has been widely adopted within the industry and has become the *de facto* standard for design of film scanners.

The associated SMPTE Recommended Practice RP-180 [18] defines spectral charac-teristics for use with colour negative densities. However RP-180 has not been uniformly implemented, and there remains an inconsistency between different film scanners with regards to their spectral response. Given the role of the negative and the print film, as outlined in the previous section, this issue will require further attention if consistent results are to be obtained.

In practice, the *Cineon* encoding scheme has served very well in characterising the greyscale responses of the input and output devices involved. In particular, it encapsulates the special characteristic of the cinema viewing environment without a reference white and fixed viewer adaptation. The 100% white value is only two-thirds of the way up the 10-bit scale, allowing for the representation of 'whiter than white' luminances of specular highlights and glare.

The logarithmic scale is not a suitable form in which to calculate psychophysical attributes such as lightness and brightness, which are functions of luminance. The com-bination of computer graphics elements with scanned images, the compositing of images from different sources, and editorial filtering and painting operations can give inaccurate results if this fact is ignored, as is commonly the case.

To perform an accurate characterisation of the input scanner for a colour management system, the following factors must be considered:

- *Densitometry*. Densitometric encoding equates to using the scanner as a densitome-ter, the accuracy of which must be included in the modelling process. Cross-colour effects between the different colour channels in the negative are used as a means of compensating for unwanted dye absorptions in the positive.
- *Processing*. Variations in the day-to-day processing in the film laboratory.
- *Density transform*. Modelling the transformation from the density of the negative to the density of the positive. Given the number of intermediate stages between the original

camera negative and the final colour print, it is more sensible to analyse the transformation between the physical properties of the input and output media, allowing for variations in the print stock.

As with input device characterisation (see below), the densitometric performance of the scanner can be made more accurate by using modelling techniques and look-up tables (LUTs) to compensate for the discrepancies between film density measurements and digital values generated by the scanner.

16.3.3 Matching the original negative

A much more straightforward method of controlling colour in post-production is to restrict the operation of the digital visual effects system to reproducing an identical copy of the original negative (with the exception of the visual effect itself, of course). This simply requires a closed-loop calibration of the film scanner and the film recorder, using a suitable set of look-up tables and colour compensation matrices. This holds true even if the film recorder is not capable of reproducing the full dynamic range of the camera negative, as is commonly the case with CRT-based film recorders typically generating maximum densities less than 1.3. In this case, the anticipated print exposure can be factored in to minimise the effect of the limited range in the highlights, since most of the useful dynamic range of print film can be reached with a 1.3 log exposure range [19]. In conjunction with a reasonable approximation of the film viewing experience on the computer monitor, it is possible to achieve a productive setup for a large number of digital visual effects. This is, in fact, how the majority of visual effects studios operate today, where in-house scanning or film recording is performed. It is particularly suitable for the more technical visual effects like wire removal, rotoscoping, etc. Unfortunately, it is not sufficient to give a realistic preview of operations that require a change of mood of the scenery, like day-for-night, sky replacements or critical blue-screen composites.

16.3.4 Colour management: a potential solution

Colour management cannot solve all of the problems involved in film grading while the process remains dependent on analogue photographic technology for both the initial image capture and final projection in the cinema. But it can greatly improve the productivity in the lengthy process of recombining the digital effect with the surrounding scenes in a film (see Figure 16.2).

Colour management introduces the new concept of *device-independent* colour representation to motion picture production. In general, colour management consists of a number of tasks, with the following two areas of particular importance [4, 5, 13]:

- *Characterisation of input devices.* Establish a mapping from the input film densities, measured by a suitable instrument, to digital signal values generated by the input device, specified in a device-independent colour space such CIE *XYZ*. In the particular case of digital film post-production the input device will be a film scanner, although future developments in the industry may also require the characterisation of digital cine cameras [10, 11].

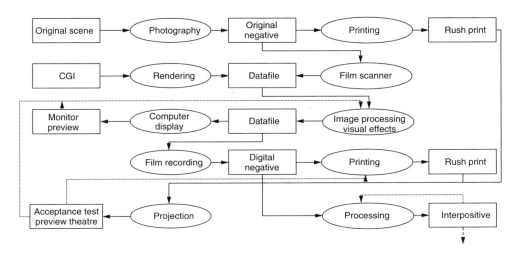

Figure 16.2 Digital visual special effects process chain.

- *Characterisation of output devices.* Establish a mapping from device-independent colour co-ordinates to the device values necessary to produce that colour in the output medium. For a CRT monitor, this means the relationship between RGB signals and the displayed colour [1, 7]. For a film recorder it means the relationship between CMY signals and the appropriate Status A densities of the print film, possibly including an intermediate negative and its appropriate Status M densities.

The characterisation of the input and output devices using device-independent colour spaces does not necessarily imply that the output will exactly match the input of a colour management system in terms of perception [4]. Three further issues also need to be considered:

- *Colour appearance.* The environments in which colours are viewed can affect their colour appearance in different ways due to the adaptation state of the viewer, viewing geometry and the effect of ambient light [4]. A colour displayed on a monitor may match a colour projected onto a cinema screen when viewed in one environment but look quite different in another.
- *Rendering intent.* It is not always necessary that the output from a colour management system should be colorimetrically identical to the input. Subjective criteria may dictate how the output should look, termed the rendering intent, and in fact this embodies important creative decisions by the cinematographer. Qualities such as pleasantness and naturalness may be appropriate [5].
- *Colour gamut.* The gamut of the output device may not allow the display of all colours of the gamut of the input medium and/or image. Possible solutions include finding the nearest, perceptually similar, colour within the output device gamut, adjusting all colours in the image, or simply indicating which colours are out of gamut [14].

16.4 THE DIGITAL STUDIO

The film industry is preparing itself for the next generation of digital film post-production: the digital mastering of complete feature films. With the arrival of digital high-definition technology, the first digital cine camera and digital projection prototypes, the requirement for a digital master has become widely accepted.

The possibilities of digital film cameras, the ever increasing volume of computer-generated material and visual effects are shifting the balance from a minority of special effects being inserted into a conventionally post-produced feature film towards a minority of live-action shots inserted into an otherwise digitally produced feature film. More importantly, the tools available for the grading and colour correction of scenes in post-production will be too tempting for any self-respecting cinematographer to ignore. The digital mastering of complete feature films offers real benefits with regards to the tortuous path of film conventional film grading (see Figure 16.3).

In current practice the critical grading decision on both conventional and digitally treated material is made by viewing a projected film print. The digital studio tries to enable the creative team to make the final colour judgement on a different medium, i.e. a computer monitor or an electronically projected image. To make this possible, a number of factors have to be taken into consideration. What is already clear is that the film makers will have to accept a number of compromises if they want to have the benefit of digital colour control for film mastering. It will take some time for the people involved to gain the confidence to make critical colour judgements using a medium that is not capable of displaying the same contrast range and colour gamut of a film print.

16.4.1 Telecine processing

When feature films are prepared for broadcasting on television, a device called a *telecine* is employed, which converts cine film to a video representation that will provide a suitable appearance on broadcast television. Adjustments are made to the colour balance of

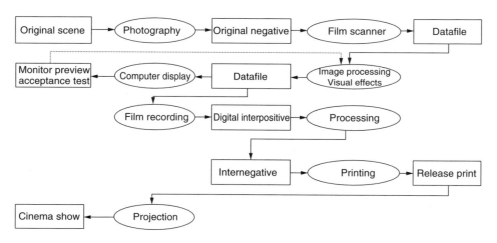

Figure 16.3   Digital lab process chain.

films intended to be displayed in darkened cinemas, so that people watching television in somewhat lighter surroundings will have an accurate portrayal of the colour. Telecine machines are operated by colourists, who make adjustments according to directorial criteria specified by either the original film-maker or the television station management. Telecine machines offer considerable flexibility in the way colour can be presented on a screen. The potential for telecine colour adjustment is illustrated by the penchant for commercial sponsors of feature film presentations on television to enhance the appearance of their product or brand [5].

What is currently common practice in telecine processing for the post-production of broadcast material will also become commercially viable for feature films. Yet there is a significant difference between those two environments. The colourist in the telecine suite has full control (in a technical if not executive sense) over all aspects of how the image is going to look. In effect the colourist has 'What you see is what you get' (WYSIWYG) control, sitting in front of a broadcast monitor. It would require remarkable advances in colour processing and display technology for this to be possible with film. Telecine is not an ideal film replication medium since, despite its flexibility, it is effectively used to tailor the visual experience of film, to fit into the narrower constraints of television. The very visual feedback in the telecine suite has created its own idiosyncratic culture of a 'colourist's control interface' in the absence of any colour science whatsoever. This is not possible with film since it is targeted at a cinema environment, which is constrained to give the best possible viewing conditions for projected film.

16.4.2 The cinematic experience

The cinematic experience is ill-defined for a number of reasons. There is no clear evidence about what we as the audience like about the visual quality of film. There are people who believe it is the fact that half the time the image on the screen is black (the time between the frames when the gate of a conventional film projector is closed), giving the brain time to think about the story. The term "Cinematic Experience" could describe many factors extraneous to the issue of film viewing, but in this chapter it means the viewing conditions, and the look of film.

The most striking phenomenon in the cinema is the absence of any visual reference outside the projected image itself. In the dark surround our visual system is constantly adapting to the picture. This makes it possible to be momentarily blinded by the relatively dim projected light after a long sequence in the dark. This capability of reproducing a sequential contrast exceeding the 'comfort level' of human vision is not matched by any other display medium. In our opinion it is one of the prime factors for the illusion of reality that a film in the cinema can provide.

A beneficial side-effect of this adaptation is that the overall colour balance of a film print is relatively uncritical. An astonishing amount of inconsistency between different film laboratories and different print runs is disguised by the fact that a feature film release print is usually produced in a single process. The audience adapts to the colour cast of the individual print, taking visual cues from the film itself.

Apart from the wide range of colours of a projected film print, what stands out is the vast dynamic range of the film material, typically having a contrast range in excess of

10 000:1. It is often argued that it is impossible to display an image in a public cinema with a simultaneous contrast ratio of much more than 100:1, due to projector lens flare, screen flare, emergency lights and diffuse reflections from the audience. The conclusion is then drawn – incorrectly – that an electronic projection system does not need to have a much higher contrast ratio to be comparable to film. Sequential contrast from one scene to the next can be much greater than 100:1 and the human visual system adapts dynamically to perceive detail in both highlight and shadow regions of the image beyond the usual static adapted range of approximately 160:1. All of this suggests that the cinematic experience is difficult to specify in terms of colour appearance and that, in order to preview film on a computer monitor, the limitations of the display must be considered.

The 'filmic look' consists of a combination of spectral, spatial, textural and temporal phenomena. In particular, flicker is caused by images changing more slowly than the temporal frequency response of the visual system, and graininess is caused by the nature of the silver halide film medium. The visual interpolation of both flicker and film grain increases the subtlety of the range of colours perceived in the cinema, relative to other electronic media. In addition, fine detail on a large screen has an elusive, transient quality that is difficult to emulate on other display systems.

16.4.3 Colour appearance models

The basis of colorimetry is that two colour stimuli with identical tristimulus values will appear to be identical under identical viewing conditions when viewed by a specified observer. When viewing conditions differ, it is almost certain that the colour percept will differ also. In order to predict how a colour stimulus will appear under different viewing conditions, colour appearance models (CAMs) [4, 20] are used.

A typical CAM takes into account such factors as the immediate area around a stimulus (the proximal field), the background to the stimulus, the area around the background (the surround), and in some cases other spatial and temporal factors. In addition to the conventional dimensions associated with device-independent colour spaces (Lightness, Chroma and Hue), there are the perceptual dimensions of Brightness and Colourfulness. Brightness refers to the attribute of a visual sensation whereby an object appears to emit more or less light. Colourfulness is the extent to which an object is perceived to have colour, which can vary with lightness, and generally increases with the luminance level. The psychophysical attributes of colour appearance, indicative of higher neuronal processing, are modelled as transformations of the device-independent colour data. Parameters allow for differences between media, based on fitting of experimental data.

Various colour appearance models are described in the literature, of varying degrees of complexity [4], but none has been specifically designed for viewing cinema. Of the existing models, the Hunt model [20] is probably most suitable, as it can be used for viewing of projected transparencies in dark or dim surrounds.

The need for colour appearance modelling is evidenced by the need to preview digital treatments of film on a CRT monitor, in order to assess how the images will look in a theatre setting. The image on a CRT must be corrected in order to simulate the viewing conditions and visual phenomena that pertain in a cinema. One of the major problems with using colour appearance modelling to compare an image on a CRT with that on another medium is the issue of adaptation mechanisms. Fairchild [4] described the problems associated with comparing hard copy with soft copy representations of images. Since CRT images are self-luminous, it is less easy to interpret them as illuminated reflective objects, and hence for the cognitive mechanism of colour constancy (associated with adaptation) to be applied [9]. Projected media are less easy to characterise. Although essentially reflective, they are also in a sense self-luminous, since light is shone through the celluloid film before it strikes the screen to create an image. This suggests that if the viewing environment of the CRT could be similar to that prevailing in a projection theatre (a *dark surround* environment in CAM parlance), one might assert that the adaptation mechanisms are similar in both cases.

The digital film grading suite is usually designed to approximate a cinema or preview theatre as closely as possible. But a small darkened room, fitted with a large computer monitor, might still give a significantly different viewing experience even for colorimetrically correct reproduction for a number of reasons. For instance, there are sources of light within the traditional cinema, in particular the exit lights, as well as reflections from the faces and clothes of members of the audience. Recreating accurately the levels of light visible in a theatre during projection of a film is difficult. It is possible that the ambient illumination, both static and dynamic, will affect the adaptation level of the viewer and hence the perception of the projected image, and will need to be taken into account in the colour appearance modelling.

There is more to the task of grading than predicting exactly how a film will look in the cinema – there is often also a requirement to achieve a consistent 'look' defined by colour. In order to preserve this look it is often necessary to compare frames from different parts of a feature film. To make a critical judgement of colour, the film maker needs a visual reference, which is something that the audience will not have. In this case both the visual task and the respective state of viewing adaptation differ between a digital grader and the audience. This indicates the necessity of adjusting the parameters of the colour appearance model and providing different viewing environments, in order to facilitate the judgement of colour, and thus ensuring a consistent appearance for particular scenes.

None of the issues raised above is relevant for the current method of film grading, where there is no immediate visual feedback apart from looking through the colour analyser. Nor are they relevant in the telecine suite, where the state of viewing adaptation does not change significantly during the scanning process. They would not be worth mentioning if it weren't foreseeable that the opportunities provided by a digital grading environment would give rise to a much more adventurous and bold approach to feature film grading. A methodology for quantifying colour appearance effects of this kind still needs to be developed, although existing models may be adapted for the purpose.

16.4.4 Gamut mapping

The colour gamut of print film is larger than the gamuts of most electronic display devices. The issue is more complicated, however, than the CIE chromaticity diagram might at first suggest [3]. The two-dimensional colour gamut consists of a convex, approximately polygonal, region in chromaticity space, whose vertices are the primary and secondary colours of the imaging device, such as the chromaticities of screen phosphors or film dyes.

In the case of an additive trichromatic device (with independent colour channels) such as a CRT display, this figure is a triangle. The missing dimension from the chromaticity diagram is luminance – although the gamut is maximum at low levels of luminance it shrinks to a single point at white [21]. A well-calibrated computer monitor is capable of displaying a contrast ratio of 1000:1 in a darkened room, but this is considerably less than the 10 000:1 often quoted for negative film. Film colours of luminance less than 10% of peak white occupy a considerably larger gamut than a typical CRT display, and it is in this aspect of rendering image detail in rich dark colours where film is most visually dramatic.

There are various strategies for gamut mapping, described in other chapters in this book, to minimise the visual impact of the differences between media. The overriding issue in the context of the digital studio is the requirement for real-time display of high-resolution images. A complex colour transformation requiring more than 40 ms (i.e. a refresh rate of 25 frames per second) to compute is not a useful way of performing this operation. Often the only viable solution is a three-dimensional look-up table, with all the disadvantages of a pre-computed, fixed relationship.

16.5 DIGITAL CINEMA

The next inevitable step in the development of digital film technology will be digital distribution and projection. The advantages for distribution in terms of control of the process, possibilities of simultaneous global marketing, flexibility of programming and potential for new forms of content are so compelling that there is little doubt about the eventual success of digital cinema.

As far as film post-production is concerned, digital cinema offers a further simplification of the colour grading issue (see Figure 16.4). It finally offers the prospect of a WYSIWIG editing and preview situation similar to the telecine suite. Two particular issues will drive the need for colour management based on colour science, and these are described below.

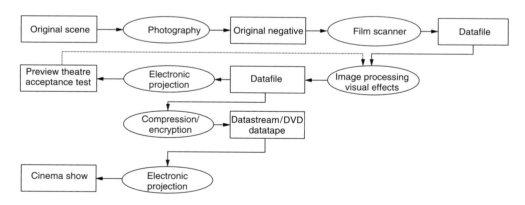

Figure 16.4 Electronic cinema process chain.

16.5.1 A standard for digital cinema distribution

It is already clear today that there will be a number of competing and fundamentally different technologies for the display of film images in the digital cinema. Moreover, they will operate in parallel with conventional film projection for many years to come.

It is obviously desirable that these competing technologies display the film images to the best of their respective capabilities, and matching as closely as possible the cinematic experience of traditional film. To this end it is necessary to standardise the response of these electronic projectors such that they can all operate from a standardised digital film distribution master and provide a consistent viewing experience similar to a distribution print today. This digital film distribution master would of course also be used to make conventional film release prints in a suitable way.

Standardisation efforts have commenced at ISO and SMPTE to define the colorimetric information of the image sequences in a device-independent manner [6]. It remains to be seen whether these efforts will be successful and sufficiently future-proof to accommodate new technologies, giving us a viewing experience one day that surpasses film as we know it.

16.5.2 An architecture for film colour management

The adoption of a standard architecture for film colour management, the means whereby any number of input and output devices could be characterised to allow accurate and consistent reproduction of colour images across a broad variety of media, would represent a significant advance for the motion picture industry. One obvious benefit would be the reduction in costs, as many of the intermediate stages would be removed. But also the flexibility of digital previewing would allow for greater artistic control over the final film print or digital projection. The existing division of labour between the artistic and the technical would be eroded, because the specification of the 'look' of a film could be dictated directly by the Director of Photography, lessening the need for specialised colourists in the laboratory. As cinema becomes increasingly digitised, such an architecture is becoming an essential requirement.

16.6 CONCLUSION

Digital film post-production is quickly changing from a specialist sector within the film industry, providing visual effects, towards a central role in mastering and distribution of complete films, replacing step by step the traditional film laboratory. As yet it is ill-prepared to meet the challenges posed by this transition, particularly with respect to colour management. There are few people within the industry who understand the underlying issues of device-independent colour representation, and even fewer people able to investigate the complex colour appearance phenomena that play such a central role in our viewing experience at the cinema.

There is no short cut on the way towards the successful implementation of a fully digital film process. The diversity of media, display technologies and viewing conditions all need to be understood to preserve the integrity of the artistic intentions.

REFERENCES

1. Berns, R. S. (1996) Methods for Characterising CRT Displays. *Displays*, **16**(4), 173–182, May.
2. DeMarsh, L. E., Firth., R. R. and Sehlin, R. C. (1985) Scanning Requirements for Motion Picture Post-Production. *J. SMPTE,* 921–924, September.
3. Fairchild, M. D., Berns, R. S. and Lester, A. A. (1996) Accurate Color Reproduction of CRT-displayed images as projected 35 mm slides. *J. Electronic Imaging*, **5**, 87–96.
4. Fairchild, M. D. (1997) *Color Appearance Models*, Addison Wesley.
5. Giorgianni, E. J. and Madden, T. E. (1998) *Digital Color Management – Encoding Solutions*, Addison Wesley.
6. Harrison, C. (2000) "A Review of Photographic Color Science, and its Application to Digital Cinema", Personal Communication, 20th February 2000.
7. Hung, P. C. (1993) Colorimetric Calibration in electronic imaging devices using a look-up-table model and interpolations. *J. Electronic Imaging*, **2**(1), January.
8. Hunt, R. W. G. (1995) *The Reproduction of Colour*, 5th edn., pp. 294–308. Fountain Press.
9. Hunt, R. W. G. (1996) Viewing parameters affecting self-luminous, reflection and transmissive colours. *Displays*, **16**(4), 157–162, May 1996.
10. Johnson, T. (1996) Methods for characterising colour scanners and digital cameras. *Displays*, **16**(4), 183–192, May.
11. Kang, H. R. (1997) Color Technology for Electronic Imaging Devices, *SPIE*.
12. Kennel, G. and Snider, D. (1993) Gray-scale Transformation of Digital Film Data for Display Conversion, and Film Recording. *J. SMPTE*, 1109–1119, December.
13. MacDonald, L. W. (1996) Developments in colour management systems. *Displays*, **16**(4), 203–212, May.
14. Morovic, J. and Luo, M. R. (1999) Developing algorithms for universal colour gamut mapping. In *Colour Imaging: Vision and Technology*, L. W. Macdonald and M. R. Luo (ed.), John Wiley, Chichester, England, pp. 253–282.
15. Pytlak, J. P. and Fleisher, A. W. (1976) A Simplified Motion Picture Laboratory Control Method for improved Colour Duplication. *J. SMPTE,* 781–786. October.
16. The CINEON encoding standard, Kodak.
17. SMPTE Standard 268M.
18. SMPTE Recommended Practice RP 180–1999. "Spectral Conditions Defining Printing Density in Motion-Picture Negative and Intermediate Films".
19. Kodak Film Data Sheets, Vision Premier Print, H-1-2383, H-1-2393.
20. Hunt, R. W. G. (1998) *Measuring Colour*, Third edition, pp. 208–247. Fountain Press.
21. MacDonald, L. W. (1997) Colour in Visual Displays. In *Colour Physics for Industry*, 2nd edn., R. McDonald (ed.), pp. 403–405. Society of Dyers and Colourists.

17

Managing color in digital image libraries

Sabine Süsstrunk

17.1 IMAGE REPRESENTATIONS IN DIGITAL IMAGE LIBRARIES

A digital image library, also called a digital image archive, can be described as a repository of digital images, deployable for various purposes. As with traditional book libraries, a digital image library may contain any number of images, from a few hundred to millions of image objects. These image files either can *represent* digital reproductions of original artwork, such as photographs or paintings, or can *be* the original digital artwork, such as an original scene captured with a digital camera or an image created on the computer. The purpose of a digital library varies: the library either contains digital image files that represent scenes or replace original artwork; it contains digital image files that should only represent a reproduction of the scene or original artwork, optimized for a given use and workflow; or it contains a combination of both. However, as with traditional libraries, the images should be accessible and usable over a long time frame, and should keep their value to the user indefinitely. The library should also be easy to manage, i.e. digital images should be deployed, processed, converted, and transmitted automatically.

Colour Engineering, edited by P.J. Green and L.W. MacDonald.
© 2002 John Wiley & Sons Ltd.

All this implies a certain strategy of initial image capture and processing techniques and a certain standardization of the workflow that should be well defined before an image library project begins.

The *usage* of the digital image files ultimately dictates image quality, and the necessary processing steps to be taken before archiving and delivering images. The higher the quality, the more expertise, time, and cost are associated to the generation, processing and storage of image files. While this chapter focuses primarily on color management relating to the capture and maintenance of digital images, the same applies, of course, to other image library issues.

Following are three most common digital image uses that define quality criteria [1]:

1. *The digital image is only used as visual reference.* The required digital image quality is low, in terms of both spatial and color resolution content. The display is limited to a screen or a low-resolution print device. Spatial resolutions do not exceed thumbnail and/or display screen resolution. Exact color reproduction is not critical; the image has only to 'resemble' the original artwork or scene. Images are usually compressed to save storage space, but primarily to save delivery time. For cases where original artwork is digitized, it suffices to use existing photographic duplicates of the originals and low-resolution scanners. Similarly, low-cost digital cameras can be employed when capturing scenes. Good examples for image libraries that primarily contain these kinds of image files are archives of on-line retailers who want to showcase a visual representation of their products, such as Amazon.com, the on-line book and CD retailer, and insurance companies that use digital cameras to create visual records of reported damages.

2. *The digital image is used for print reproduction.* The requirements for the digitizing system depend on the desired reproduction quality. Limiting output to certain spatial dimensions and a certain color encoding will define the digitizing devices and initial processing. Commercial digital image archives, such as Corbis Corporation [2, 3] and Getty Images, Inc. [4], and many cultural institutions digitize their analog image collections with this purpose in mind. Similarly, professional photographers who use digital cameras will build a collection of such images, and therefore an image library, over time. Consumer digital printing for print sizes larger than $10 \times 15\,\text{cm}$ also falls in this category. Commercially available scanners, digital cameras, and imaging software can produce image files for such applications.

3. *The digital image represents a 'replacement' of the original in terms of spatial and color information content.* This goal is the most challenging to achieve. The information content in terms of pixel equivalency varies from original to original. In terms of scanning original photographs, for example, it is not only defined by film format, but also by emulsion type, shooting conditions and processing techniques (see Table 17.1). Color information can only be captured adequately if the digitizing device supports the originals' color gamut in terms of encoding bit-depth and lack of metamerism. Such image library applications can be found in museums and other collections, where colorimetrically accurate capture is important, and/or the physical artwork is too fragile to be handled daily or is already deteriorating so that the image content as seen today needs to be preserved. The *Vasari* and *Marc* projects [5, 6], creating

Table 17.1 Sampling resolution for extraction of photographic film information. These values assume an excellent quality camera lens and normal processing

Film speed	ISO	Pixels per inch required
Very low	<64	3500–5000
Low	64–200	2500–3500
Medium	200–320	2000–2500
High	400–800	1500–2000
Very high	>800	800–1500

Source: Holm, J. (1996) The photographic sensitivity of electronic still cameras, J. Photographic Science and Technology, Japan 59(1), 117–131.

an image library of paintings for the National Galleries in London, are an example of such an application. Producing these files usually requires specialized hardware and software.*

Each of the usages outlined above will determine the quality of the *digital master* to be archived. It represents the highest quality file that has been digitized [7]. Since it represents the information that is supposed to survive long term, the encoding has to be appropriate for all current and future – as yet unknown – usage. The choices made in the initial digitization and processing of images are final and can usually not be reversed, and therefore great care should be given in designing the technical specifications of the initial capturing process. For daily use, *derivatives* of digital master files can be created, which are encoded according to their purpose. In most image libraries, various image derivatives are found, each appropriate for a different application [8]. See Table 17.2 for a comparison of different image library usage and recommended encoding specifications.

17.2 DIGITAL IMAGE COLOR WORKFLOW

The color workflow of a digital image can be described as follows (see Figure 17.1). An image is captured and encoded into a *sensor* or source device space, which is device and image specific. It may then be transformed into an *input-referred* image representation, i.e. a color encoding describing the scene's or original's colorimetry. In most workflows, however, the image is directly encoded from source device space into an *output-referred* image representation, which describes the image appearance on some real or virtual output.

---

* The MARC project, for example, used a Kontron ProgRes 3012 camera that could be moved to seven horizontal and nine vertical positions, resulting in an image file of 19 704 × 19 984 pixels. Seven interference filters were placed in front of the light source to control the spectral power distribution of the illuminant during exposure. The resulting seven-channel image files were converted to *XYZ* tristimulus values. For archiving, the master files were perceptually encoded to CIE $L^*a^*b^*$ values to save bit-depth. The size of the largest archive files is ∼1.6 gigabytes (GB).

Table 17.2 Comparison of different image library purposes and recommended encoding specifications

Master file	Visual representation	Print reproduction	Replacement of the original
Purpose of image library	• Consumer (electronic sharing) • On-line visual representation (electronic catalogs, image kiosks) • Low-end print representation (prints up to 8 × 10 inch)	• Commercial image libraries (picture agencies, publishing houses, news agencies) • Professional photographers • Museums, cultural institutions • Consumer printing (>10 × 15 cm)	• Museums, cultural institutions with high quality demand and/or fragile original artwork
Derivatives	• Thumbnails	• Lower print resolutions • Screen resolutions • Thumbnails	• Print resolutions • Screen resolutions • Thumbnails
Spatial resolution[a,b,c] (for one dimension)	• Thumbnails: ⩽250 pixels • Screen resolution: ⩽1600 pixels • Low-end printing: ⩽10 × 15 cm	• Dependent on the reproduction intent, the size and quality of the original artwork, and/or limited by the maximum spatial resolution of the digitizing device • ~3000 to 7000 pixels	• Dependent on the size and quality of the original artwork, and/or limited by the maximum spatial resolution of the digitizing device
Image encoding of master file	Output-referred	Output-referred	Sensor Input-referred
Compression of master file	None, lossless, or lossy (JPEG, JPEG2000)	None or lossless	None or lossless
File format of master file	EXIF, TIFF, JFIF, JP2	EXIF, TIFF, JPX	TIFF, JPX

[a]ISO 12233: 2000 Photography – Electronic still picture cameras – *Resolution measurements*.

[b]ISO/CD 16067-1 (2001) Photography – Electronic scanners for photographic images – *Spatial resolution measurements – Part 1: Scanners for reflective media.*

[c]Note that pixel dimension is only one parameter describing the spatial resolution of a digital image file. Other factors, such as the size, resolution and sharpness of the original, the spatial frequency response of the digitizing system, the quality of pixel reconstruction and interpolation algorithms, and the resolution of the output device, all influence the resolution and apparent sharpness of the image reproduction.

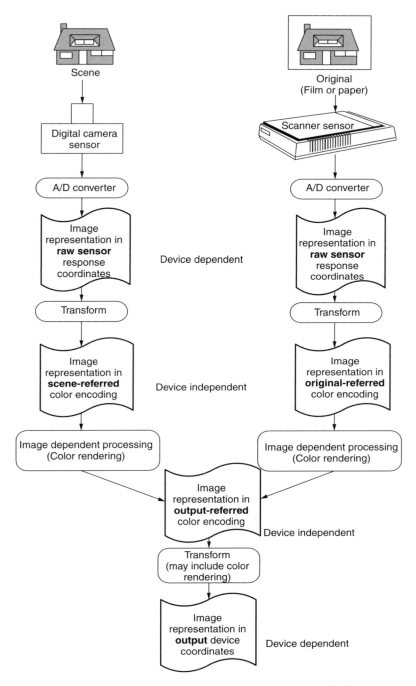

Figure 17.1   Color image workflow from capture to display.

If the output-referred color space describes a virtual output, then additional transforms are necessary to encode the image into *output* coordinates, which are dependent on the specific device [9, 10].

According to the CIE, a *color space* is a geometric representation of colors in space, usually of three dimensions [11]. The basis functions are color matching functions, usually CIE color matching functions. Spectral spaces are spanned by a set of spectral basis functions. The set of color spaces is therefore a subset of the set of spectral spaces. However, in practice, the difference is often neglected, and all representations of color in space are called a 'color space'.

From the point of view of color imaging, the term *color encoding* needs also to be defined. Color encoding refers to a quantized numerical representation of an image in a color space, including any associated data required to interpret the color appearance of the image [12]. A color space can have more than one color encoding associated with it. For example, CIEL*a*b* is a color space, but the encoding of images within the color space, and the image appearance associated with it, is determined by the color encoding specifications such as the ICC PCS specifications [13]. These color encodings are also often called 'color spaces', which leads to some confusion in the imaging literature and industry. Note also that color reproduction quality is not necessarily defined by the color space an image is encoded in, but by the quality of the transformation applied to the image data to encode the image into a given color space.

17.2.1 Sensor encoding

When a scene or original is captured, either by a scanner or by a digital camera, its first color encoding is device and scene specific, defined by illumination, sensor, exposure, and filters (see Figure 17.2). In the case of scanners, the illumination should be nominally constant for each image. With digital cameras, the illumination can vary from scene to scene, and even within a scene.

When a high quality master file is desired, storing 'raw image' data encoded in sensor space is most appropriate. Any further processing can degrade the image, as algorithms are based on today's know-how and might be improved in the future. Additionally, these transforms are not always reversible, due to re-quantization or clipping, even when stored with the image files. In most cases, the hardware and software necessary to create and archive such master files is not commercially available and has to be developed in-house or with academic or industrial partners [6, 14].

Images encoded in raw-data, or sensor space, are not viewable without further processing.

Encoding

When images are archived in sensor space, camera or scanner characterization data, such as device spectral sensitivities and illumination data, have to be retained so that further color and image processing is possible. Additionally, it has to be clearly noted which processing has already been applied to the image. The files can be stored before or after linearization, for example. If stored before linearization, the focal plane opto-electronic conversion function (OECF) and the scene flare characteristics have to be retained for

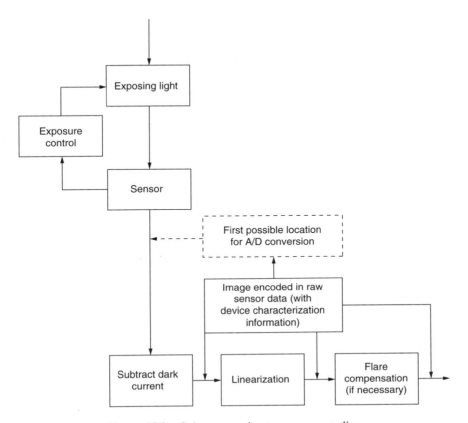

Figure 17.2   Color processing to sensor encoding.

further processing. The image should be saved in a standard file format, such as TIFF [15], whose encoding specifications are widely published, and libraries are available. The International Organization for Standardization (ISO) has developed a specific TIFF format called TIFF/EP [16], which has defined tags for the necessary information that has to be stored to further process images from digital cameras.

If the capturing conditions are always the same, it might be advantageous to store the color metadata not in the file, but in a separate database, which allows the creation and definition of the tags necessary for an individual application. However, in real situations it can occur that the image data and metadata are de-synchronized, usually due to human error. Storing image metadata in a database and in the master file can prevent this from having a negative effect on future automated processing.

It is also recommended to store as much information as possible about the original artwork, such as original size, reflection characteristics, and film emulsion type in case of photographs. In general, it is always advisable to archive more, rather than less, information about the original scene or artwork, the digitizing system, and the illumination conditions.

Trichromatic and multi-spectral imaging

The spectral sensitivities of the filters for a digitization device should be chosen, if possible, according to the original. When scanning color negatives, for example, the filters used should be optimized to the film dye densities. Filters for scanning color positives, as well as digital camera filters, are usually broader and overlap. Out of image system considerations such as noise and speed, it is rare to find filters in commercially available capturing devices that match the color matching functions of the human eye. As a result, device metamerism can occur, i.e. two different colors that can be distinguished by the human eye can be encoded to the same digital values, or vice versa. Depending on the application and originals, that effect is more or less negligible. If very precise color reproduction is required, the use of a multi-spectral capturing system is appropriate. Such systems are usually employed in a museum environment where original paintings are scanned, such as for the Vasari and Marc projects. Depending on the system, they consist of six to eight filters with varying bandwidth and peak spectral sensitivities [17–20]. XYZ tristimulus values can then be calculated from the different channels for a colorimetric representation of the original.

Scanners versus digital cameras

In many cultural institutions, the size and/or fragility of the artwork necessitate that a digital camera and not a scanner is employed, even if the artwork is two-dimensional. A monochromatic digital camera system can more easily be adapted with different filters, either in front of the camera or in front of the light sources, than a scanner.

17.2.2 Input-referred image encoding

The transformation from raw data to *input-referred*, device-independent color encoding is image and/or device specific: linearization and flare correction, pixel reconstruction (if necessary), white point selection, followed by a transformation to the input-referred color space (see Figure 17.3).

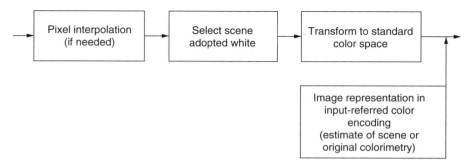

Figure 17.3   Color processing from sensor encoding to input-referred encoding.

The purpose of input-referred image encoding is for the image to represent an estimate of the scene's or the original's colorimetry. An input-referred encoding maintains the relative dynamic range and gamut of the scene or original. A *scene-referred* encoding describes the input-referred encoding of an original scene, whereas an *original-referred* encoding describes the input of original artwork.

Input-referred image encodings can be used for archiving master image files when it is important that the original colorimetry is preserved so that a facsimile can be created at a later date, and the transformation from sensor encoding to input-referred encoding is unambiguous. Another advantage of input-referred encodings, especially if the images are encoded in higher bit-depth, is that they can always be tone and color processed for all kinds of different rendering intents and output devices at a later date. A generic rendering algorithm can then be applied without having to consider from which device the data is coming from.

Examples of color spaces that are used to encode an estimate of the scene's or original's colorimetry are CIEXYZ, CIEL*a*b*, CIELuv, and Photo YCC [21]. Specific scene-referred color encodings are RIMM RGB [22], scRGB [23], and ISO RGB [24]. Input-referred encoded images will need to go through additional transforms to make them viewable or printable.

Encoding

The quality of the colorimetric estimate depends on the ability to choose the correct scene illuminant, and the correct transformations from device RGB to device-independent coordinates.

Estimating the illuminant

In case of imaging natural scenes, where the scene illuminant is unknown and/or mixed, the scene adopted white has to be estimated. The most common method is the *gray world technique* [25], where the illuminant estimation is taken as the average of the image's R, G, and B values. Another popular method is the *maximum RGB* [26] technique, where it is assumed that the image contains a white Lambertian surface. The illuminant's tristimulus values are then calculated from the maximum R, G, and B values of the image file. Finlayson *et al.* [27] proposed a method called *color by correlation*, where the illuminant with the highest correlation is chosen from a comparison of plausible illuminant matrices containing image colors.

Device RGB to device-independent coordinates

The transformation is usually based on minimizing an error criterion, such as CIE XYZ, CIE ΔE_{Lab}, CIE ΔE_{94}, or CIE ΔE_{Luv} [28]. The calculation of the transform is input system and application dependent. If the illuminant spectral power distributions, the reflection spectra of the original artwork, the spectral sensitivities of the filters and CCD, and the OECF of the digitizing system are known, the transform can theoretically be calculated. However, because digitization devices have spectral sensitivities that are usually not 'equivalent' (i.e.

within a linear transformation) to the CIE XYZ color matching functions, this transform cannot be perfect for all possible original stimuli. In practice, it is therefore advisable to image a test chart that has similar or identical reflection or transmission spectra as the original's to be digitized. For many applications, the commercially available MacBeth ColorChecker™ [29] is used as target. For photographic slide and print scanning, an IT-8 [30] target on the same film or print material can be imaged. When paintings are digitized, a target with representative color samples should be used so that the transform is based on realistic reflection spectra.

A simple transform from linear device *RGB* sensor data to device-independent values under the same illuminant is as follows [31]:

$$\mathbf{S} \approx \mathbf{PT} \tag{17.1}$$

where $\mathbf{S}(m \times 3)$ are the target color's (normalized) device-independent color values, $\mathbf{P}(m \times 3)$ are corresponding linear (and normalized) RGB device code values, $\mathbf{T}$ is a 3×3 matrix, and m refers to the number of target color patches. The matrix $\mathbf{T}$ that best maps $\mathbf{P}$ to $\mathbf{S}$ minimizing least-squares error is calculated as follows:

$$\mathbf{T} = (\mathbf{P}^T \mathbf{P})^{-1} \mathbf{P}^T \mathbf{S} \tag{17.2}$$

The above algorithm has been extended to a white-point preserving least-squares fit where all achromatic colors are mapped without an error [32], as the human visual system is especially sensitive to slight colorcasts in achromatic colors.

When minimizing an error criterion based on CIE ΔE_{Lab}, CIE ΔE_{94}, or CIE ΔE_{Luv}, a method to find transform $\mathbf{T}$ is to use an optimization routine for finding the entries of $\mathbf{T}$ by minimizing mean ΔE between the actual target tristimulus values ($\mathbf{S}$) and the predicted tristimulus values, such as:

$$T = \arg\min \frac{1}{m} \sum_{i=1}^{m} \Delta E_i \tag{17.3}$$

Basing the colorimetric estimate, and therefore the transformation matrix $\mathbf{T}$, on minimizing CIE L*a*b* or CIE XYZ error is not necessarily the right approach when imaging natural scenes. Natural scenes can contain very saturated colors that are close to the spectral locus, such as a colored light source. The CIE L*a*b* perceptual color space was developed based on surface colors that span a smaller gamut than all visible colors. Holm *et al.* [33] have shown that the best minimization criterion is dependent on which assumption was made about the illuminant's spectral power distribution and the spectral radiance distributions within the scene.

Chromatic adaptation transforms

Some input-referred color encodings specify a fixed observer adapted white-point. As a consequence, the sensor code values need to be mapped, using a chromatic adaptation transform (CAT), if the original/scene adapted white differs. Several chromatic adaptation transforms are described in the literature, most based on the von Kries model [34]. CIE tristimulus values are linearly transformed by a 3×3 matrix $\mathbf{M}_{CAT}$ to derive $R'G'B'$ responses under the first illuminant. The resulting $R'G'B'$ values are independently scaled to get $R'G'B'$ responses under the second illuminant. The scaling coefficients are based on the illuminants' white-point $R'G'B'$ and $R'G'B'$ values. If there are no non-linear coefficients, this transform can be expressed as a diagonal matrix. To obtain CIE tristimulus values $X'Y'Z'$ under the second illuminant, the $R'G'B'$ are then multiplied by $[\mathbf{M}_{CAT}]^{-1}$, the inverse of matrix $\mathbf{M}_{CAT}$. Equation (17.4) describes a matrix notation of this concept:

$$
\begin{bmatrix} X'' \\ Y'' \\ Z'' \end{bmatrix} = [\mathbf{M}_{CAT}]^{-1} * \begin{bmatrix} R''_w/R'_w & 0 & 0 \\ 0 & G''_w/G'_w & 0 \\ 0 & 0 & B''_w/B'_w \end{bmatrix} * [\mathbf{M}_{CAT}] * \begin{bmatrix} X' \\ Y' \\ Z' \end{bmatrix} \tag{17.4}
$$

Quantities R'_w, G'_w, B'_w and R''_w, G''_w, B''_w are computed from the tristimulus values of the first and second illuminants, respectively, by multiplying the corresponding XYZ vectors by $\mathbf{M}_{CAT}$.

The currently most popular chromatic adaptation transforms are the linearized Bradford CAT [35], the Sharp CAT [36] and *CMCCAT2000* [37]. The different transformation matrices $\mathbf{M}_{CAT}$ are as follows:

$$
\mathbf{M}_{BFD} = \begin{bmatrix} 0.8951 & 0.2664 & -0.1614 \\ -0.7502 & 1.7135 & 0.0367 \\ 0.0389 & -0.0685 & 1.0296 \end{bmatrix}
$$

$$
\mathbf{M}_{Sharp} = \begin{bmatrix} 1.2694 & -0.0988 & -0.1706 \\ -0.8364 & 1.8006 & 0.0357 \\ 0.0297 & -0.0315 & 1.0018 \end{bmatrix}
$$

$$
\mathbf{M}_{CMCCAT2000} = \begin{bmatrix} 0.7982 & 0.3389 & -0.1371 \\ -0.5918 & 1.5512 & 0.0406 \\ 0.0008 & 0.0239 & 0.9753 \end{bmatrix}
$$

However, new research has shown that these transforms are not unique, assuming a von Kries adaptation model [38].

Considering that research has not yet progressed to the point where a unique answer can be given on how to estimate scene adopted white, which error to minimize when deriving the transform, and which chromatic adaptation transform to use, it is recommended that when input-referred master files are archived, the encoding transforms are retained.

17.2.3 Output-referred image encoding

Output-referred image encodings refer to image representations in color spaces that are based on the colorimetry of real or virtual output device characteristics. Images can be transformed into output-referred encodings from either source or input-referred image encodings.

Output-referred color encodings, based on virtual output devices, are the perceptual intent ICC PCS, ROMM RGB [39], Adobe 98 RGB, sRGB [40], e-sRGB [41], and SWOP CMYK [42]. Output-referred image representations can also be encoded in other color spaces, such as CIE L*a*b*, CIE Luv, PhotoYCC, YUV, YC_bC_r [43], etc. These color spaces can accommodate both input- and output-referred encodings.

Archiving in output-referred encoding is appropriate for image libraries that archive primarily master files whose purpose is to be a reproduction of the original, and not a representation. Most current image libraries fall into this category, as commercial hardware and software are available for this task. The choice of which output-referred color encoding is appropriate for the master file depends on the application. However, whenever possible, the same encoding should be chosen for the master files, especially when large number of images are processed to derivatives at a later date. In general, all image library derivatives are encoded in output-referred representations.

Encoding

The complexity of these color rendering transforms varies: they can range from a simple gamma-function, such as is employed by video encoding, to complicated image-dependent algorithms that also consider spatial image information (see Figure 17.4). The transforms are usually nonreversible, as some information of the original scene encoding is discarded or compressed to fit the dynamic range and gamut of the output. Additionally, if the rendering intent of an image has been chosen, it may not be easily reversed. For example, an image that has been pictorially rendered for preferred reproduction cannot be re-transformed into a colorimetric reproduction of the original without knowledge of the rendering transform used. If image information has been clipped, a reverse transform is not possible.

Most commercially available scanners and digital cameras, except for the high-end models, automatically apply some image-dependent color rendering and deliver files

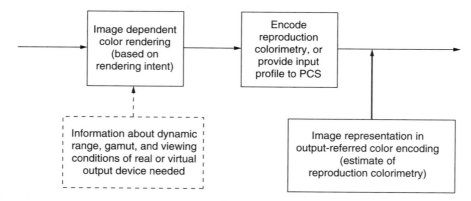

Figure 17.4 Color processing from input-referred encoding to output-referred encoding.

in an output-referred encoding. With scanners and high-end digital cameras, the user can additionally 'manipulate' the rendering by changing color, saturation, contrast, etc. As the transform algorithms – and the resulting image reproduction quality – are the differentiating factors between manufacturers and their devices, it is hard to obtain much information about them. Additionally, these transforms can never be standardized, as most images are subjectively evaluated, and preference varies between observers.

Today, it is difficult commercially to obtain software that automatically applies color rendering to sensor or input-referred encoded images, especially if the images are encoded with higher than 8 bits per channel. The closest to such applications are high-end scanner and digital camera drivers that allow importing high bit-depth data files. However, they can usually only interpret the image files coming from their own devices, as the device metadata is stored in the driver and not in the image file. This is the main reason why commercial image archives, such as Corbis and Getty, as well as many cultural institutions archive their master files in an output-referred encoding. It is to be expected that the imaging industry will soon address this missing piece in the image production pipeline.

Some output-referred color encodings are designed to closely resemble the anticipated output device characteristics, ensuring that there is little loss when converting to the output specific device coordinates. Most commercial image applications support only 24-bit image encoding, making it difficult to make major tone and color corrections at that stage without incurring visual image artifacts. Some output-referred RGB color encodings are even designed so that no additional transforms are necessary to view the images; in effect, the output-referred RGB color space is the same as the monitor output space. For example, sRGB is an output-referred color encoding that describes a specific output and as such can be equivalent to an output device encoding. It should be noted that there are no 'real' sRGB monitors, but the encoding specifications 'resemble' the characteristics of most of today's CRT monitors so that the actual image appearance is approximately the same as the encoded image appearance.

17.2.4 Output device specific encoding

Encoding

Transforms from output-referred encodings to output coordinates are device and media specific (see Figure 17.5). If an output-referred color space is equal or close enough to the real device characteristics, such as 'monitor' RGBs, no additional transformation to device-specific digital values is needed. For other output applications, such as print, there is a need for additional conversions. This can be accomplished using the current ICC color management workflow. An 'input' profile maps the reproduction description in the output-referred space to the (perceptual intent) profile connection space (PCS), and the output profile maps from the PCS to device- and media-specific values.

However, if the gamut, dynamic range, and viewing conditions of the output-referred encoding are very different from those of the actual output, it might be more advantageous to use color rendering that allows image specific transforms than to use the current ICC color management systems that contains only 'device-to-device' mapping.

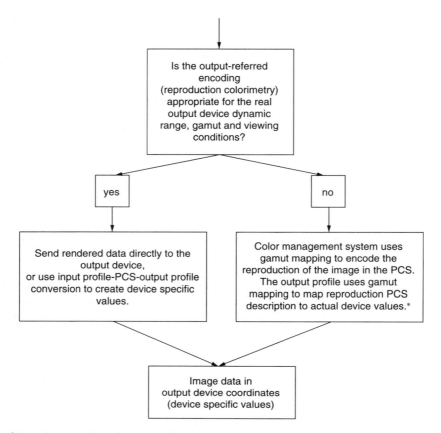

Figure 17.5 Color processing from output-referred encoding to output device coordinates.

Rendering for another encoding

Apart from some graphic arts applications, it is rare today that images are archived in output coordinates, such as device- and media-specific RGB, CMY, or CMYK. However, there are many legacy master files, such as CMYK separations and RGB monitor specific images, that need to be rendered so that they can be viewed and printed on other devices. If the color encoding is known and sufficiently characterized, it is possible to 'reverse'-transform these images to another, better suited color encoding, such as one of the standard output-referred color encodings discussed elsewhere in this chapter. If the two color encodings are similar and the viewing conditions assumed to be the same, for example Apple RGB and sRGB, the reverse-transform can simply be executed by applying the inverse of the first non-linear transfer function, followed by a chromatic

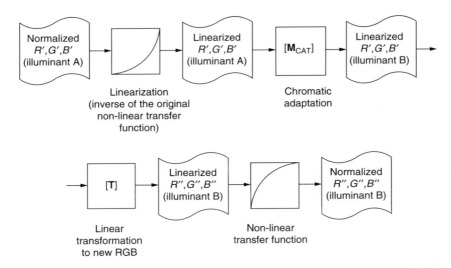

Figure 17.6 Schematic representation of reformatting from one output-referred color encoding to another. Using these transformations is only appropriate if the two color encodings have similar gamuts and viewing conditions.

adaptation transform (if necessary), followed by a linear matrix conversion to the new color space, and then again perceptual encode using the new non-linear transfer function. See Figure 17.6 for a schematic representation of such a reverse-transform.

This simple transform will clip colors that are outside the destination gamut. Therefore, if the gamuts of the two encodings have a different shape, gamut mapping should be applied to map from one encoding to another. This can be accomplished using an ICC profile. However, if the gamuts, white-point, and viewing conditions are very different, an image-dependent reproduction model might be more appropriate.

Each transformation and resulting new quantization can introduce visible artifacts. It is therefore recommended that only new derivatives are reverse-transformed, and a copy of the original master file is kept. If, in a few years, other 'standard' output-referred encodings based on new output devices become popular, then new derivatives can be created from these master files.

17.3 THE USE OF ICC COLOR MANAGEMENT IN IMAGE LIBRARIES

The International Color Consortium [44] publishes a profile specification that includes the transformation from device-specific encoding values to a standard color encoding, called the profile connection space (PCS). The device-specific values are either from actual devices, such as input devices (scanner, digital camera), displays (monitor), and output devices (printer, film recorder), or from virtual devices such as a standard output-referred encoding. The profiles are intended to facilitate the interchange of color information between different operating systems, applications, and devices.

17.3.1 ICC profiles and master files

It is not recommended for image libraries to archive master files with only ICC profiles to characterize the color encoding. Current ICC profiles do not contain device characterization data as a matter of course; they contain only the *transformation* from a device-specific space to the PCS. For example, it is impossible to determine the sensor spectral characteristics of a scanning system from a scanner profile, as the sensor data was transformed to an output-referred encoding and the specific transforms not retained. The image encoding is therefore determined by the device and is nonstandardized, and the only clue about its characteristics is given by the profile.

Additionally, the profile and PCS specifications have evolved, and will continue to evolve. While it is reasonable to assume that applications are capable of reading older-version profiles for two to three generations, accurate processing of these older profiles in combination with newer profiles is not guaranteed. In an image library environment where color transformation accuracy and longevity are important, updating all the different profiles associated with different devices and originals can be difficult to manage. It is therefore recommended that output-referred master files are archived in a well-defined, if not standard color encoding. In that case, only one profile needs to be updated when profile specifications change.

17.3.2 ICC profiles and derivative files

The ICC profile associated with an output-referred master file encoding can be used to create derivative files in other output-referred encodings, or it can be used to directly map to the output device coordinates. Practically, ICC profiles certainly facilitate the production workflow of images, such as for previewing and soft-proofing on a monitor and printing. Therefore, all derivative image files intended for print reproduction should be deployed with an ICC profile. Most prepress and professional image manipulation programs can read and process them. While using ICC color management does not always guarantee perfect color reproduction, it does communicate an encoding intent. It is up to the user of the derivative files to process the image additionally if so desired.

Thumbnail and screen resolution images should be encoded so that they do not need to be color managed, i.e. their encoding should closely resemble the intended output device. It is not reasonable today to assume that all computer systems and applications are color managed. Indeed, many viewing applications assume that if no profile is attached to an RGB image file, the file is encoded in sRGB and is displayed as such. If the applications are color management enabled, i.e. most printer drivers for desktop printers, they will transparently transform the image to device-specific values, assuming sRGB encoding.

When file size is important, such as for limited bandwidth transmission, the additional overhead of an ICC profile, even if it is a small matrix-based profile, might already be unacceptable. For these applications (web, PDA, cellphones), the image should be encoded in an output-referred space that resembles the actual output characteristics.

17.4 COLOR SPACES AND COLOR ENCODINGS

As outlined above, the choice of color space and resulting color encoding depends on the purposes of the image library's master and/or derivative files. Archiving in device-specific sensor space is appropriate for very high quality applications where the original data needs to be preserved. However, it requires a consequent storing of device and image metadata so that the image can be processed in the future. Archiving in a device-independent input-referred color space is also appropriate, provided the transform from sensor to input-referred encoding is straightforward, i.e. the white-point is known and does not have to be estimated, or the scene colorimetry estimate is sufficiently accurate for further image representations. It might also be expedient to represent images that are acquired with different capture devices, such as digital cameras, film and print scanners, in the same input-referred color space so that later the images can be processed as if they had all come from a common source. Storing in an output-referred color space is recommended for all derivative images that have been rendered for a given intent, or where the original appearance is not important and will never have to be rendered. Archiving output device-specific image data is not recommended for image libraries.

17.4.1 Color space and color encoding characteristics

To determine the correct color space and the correct color encoding parameters for a given image application, the following points need to be considered [15].

Extent of color gamut

Depending on the original scene and applications, the color gamut necessary for an input-referred color space can vary tremendously. For print scanning applications, it is sufficient to define a color gamut that includes all the reflective surface colors, while in digital photography applications a more extended gamut is necessary to allow for extreme scene luminance ratios encountered in real scenes, and to encompass spectral radiances given by transmissive and emissive surfaces that are on the border of the spectral locus. For output-referred color encodings, it is sufficient to define gamut boundaries that encompass all soft- and hardcopy output devices.

The size and shape of the color gamut and the degree of perceptual uniformity of the color space will affect quantization efficiency, i.e. how many bits are needed to encode colors. In general, larger color gamuts will need larger bit-depth encoding to prevent visual quantization artifacts when a smooth ramp is imaged.

Perceptually uniform transfer function

A perceptually uniform tone-scale encoding minimizes the bit-depth needed to encode an image. These transfer functions encode the scene or original's intensity or luminance nonlinearly so that equal quantization steps correspond to approximately equal perceptual differences of lightness. A 'traditional' transfer curve to perceptual uniform encoding is the transform from luminance values to lightness values defined in colorimetry:

$$L^* = 116(Y/Y_n)^{1/3} - 16 \quad \text{for } Y/Y_n > 0.008856$$
$$L^* = 903.3(Y/Y_n) \qquad \text{for } Y/Y_n < 0.008856 \tag{17.5}$$

where Y is equal to the luminance of the sample, and Y_n is the luminance value of the reference white.

For imaging applications, however, other transforms might be considered. For input-referred image encoding involving a large scene luminance range, it may be desirable to employ fewer bits when encoding highlights and shadows, and more bits when encoding midtones. In effect, such a transfer curve would be analogous to the one employed with conventional film, and can be modeled by an exponential or polynomial function. For output-referred color encodings, the transfer function should fit the expected tone reproduction characteristics of a certain class of output devices. CRT monitors, for example, have an inherent non-linear response that can be characterized with a power function as follows:

$$L = (V + e)^\gamma \tag{17.6}$$

where L is the luminance in cd/m^2, V is the voltage (mV), e is equal to the offset, and γ usually varies between 2.35 and 2.5. Many output-referred color encodings therefore define a non-linear transform that is equal to a power function of 1/2.2 or 1/2.4, which has the added advantage of being close to the nonlinearity we perceive (see Figure 17.7).

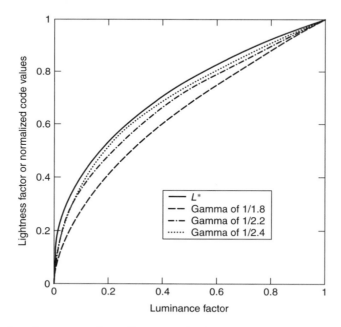

Figure 17.7 Popular perceptual non-linear transfer functions used in color image encodings.

Apple Macintosh early on defined a transfer function of 1/1.8 that corresponds well to the dot-gain function usually found in the printing process. As the Macintosh platform dominated graphic arts and prepress, a 'gamma of 1.8' became a *de facto* standard for still image encoding. Apple RGB and ROMM RGB encodings specify such a transfer function.

Following is an example of a specific non-linear transfer function, encoding linear sRGB values to non-linear sRGB values:

$$C' = \begin{cases} 12.92C & C \leqslant 0.0031308 \\ 1.055C^{1/2.4} - 0.055 & 0.0031308 < C \leqslant 1 \end{cases} \tag{17.7}$$

where C denotes the R, G, or B value.

Dynamic range

Scene luminance ratios average about 160:1, but can be significantly higher for real scenes, up to 50 000:1. Scene-referred encodings usually need to support a variety of dynamic ranges and colorimetric values larger than white, while original-referred color encodings need to accommodate the dynamic range of the original captured ($\sim$1000:1).

Output-referred color encodings can be limited to the maximum dynamic range of display and hardcopy output (max. 1000:1). The higher the dynamic range, the more bits will be necessary to represent it.

White-point

Scenes and originals can be illuminated in many different ways. Natural outdoor scenes can include multiple light sources. It is therefore necessary for input-referred color encodings to accommodate a variety of white-points, either by specifying a chromatic adaptation transform, or by employing a floating white-point. For output-referred color encodings, a white-point can be chosen that is consistent with the viewing conditions of the output, such as CIE illuminant D50 for printing applications or CIE illuminant D65 for CRT monitor applications.

Viewing conditions

Color spaces relate image code values to image colorimetry. However, in order to interpret image colorimetry in terms of color appearance, it is necessary to define the viewing conditions associated with each encoding. Observer adaptive white, adaptive white luminance level, viewing surround, and flare are typically defined. While the types of conditions remain the same, the viewing environment will naturally differ for input-referred and output-referred encodings. For example, the RIMM RGB (input-referred) and ROMM RGB (output-referred) encodings define the following viewing conditions:

RIMM RGB:

- *Observer adaptive white*: CIE standard illuminant D50
- *Luminance level for observer adaptive white*: >1600 cd/m^2

- *Viewing surround*: average (overall luminance level and chrominance of the surround is assumed to be similar to that of the scene)
- *Flare*: no viewing flare, and the scene colors are assumed to be encoded using flareless (or flare corrected) colorimetric measurements.

ROMM RGB:

- *Observer adaptive white*: CIE standard illuminant D50
- *Luminance level for observer adaptive white*: 160 cd/m^2
- *Viewing surround*: average (overall luminance level and chrominance of the surround is assumed to be similar to that of the reproduction)
- *Flare*: 0.75% viewing flare, referenced to the observer adaptive white, and the image colors are assumed to be encoded using flareless (or flare corrected) colorimetric measurements.

Quantization efficiency

The number of bits per channel necessary for an acceptable quality image representation depends on several factors, namely the size of the gamut and dynamic range, the perceptual uniformity of the color space, the need for color rendering and/or image manipulation, and compression. The first two points have already been addressed. If an image is going to be color rendered or 'manipulated', i.e. adjustments to its color balance and tonality are made, then more bits will be needed so that artifacts introduced due to repeated quantization will not be visible. The capturing device also limits bit-depth.

Due to all these variables, the following recommendations should be regarded as very general practices. Sensor-encoded master files should certainly be archived in the device bit-depth, which ranges today from 10 to 16 bits. Input-referred encodings with no perceptual transfer curve applied (such as CIE XYZ) should also be archived in the initial encoding bit-depth. Input-referred encodings that are perceptually linear should still be encoded with more than 8 bits per channel (10, 12 or 16 bits) when image-dependent color rendering is applied to render them to output-referred encodings, and/or when the output-referred encoding or output device-specific encoding specifies a large gamut and dynamic range. Output-referred encodings that are already rendered and are close to the gamut of a real output device can be encoded in 8 bits per channel. That is especially true for output-referred derivatives that need to be interpreted by today's software applications and devices.

Compression

Image compression schemes, such as JPEG and JPEG2000, usually convert the image to a luminance–chrominance–chrominance (YCC) representation and quantize high spatial frequency image information and chrominance information with fewer bits. They take advantage of the lower contrast sensitivity of the human visual system to high frequency and color information. It is important that when lossy compression is applied, the color encoding supports transforming to YCC and back without introducing visual artifacts due to re-quantization.

Table 17.3 Attributes of input-referred and output-referred color encodings

	Input-referred encoding	Output-referred encoding
Image representation	Colorimetric estimate of a scene/original	Colorimetric estimate of a reproduction
Color gamut	Large enough to encompass most scene and/or original colors	Large enough to encompass most output device colors
Perceptual uniformity (transfer function)	Data is optionally encoded using a transfer function for approximate perceptual uniformity (invertibility desired)	Data is optionally encoded using a gamma-type power function to approximate perceptual uniformity on the output device (invertibility desired)
Dynamic range	Must handle typical scene luminance ratios and/or original luminance ratios	Must handle a reproduction device/media luminance ratio
White-point	Should accommodate floating white-points or chromatic adaptation to a fixed white-point	Fixed white-point determined by reproduction viewing conditions (D50, D65)
Viewing conditions (linkage to color appearance)	Luminance level, viewing surround, adapted white-point, and viewing flare, as typical of outdoor environments	Luminance level, viewing surround, adapted white-point, and viewing flare as typical of indoor environments
Quantization/encoding	Quantization errors not visible on smooth, noiseless ramps; extended bit-depth encoding desired (10, 12, 16 bits per channel)	Quantization errors not visible on smooth, noiseless ramps; 8, 10, 12 or 16-bit encoding (8-bit for applications)
Compressibility	Not very important	Importance dependent on the imaging application (easy conversion to YCC color encoding)
Usual color encodings	CIEXYZ, CIELab, CIELuv, RIMM RGB, Photo YCC, ISO RGB, scRGB	e-sRGB, s-RGB, ROMM RGB, Adobe RGB 98, YC_bC_r (legacy: Apple RGB)

Table 17.3 summarizes the most important attributes of input-referred and output-referred color encodings.

Examples of color spaces and color encodings

In theory, every image library can design its own color encodings optimized for its applications, using the guidelines outlined above. The encoding would be based on either CIEXYZ, CIEL*a*b*, CIEL*u*v*, RGB, YCC or an *n*-channel space. The encoding specifications depend on the desired image representation and the application. These specifications need to be stored so that future processing can always be guaranteed. However, while an application-specific encoding might be desirable for the master archive file, it is not usually reasonable to use non-standard image encodings for derivatives files. In practice, commercial imaging applications drive most image production workflows and display, and are therefore the recipients of image library derivative files. They understand only a limited number of color encodings, and it is advantageous considering productivity and quality to keep derivative files in a standard color encoding. If a (higher quality) master file is available, new derivatives with different encodings for different output-driven purposes can always be generated from these archive files.

The following paragraphs outline some of today's color encoding practices found in image libraries and discuss their appropriateness for different applications. Note that the discussion is limited to existing practices and newly standardized color encodings whose specifications are published. Proprietary color encodings are not considered. See Table 17.4 for a comparison of the discussed color encodings.

Sensor space encoding

Sensor space encodings are usually found in two very different applications. Image libraries that employ multi-spectral image capturing systems archive n-channel files of the original data, and conversions to output-referred spaces or output device spaces are usually done directly from these files. These transformations are usually based on minimizing colorimetric errors to the original.

Sensor space encoded images are also archived in many commercial image libraries that use ICC enabled scanners and digital cameras. As discussed earlier, that approach does not truly consist of sensor space encoding, but of a nonstandard output-referred encoding that is characterized by the profile inherent transform to the PCS. That strategy is not recommended for master files.

Input-referred encoding

Many input-referred encodings are based on CIEXYZ, CIEL*a*b*, or CIELuv color spaces. The transformation from sensor to input-referred encoding is based on minimizing colorimetric errors. The encoding specifications, such as bit-depth, white-point, luminance ratio, viewing conditions, etc., depend on the application.

RIMM RGB

Eastman Kodak has proposed an input-referred RGB color encoding, called Reference Input Medium Metric RGB (*RIMM RGB*). The reference primaries describe a wide gamut RGB color space that allows possible color encodings outside the spectral locus (see Figure 17.8). The reference viewing conditions used to encode scene color values are typical of outdoors viewing environments. The observer adaptive white-point has the chromaticity values of CIE standard illuminant D50, necessitating that a chromatic adaptation transformation must be applied to the image data in cases where the chromaticity of the observer adaptive white for an actual scene differs from that of the reference

Table 17.4 Attributes of standard color encodings

Color encoding	Type	Encoding	Gamut	White-point	Primaries		Specified dynamic range and viewing conditions
						x $\quad$ y	
RIMM RGB	Input-referred	8-bit non-linear ($\gamma = 0.45$), 12-, 16-bit optional	Extended	D50	R 0.7347 0.2653 G 0.1596 0.8404 B 0.0366 0.0001	Typical luminance and viewing conditions for outdoor viewing defined Extended encoding defined for high dynamic range images
Photo YCC	Input-referred	8-bit non-linear ($\gamma = 0.45$)	Extended	D65	R 0.64 0.33 G 0.30 0.60 B 0.15 0.06	Typical luminance and viewing conditions for outdoor viewing defined
sRGB	Output-referred	8-bit non-linear ($\gamma = 0.417$)	CRT	D65	R 0.64 0.33 G 0.30 0.60 B 0.15 0.06	Typical CRT monitor reference viewing environment defined, with D50 as ambient white-point
e-sRGB	Output-referred	10-, 12-, 16-bit non-linear ($\gamma = 0.417$)	Extended	D65	R 0.64 0.33 G 0.30 0.60 B 0.15 0.06	Typical CRT monitor reference viewing environment defined, with D50 as ambient white-point
ROMM RGB	Output-referred	8-bit non-linear ($\gamma = 0.556$) 12-, 16-bit optional	Extended	D50	R 0.7347 0.2653 G 0.1596 0.8404 B 0.0366 0.0001	Reproduction viewing environment defined
Adobe RGB 98	Output-referred	8-bit non-linear ($\gamma = 0.45$)	Extended CRT	D65	R 0.64 0.34 G 0.21 0.71 B 0.15 0.06	No
Apple RGB	Output-referred	8-bit non-linear ($\gamma = 0.556$)	CRT	D65	R 0.625 0.34 G 0.28 0.595 B 0.155 0.070	No

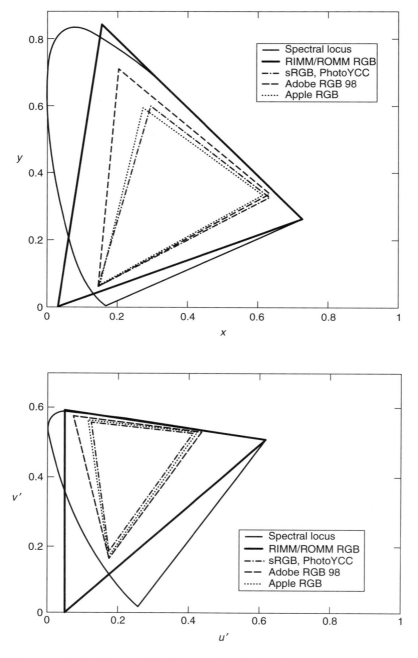

Figure 17.8 x, y and $u'v'$-Chromaticity diagrams for sRGB, ROMM RGB, Adobe RGB 98 and Apple RGB.

conditions. The non-linear transfer function is based on the ITU-R BT.709 recommendations [45] and is equal to 0.45 (1/2.2).

RIMM RGB master file encoding can be considered if the corresponding output-referred derivative encoding is ROMM RGB see Chapter 14.

PhotoYCC

Eastman Kodak also defined PhotoYCC, which is an encoding that was primarily developed for the *Kodak PhotoCD* system. However, the specifications have been published, and many image libraries that have used – or still use – the PhotoCD system have master files archived in PhotoYCC.

PhotoYCC was originally designed as an input-referred encoding, but could also be used as an output-referred color encoding. The viewing conditions are the same as RIMM RGB, but the scene adaptive white is assumed to be CIE illuminant D65. The encoding is based on a reference image capturing device that has the sensitivity of the ITU-R BT.709 color matching functions. Linearized device-specific *RGB* values can be mapped to RGB_{709} values using Equation (17.1). These RGB_{709} values are nonlinearly encoded with a gamma function of 0.45 (1/2.2) to $R'G'B'_{709}$. $R'G'B'_{709}$ is linearly transformed to Luma and chroma values, and the encoding is usually quantized to 8-bit. PhotoYCC can easily be transformed to a monitor-based RGB space for viewing, or to YC_bC_r for compression.

Output-referred color encoding

There are several output-referred color encodings that have been standardized, or have become *de facto* industry standards. All have been designed with certain applications in mind.

sRGB

sRGB is defined by the IEC 61 966-2-1 standard as a default color encoding for multimedia applications. It is an output-referred encoding, based on the characteristics of a CRT reference display. In its normative part, the standard defines the relationship between 8-bit sRGB values and CIE 1931 XYZ values as measured at the faceplate of the reference display. The reference display white-point and primaries are defined according to ITU-R BT.709 (see Table 17.4). The colorimetry seen by an observer looking at an image on a reference display in the reference viewing conditions is described in an informative annex.

The purpose of sRGB is to define an output-referred color space for data interchange in multimedia where ICC color management is not implemented. Due to similarities of the defined reference display to real CRT monitors, often no additional color space conversion is needed to display images. However, conversions are required to transform image data to devices with different dynamic ranges, gamuts and viewing conditions. As mentioned before, if these parameters are close enough to those of sRGB, an ICC profile can be used. If they are very different, an image-dependent mapping has to be applied for good results.

sRGB has also become the *de facto* encoding of consumer digital camera images. Most consumer desktop color printers today assume for the transformation to device-specific coordinates that an image is encoded in sRGB unless otherwise specified. sRGB encoding is therefore ideal for all image library derivatives that involve either monitor viewing, such as publishing to the web or electronic kiosk applications, or consumer applications.

e-sRGB

An extended gamut and bit-depth encoding, e-sRGB, has recently become an ANSI standard. Most encoding parameters, such as transfer function, viewing conditions, and primaries are similar to sRGB. However, e-sRGB defines an offset and over range to encode a larger gamut and dynamic range than sRGB. The gamut of e-sRGB is very close to the gamut of all visible colors.

e-sRGB is designed as a storage and interchange space for photographic applications, and is well suited to encode output-referred master files. Because of its similar encoding specifications, the conversion from e-sRGB to sRGB is simple to implement, so that derivative files intended for monitor viewing are easily created. Transformations to ROMM RGB, YCC and ICC PCS are also specified.

ROMM RGB

ROMM (Reference Output Medium Metric) *RGB* is a wide-gamut, output-referred RGB color encoding. It was designed by Eastman Kodak and is intended as an RGB color space for manipulating and editing images after the initial rendering has been applied. The ROMM RGB primaries are not tied to any monitor specification. Rather they were selected to wholly enclose an experimentally determined gamut of surface colors, so that there would not be any loss of color information when representing reflectance colors that had been captured in an input-referred color space, such as RIMM RGB. ROMM RGB uses a CIE illuminant D50 white-point, which is a standard for viewing and evaluating graphic arts reproductions, as well as the ICC PCS white-point.

ROMM RGB has the largest gamut of the output-referred RGB spaces described here. By selecting a gamut that wholly encloses most real-world surface colors, many ROMM RGB values are wasted in the production of reflection hard-copy, in that they do not correspond to reflectance colors and are never used. ROMM RGB defines therefore 12- and 16-bit encoding to allow for greater precision in addition to the usual 8-bit encoding.

ROMM RGB encoded images do not display well without additional transformations to device-specific values. The encoding parameters have rather been chosen to fit well into an ICC color managed graphic arts workflow. Therefore, ROMM RGB encoding is most appropriate for master files or derivative files that will undergo further color transforms and/or be printed on hard-copy output.

Adobe RGB 98

Adobe, Inc. introduced the concept of a working space that is device independent in their currently most popular commercial image manipulation application, Adobe Photoshop©.

The goal is to make the image data more portable and not tied to a RGB display. It is also the space to which the user will import images from different sources and in which the user will make editing decisions.

Adobe RGB 98 was intended to provide a larger gamut than previous monitor spaces, so prepress users can set it as the default working space in Adobe Photoshop. It is loosely based on the SMPTE-240M standard and was later renamed Adobe RGB 98.

The Adobe RGB 98 encoding gamut is smaller than the ROMM RGB gamut, but still encloses most device colors of current CMYK printers. It is close enough in gamut to halftone printing device gamuts that applying ICC color management usually results in good reproduction quality. It is therefore very popular in graphic arts and professional photography applications. Commercial image libraries, such as Corbis and Getty, distribute their print-intended images in a color space that has similar characteristics. Images encoded in Adobe RGB 98 also display reasonably well on an uncalibrated monitor, so that image buyers that do not use color management get a reasonable good representation of image appearance.

While the encoding is reasonable for derivative files that are distributed today to graphic arts applications that might or might not be color managed, it is worth considering archiving output-referred master files with a larger gamut so that more scene and original colors can be retained. It is clear that future display and print applications will be able to reproduce a larger gamut and dynamic range.

Apple RGB

Apple RGB color encoding is based on the classic Apple 13″ RGB monitor. Because of its popularity and similar Trinitron-based monitors that followed, many key publishing applications used it as the default RGB space in the past, and therefore many legacy images are encoded in this color space. The gamut of Apple RGB and sRGB are similar, but the non-linear transfer function differs (1/1.8 for Apple RGB, 1/2.4 for sRGB). Therefore, Apple RGB images will appear too dark on noncalibrated CRT monitors. This effect is especially visible in web applications where thumbnails with different color encodings are displayed together. However, it is not appropriate to transform Apple RGB master files to another output-referred color encoding, as the occurrence of visual quantization artifacts increases with the number of color transformations applied to an image, and it is likely that a new default encoding for display viewing will be defined once the predominant display technologies change. Rather, new derivatives should be created for the application desired.*

Future display considerations

If possible, one should avoid archiving master files with an output-referred color encoding that is based on today's CRT monitor specifications. Most flat panel displays (LCD,

---

* Note that when transforming from Apple RGB to sRGB for monitor viewing, it is usually sufficient just to adjust the non-linear transfer function. The primaries are close enough so that in practical viewing applications the slight color shift from one encoding to the next is not noticeable.

plasma) have an S-shaped transfer function, which differs from display model to display model. Displaying CRT encoded images on these displays without correction results in significant errors in either highlights or shadows. Additionally, the viewing conditions in future work environments could be very different from those of today. Most output-referred RGB monitor encodings define viewing conditions (dim surround) and dynamic range ($\sim$80:1) based on the limitations of CRT monitors. The latest desktop flat panel displays can output 230+ cd/m^2 with a dynamic range over 300:1, which will reduce the effect of the ambient luminance level. Personal digital assistants (PDA) and some laptop LCDs, however, usually have a lower dynamic range ($\sim$40:1) to minimize power consumption. Considering the increase of portable devices that can display color images, such as PDAs, DVD players, and cellphones, defining a 'normal' display viewing environment will become more difficult.

It is easy to see that one standard output-referred encoding based on display technology, such as the current sRGB, will not be feasible in the future. Therefore, output-referred master files should be archived in an output-referred encoding that is not limited to current technology, with the exception of 'visual representation' files captured with current consumer scanners and digital cameras that support only sRGB encoding.

17.5 COMPRESSION AND FILE FORMATS

17.5.1 Compression

For many applications, the size of an image file needs to be reduced to minimize transfer time over limited bandwidth. The cost of storage is usually insignificant for an image library compared to the cost of production and maintenance. As a result, it is more common to find derivative files that are compressed than master files.

Lossless compression

Lossless compression schemes use different bit-representation to minimize the number of bits encoded. When the image is decompressed, it is identical to the original. The *Huffman Coding*, for example, reduces the average number of bits to represent a set of values by using shorter codes for more probable values. *LZW* (Lempel–Ziv–Welch) is a lossless coding implemented in the GIF file format and maps variable-size blocks of symbols to fixed-length codes via table look-up [46]. Generally, the size of the file of losslessly encoded images is about 0.5 to 0.7 the size of the uncompressed file.

Lossy compression

JPEG (Joint Photographic Expert Group) [47] is the most popular lossy image compression scheme today. It is supported by all applications able to display images and has the following structure. The input image passes first through a color conversion. Then it is mapped to the frequency domain with a discrete cosine transform (DCT). The resulting coefficients are quantized and entropy encoded. The rate-distortion (RD) unit controls the quantization step size as a function of bit rate and distortion, which the user controls by determining the 'quality' (Q) factor. High frequency image information and chrominance information are encoded with fewer bits to take advantage of the lower contrast sensitivity of the human visual system to high frequency and color information [48].

Because compression in the RGB domain is very inefficient, images are usually converted to an opponent (luminance–chrominance) color space. While there are several encoding possibilities, such as converting the image to CIE L*a*b*, YC_bC_r, based on the ITU-R BT. 601 recommendations [49], is the most often applied opponent color encoding. Its transformation from RGB is as follows:

$$\begin{bmatrix} Y \\ C_b \\ C_r \end{bmatrix} = \begin{bmatrix} 0.299 & 0.587 & 0.114 \\ -0.169 & -0.331 & 0.500 \\ 0.500 & -0.419 & -0.081 \end{bmatrix} \begin{bmatrix} R \\ G \\ B \end{bmatrix} \tag{17.8}$$

where Y describes the luminance, and C_b, C_r the chrominance channels.

What is commonly called a 'JPEG file' is a JPEG compressed file saved in the JFIF file format, which was designed by C-Cube Microsystems and is in the public domain. However, JPEG compression is also supported by SPIFF, EXIF and some TIFF file formats, such as TIFF/FX, the standard fax format, which defines a CIE L*a*b* opponent color encoding.

ISO/IEC JTC1 SC29 is currently working on JPEG2000, a compression scheme that uses a discrete wavelet transform (DWT) and offers a better quantization efficiency and more features compared to JPEG, such as lossless compression and random access within the image. The baseline decoder [50], the first JPEG2000 part published, assumes an sRGB encoding when no color encoding is specified, but will accept 'simple' (matrix based) ICC profiles. The corresponding file format is JP2. The extended version will support full ICC profiles and up to 256 color channels. The corresponding file format for the extend version is JPX.

Depending on the level of compression and the display conditions, there will be no artifacts visible due to compression. The compression is then considered to be *visually lossless*. However, compressed images should not be further rendered or color adjusted, as they are very sensitive to quantization artifacts. Therefore, compression should only be applied as the last image processing step before delivery for derivative files. Images that need to be further processed, such as input-referred master files or output-referred derivative files encoded in an output-referred color space that will undergo further processing, should not be compressed or only be compressed with lossless coding.

17.5.2 File formats

To ensure longevity, master files should be encoded in file formats whose specifications are freely available and can be archived with other metadata. A file format recommended for master files is TIFF (Tagged Image File Format). Its specifications are readily available and no licensing fees have to be paid. Another possibility is the use of one of the standardized TIFF versions like TIFF/EP, TIFF/IT, or EXIF [51]. The latter is primarily used in digital

Table 17.5 A comparison of file formats, color encoding, compression and image library applications

File formats	TIFF	JPX	PhotoCD	EXIF	JFIF	JP2	GIF	PNG
Most common image representation	All	All	Input-referred	Output-referred	Output-referred	Output-referred	Output-referred	Output-referred
Color encoding[a]	All	All	Photo YCC	All (usually sRGB)	YC_rC_b, ICC profiles	sRGB, simple ICC profiles	RGB	sRGB, ICC profiles
Master file	Yes	Yes	PhotoCD system	Consumer, digital camera output	Consumer, digital camera, scanners	Not recommended	Not recommended	Not recommended
Derivative files	Print	Print	No	No	Low-end print, monitor	Low-end print, monitor	Monitor	Monitor
Compression	None, lossless	JPEG 2000, lossless and lossy	Lossy	None, JPEG	JPEG	JPEG2000	Color palette, LZW	Flate

[a]Either natively or with ICC profiles.

Sources:

ISO 12233: 2000 Photography – Electronic still picture cameras – *Resolution measurements.*

ISO/CD 16067-1 (2001) Photography – Electronic scanners for photographic images – *Spatial resolution measurements – Part 1: Scanners for reflective media.*

cameras to store image data. Once the specification has been published, JPX, the file format associated with the JPEG2000 extended version, can also be considered.

The original file format associated with the Photo YCC encoding was specifically designed for scanning photographic film, and specified a pyramid structure with 5–6 resolutions (base/16 to 16 base, optionally 64 base). The maximum resolution is 2048 × 3072 for 16 base and 4096 × 6144 for 64 base, which makes it unsuitable for many image library applications where the originals are scanned at higher resolution and/or with a different aspect ratio than the format supports. The image is slightly lossy compressed by sub-sampling the chroma channels. The file format is used only in Kodak PhotoCD scanning systems. Plug-ins are available to open the files in other applications.

GIF (Graphic Interface Format) was developed by CompuServe, Inc. as a protocol for on-line transmission and interchange of raster graphic data. Colors are specified in uncalibrated device-dependent RGB and palettized, either image dependent or using a standard palette. The raster data is then compressed using LZW. The encoding of GIF is very efficient, and almost all web browsers support the format.

PNG (Portable Network Graphics) is a patent-free replacement of GIF that also supports palettized color, grayscale, and RGB color. Additionally, ICC Profiles can be embedded. PNG is a recommended file format for derivatives that will be used in web and portable device applications were the transmission bandwidth is low.

The color encoding and file formats (see Table 17.5) appropriate for derivative files are dependent on the applications the images are intended for. Depending on what the specific software application supports, formats other than those listed above can be used, such as VRML, PDF, or QuickTime. A number of file formats support multiple images (animated GIF, MNG, FlashPix, QuickTime), multi-page images (TIFF/FX, PDF), compound images (TIFF-FX Profile M, PDF), and compound documents (XML, HTML, SVG, PDF) [52].

17.6 OTHER ISSUES IN BUILDING AND MAINTAINING A DIGITAL IMAGE LIBRARY

There are, of course, many other issues to consider when building and managing a digital library besides the color encoding and management described in this chapter. Some of the more important issues are briefly outlined below.

17.6.1 Storage

The longevity of digital image files is dependent not only on the initial encoding, but also on the file format, the storage media and the media format, i.e. the software used to write the image files onto the media. Contrary to analog libraries and archives, where longevity decreases when objects are handled often, digital image libraries need an active migration policy to ensure digital image permanence [53, 54].

17.6.2 Text metadata

As opposed to an analog library, the digital image object is not visible without some technological intervention. Browsing the library is only feasible when the number of images is small. Consequently, cataloging information has to be associated with the images in order for a search and retrieval system to find the correct images. Such information includes author, copyright, date, caption, image content, etc., as well as administrative metadata. For describing image content, a *thesaurus* commonly controls the indexing terminology so that identical content is described with the same vocabulary, and the indexing and search terms can be translated to other languages.

The indexing and managing of such text metadata can be as involved as the capturing and managing of the pixel data. To date, there is no agreement between the different image library application communities as to exactly which text data needs to be captured. The two large commercial libraries, Getty and Corbis, use their own indexing scheme and thesaurus. Some efforts at standardization have been made for individual communities: The *Dublin Core Metadata Initiative* [55] has published standard categories for cultural objects, intended for museum and library archives, and the *International Press Telecommunications Council* [56] has developed a set of standard indexing fields for digital news photos.

17.6.3 Content-based image retrieval (CBIR)

Content-based image retrieval systems automatically index images according to their visual content. CBIR was primarily developed to be an alternative to the manually intensive labor of capturing textual information about an image, and misclassification due to subjective image interpretation. The goal is to design image representations that correlate well with the human visual perception, so that search results 'resemble' the original query.

Visual features can be categorized with general features, such as color, textures, and shape, or domain-specific features such as human faces, fingerprints, etc. Color features are mostly extracted by analyzing the color or cumulative color histogram of an image [57, 58]. Distances are calculated by intersecting the histograms or by using a Euclidean distance. The resulting *feature vectors* are then used to retrieve 'similar' images, either by analyzing the original image of a query, or by allowing the user to sketch an original image [59, 60].

17.6.4 Database architecture

The architecture of a digital library depends on its users and usage. However, there are some elements that are common to all [61].

- A *graphical user interface* (GUI) that lets the user browse and query the database
- A *client/server* interface that interprets the queries and sends them to an access server
- An *access server* that manages the data retrieval from the archive servers and controls user sessions, user profiles and access rights
- An *archive server* that contains pixel and text data.

17.7 SUMMARY

The imaging industry and international standards bodies continue to develop standard color encodings for different imaging applications, and for the storage and interchange of images. It is to be expected that in a few years, such specifications will be adopted by the imaging industry, and specific recommendations for color encodings applicable to image libraries can be developed. In the meantime, however, each library has to choose its own encoding strategies. These strategies should be developed according to not only current imaging technologies, but also to future, as yet unknown, imaging devices with extended color gamuts and nonstandard transfer curves.

In general, the advantages of encoding master files in higher bit-depth in sensor or input-referred color spaces will require extra steps in the production workflow, and non-commercial software might have to be developed. However, considering the possibility of creating new derivatives with different color encodings in the future, having more image 'information' available will prevent master files from becoming obsolete.

REFERENCES

1. Frey, F. and Süsstrunk, S. (1996) Image Quality Issues for the Digitization of Photographic Collections. *Proc. IS&T 49th Annual Conference*, 349–353.
2. Corbis Corporation, www.corbis.com.
3. Süsstrunk, S. (1998) Imaging Production Systems at Corbis Corporation. *Proc. ICPS'98*, 232–234.
4. Getty Images, Inc. www.gettyimages.com.
5. Saunders, D. and Cuppit, J. (1993) Image Processing at the National Gallery: The Vasari Project. *National Gallery Technical Bulletin*, **14**, 72–85.
6. Granger, S., Saunders, D., Martinez, K. and Cupitt, J., 'From Vasari to Viseum: the development of methods and systems for the acquisition, display, publishing and transmission of high resolution colorimetric digital images of works of art,' available at http://ds.dial.pipex.com/stewartg/vasari.html.
7. Holm, J. and Judge, N. (1995) Electronic Photography at the NASA Langley Research Center. *Proc. IS&T 48th Annual Conference*, 436–441.
8. Frey, F. S. and Reilly, J. M. (1999) *Digital Imaging for Photographic Collections*, Image Permanence Institute, RIT. Also available at http://www.rit.edu/ipi.
9. Holm, J. (1999) Integrating New Color Image Processing Techniques with Color Management. *Proc. IS&T/SID 7th Color Imaging Conference*, 80–86.
10. Süsstrunk, S., Buckley, R. and Swen, S. (1999) Standard RGB Color Spaces. *Proc. IS&T/SID 7th Color Imaging Conference*, 127–134.
11. CIE (1987) Publication No. 17.4 *International Lighting Vocabulary*.
12. ISO/WD 22028-1.15 (2001) *Photography and graphic technology – Extended color encoding for digital image storage, manipulation and interchange – Part 1: Architecture and Requirements*.
13. ICC.1: (2001) *File Format for Color Profiles*, International Color Consortium. Available at www.color.org.
14. Mintzer, F. (1999) Developing Digital Libraries of Cultural Content for Internet Access. *IEEE Communications Magazine*, **37**(1), 72–78.
15. *Tagged Image File Format*, specifications 6.0, Adobe (1995). Available at http://partners.adobe.com/asn/developer/pdfs/tn/TIFF6.pdf.
16. ISO 12234-2: (2001) *Photography – Electronic still picture cameras – Removable memory – Part 2: Image data format – TIFF/EP*.
17. Martinez, K., Cuppit, J. and Saunders, D. (1993) High resolution colorimetric imaging of paintings. *Proc. SPIE*, **1901**, 110–118.
18. Burns, P. D. and Berns, R. S. (1996) Analysis Multispectral Image Capture. *Proc. 4th Color Imaging Conference*, 19–22.

19. Sugiura, H., Kuno, T., Watanabe, N., Matoba, N., Hayashi, J. and Miyake, Y. (1999) Development of Highly Accurate Multispectral Cameras. *Proc. International Symposium on Multispectral Imaging and Color Reproduction for Digital Archives*, Society of Multispectral Imaging of Japan, pp. 73–80.
20. Brettel, H. and Schmitt, F. (2001) Multispectral Internet Imaging. *Proc. SPIE*, Vol. 4311, pp. 17–26.
21. Giorgianni, E. J. and Madden, T. E. (1998) *Digital Color Management – Encoding Solutions*, Addison-Wesley.
22. I3A 7466: (2001) *Photography – Electronic still picture imaging – Reference Input Medium Metric RGB Color encoding (RIMM-RGB)*.
23. IEC/DIS 61966-2-2 (2001) *Multimedia systems and equipment – Color measurement and management – Part 2-2: Color management – Extended RGB color space – scRGB*.
24. ISO/WD4.0 17321 (1999) – *Color characterization of digital still cameras*.
25. Hunt, R. W. G. (1995) *The Reproduction of Color*, 5th ed. Fountain Press.
26. Land, E. H. (1977) The retinex theory of color vision. *Scientific American*, 108–129.
27. Finlayson, G. D., Hordley, S. D. and Hubel, P. M. (1999) Color by correlation: a simple unifying theory of color constancy. *Proc. IEEE International Conference on Computer Vision*, 835–842.
28. Hunt, R. W. G. (1998) *Measuring Color*, 3rd ed. Fountain Press.
29. McCamy, C. S., Marcus, H. and Davidson, J. G. (1976) A Color Rendition Chart. *J. Applied Photographic Engineering*, **2**, 95–99.
30. ISO 12641: (1997) *Graphic technology – Prepress digital data exchange – Color targets for input scanner calibration*.
31. Wandell, B. A. and Farrell, J. E. (1993) Water into Wine: Converting Scanner RGB to Tristimulus Values. *Proc. SPIE*, **1909**, 92–101.
32. Finlayson, G. D. and Drew, M. S. (1997) White-point preserving color correction. *Proc. IS&T/SID 5th Color Imaging Conference*, 258–261.
33. Holm, J., Tastl, I. and Hordley, S. (2000) Evaluation of DSC (Digital Still Camera) Scene Analysis Error Metrics – Part 1. *Proc. IS&T/SID 8th Color Imaging Conference*, 279–287.
34. von Kries, J. (1902) Chromatic Adaptation. *Festschrift der Albrecht-Ludwigs-Universität* [Translation: MacAdam, D. L., *Colorimetry-Fundamentals*. SPIE Milestone Series, Vol. MS, 77, 1993].
35. Lam, K. M. (1985) *Metamerism and Color Constancy*, Ph.D. Thesis, University of Bradford.
36. Finlayson, G. D. and Süsstrunk, S. (2000) Performance of a chromatic adaptation transform based on spectral sharpening. *Proc. IS&T/SID 8th Color Imaging Conference*, 49–55.
37. Li, C., Luo, M. R. and Rigg, B. (2000) Simplification of the CMCCAT97. *Proc. IS&T/SID 8th Color Imaging Conference*, 56–60.
38. Finlayson, G. D. and Süsstrunk, S. (2001) Spherical Sampling and Color Transformations,' to be published in *Proc. IS&T/SID 9th Color Imaging Conference*.
39. PIMA 7666: (2001) *Photography – Electronic Still Picture Imaging – Reference Output Medium Metric RGB Color encoding: ROMM-RGB*.
40. IEC 61966-2-1 (1999) *Multimedia systems and equipment – Color measurement and management – Part 2-1: Color management – Default RGB color space – sRGB*.
41. PIMA 7667: (2001) *Photography – Electronic Still Picture Imaging – Extended sRGB color encoding, e-sRGB*.
42. ANSI/CGATS TR 001 (1995) *Graphic Technology – Color Characterization Data for Type 1 Printing*.
43. Poynton, C. (1996) *A Technical Introduction to Digital Video*, John Wiley & Sons, Inc.
44. International Color Consortium, www.color.org.
45. ITU-R Recommendation BT.709-3: (1998) *Parameter values for the HDTV standards for production and international programme exchange*.
46. Held, G. and Marshall, T. R. (1996) *Data and Image Compression*, Wiley & Sons.

47. ISO/IEC 10918-1: (1990) (ITU-T T.81): *Information Technology – Digital compression and encoding of still images – Requirements and guidelines.* Available at http://www.w3.org/Graphics/JPEG/itu-t81.pdf).
48. Mitchell, J. L. and Pennebaker, W. B. (1993) *JPEG Still Image Compression Standard*, Chapman & Hall.
49. ITU-R BT.601-5: (1999) *Studio encoding parameters of digital television for standard 4:3 and wide screen 16:9 aspect ratios.*
50. ISO/IEC 15444-1: (2001) *Information Technology – JPEG image coding system.*
51. JEIDA-49-1997: *Digital Still Camera Image File Format Standard*, Version 2.0.
52. Buckley, R. and Beretta, G. B., (2001) *Color Imaging on the Internet*, short-course, PICS, available at http://www.hpl.hp.com/imaging/cii.
53. Rosenthaler, L. and Gschwind, R. (1998) Long Term Preservation and Computers – a Contradiction?' *Proc. ICIPS'98*, **2**, 239–246.
54. Frey, F. and Süsstrunk, S. (2000) Digital Photography – How long will it last?' *Proc. ISCAS'2000*, **V-113**.
55. The Dublin Core Metadata Initiative, http://dublincore.org/.
56. International Press Telecommunications Council, http://www.iptc.org.
57. Swain, M. and Ballard, D. (1991) Color Indexing. *J. Computer Vision*, **7**.
58. Stricker, M. and Orengo, M. (1995) Similarity of Color Images. *Proc. SPIE*, **2420**.
59. Rui, Y., Hung, T. S. and Chang, S. F. (1999) Image Retrieval: Current Techniques, Promising Directions, and Open Issues. *J. Visual Comm.*, **10**, 39–62.
60. Gunther, N. J. and Beretta, G. B. (2001) A Benchmark for Image Retrieval using Distributed Systems over the Internet: BIRDS-I. *Proc. SPIE*, **4311**.
61. Aquarelle, *Final Report*, available at http://aquarelle.inria.fr/aquarelle/EN/final-report.html.

Standards activities for colour imaging

David Q. McDowell

18.1 INTRODUCTION

Today, we are surrounded by colour images that come to us via television, motion pictures, our computer displays, the World Wide Web, printed material, and photographs to name the most common. This chapter will focus primarily on the colour related standards of printing and publishing and still photography. Ironically, until the arrival of digital imaging in the early 1980s these industries had few colour related standards.

In the early 1980s the only graphic arts standards that touched on colour were two standards that specified the colorimetric values and provided samples of the solids of representative inks used for colour printing using letterpress and offset printing. These were ISO 2845, *Set of printing inks for letterpress printing – Colorimetric characteristics* and ISO 2846, *Set of printing inks for offset printing – Colorimetric characteristics*. Both were published in 1975 and had little application.

The photographic industry had an international viewing standard, ISO 3664, *Photography – Illumination conditions for viewing colour transparencies and their reproductions*, also published in 1975. Of course, while not colour standards, the ISO 5 series of standards

Colour Engineering, edited by P.J. Green and L.W. MacDonald.

for densitometry were, and are, used extensively for the process control of colour imaging by both the photographic and printing industries and have been kept current.

This lack of colour standards is surprising, because the graphic arts has been producing coloured images since the mid-1700s and we have had practical colour photography since the introduction of Kodachrome film in 1935. However, both colour printing and colour photography were essentially closed systems using analogue film as the interchange media. It was not until sufficient computer power was available to manipulate and exchange images as digital data that meaning had to be assigned to information being exchanged. It was this need that led to the development of colour related standards for graphic arts and photography, starting in the late 1980s.

The following describes some of the initial efforts at developing colour standards for imaging in the graphic arts and photographic industry, some of the standards that have resulted from the ongoing efforts and the standards currently in development. Because the author was directly involved in the preparation of many of these standards, in many cases the descriptions of the goals and/or contents of these standards are summaries of the introductions and scope of the documents themselves.

18.2 THE INITIAL GRAPHIC ARTS COLOUR STANDARDS

The first steps into colour data definition for digital data in the graphic arts started in the late 1980s under the auspices of the ANSI IT8 Committee (IT8 simply stands for Image Technology Committee #8). Working Group 11 (WG11) of IT8 was created to define standard targets to be used in the characterization of both scanner input of colour information and CMYK output printing and proofing devices. At about the same time ANSI CGATS (Committee for Graphic Arts Technologies Standards) started work on the definition of conditions to be used in the measurement and computation of colorimetric information for graphic arts images.

18.2.1 Scanner targets

Why a scanner target as the first step into colour standards? Pragmatically, it was obvious, relatively easy, and a model existed – the Kodak Q60 scanner calibration target.

The problem that needed to be addressed was rooted in the fact that the responsivity of a scanner does not match human vision – is not colorimetric – and each manufacturer of photographic goods uses different dyes to metamerically create the colour that we see. This means that to properly characterize (initially, before colour management, we called this calibration) a scanner, a known sample of each product to be scanned must be presented to the scanner and the scanner responsivity to that dye set determined. Without a common target definition and layout, it was not practical for either scanner manufacturers or film manufacturers to develop general purpose software tools to facilitate this characterization task.

The IT8 committee chose to colorimetrically define the colours that should appear in the target, and then leave it to each manufacturer of photographic goods to produce targets using their dye sets. The committee also specified the data to be reported and an associated data format. The combination of a target and its associated data allows the colour characterization a scanner for that particular film type and the standard definition of a layout and data format enables general purpose software development. The requirements are defined in the ANSI IT8.7/1 and ANSI IT8.7/2 standards. Their official titles are *Graphic technology – Color transmission target for input scanner calibration* and *Graphic technology – Color reflection target for input scanner calibration*. These two standards have been combined into a single ISO standard, ISO 12641:1997 *Graphic technology – Prepress digital data exchange – Colour targets for input scanner calibration*. Targets meeting these standards are being produced by Kodak, Agfa, and Fuji and in many ways are the cornerstone of today's colour management systems.

But let's look back at the development of these standards to better understand both their significance and the details of their design. The initial look at graphic arts input scanning, by WG11, revealed that the total graphic arts transmission input was encompassed by a very limited set of products. The Agfachrome, Ektachrome, Fujichrome, Kodachrome and Konicachrome film families represented over 95% of the transmission material available to be scanned. The colour gamut of these materials included a very large common area and this became the focus of the committee's approach.

Three working agreements were quickly adopted. First, target definitions would be based on the CIELAB colour space to provide a reasonably uniform visual mapping of the colour space defined by these films and targets. It was also agreed that the same colorimetric values for different products should represent the same visual appearance when the products were viewed together. This led to the second decision – that is the white point should be the illuminator rather than D_{min} of the film. Further, D50 (as defined by the ANSI and ISO viewing standards, PH2-28 and ISO 3664) was adopted as the reference illuminant and the CIE 2° observer was selected to define the colorimetry. The design of the Kodak Q60 scanner calibration target had been offered to the standards community, and the third decision was that the basic Q60 design should be accepted as a working reference for the IT8 activity.

The original Kodak Q60 target consisted of three parts. These were single- and two-colour dye scales, a neutral dye scale, and a colour gamut area. The colour gamut area included patches at 12 hue angles, each arranged with three chroma values at each of three lightness levels. A tenth patch at each hue angle was located at the lightness and chroma corresponding to the maximum chroma available at that hue angle.

The IT8 target that evolved is shown in Plate 2.2. The dye scale definitions in columns 13 through 19 were adapted directly from the Kodak target. Column 16 is specified to be a neutral scale that varies in equal steps of lightness (not density) from D_{min} to neutral D_{max}. The dye amounts used to create the neutral are used either singly or in pairs to create the cyan, magenta, yellow, red, green and blue scales in columns 13–15 and 16–19.

The colour gamut area, columns 1–12, uses the same 12 approximately equally spaced hue angles are as used in the Q60. These are listed in Table 18.1. At each hue angle three lightness levels are specified and at each lightness level four chroma values.

The determination of the lightness levels and chroma values represented an unusual level of sharing among the four film companies involved. Each company (Agfa, Fuji, Kodak, and Konica) shared the spectral dye curves of their products and achievable maximum densities. From these data, the theoretical gamut that could be created by

Table 18.1 Transmission target L^* and C^* values versus hue angle

Row	Hue angle	Column												
		1	2	3	4		5	6	7	8		9	10	11	12
		L1	C1	C2	C3	C4	L2	C1	C2	C3	C4	L3	C1	C2	C3	C4
A	16	15	10	21	31	– [a]	35	15	30	45	– [a]	60	8	16	24	– [a]
B	41	20	11	23	34	– [a]	40	17	34	51	– [a]	65	7	15	22	– [a]
C	67	30	11	22	34	– [a]	55	20	40	60	– [a]	70	9	17	26	– [a]
D	92	25	9	18	27	– [a]	50	17	35	52	– [a]	75	23	46	69	– [a]
E	119	30	11	22	33	– [a]	60	20	39	59	– [a]	75	12	25	37	– [a]
F	161	25	10	21	31	– [a]	45	17	35	52	– [a]	65	12	25	37	– [a]
G	190	20	7	14	21	– [a]	45	14	29	43	– [a]	65	11	23	34	– [a]
H	229	20	7	15	22	– [a]	40	13	25	38	– [a]	65	7	15	22	– [a]
I	274	25	14	27	41	– [a]	45	10	21	31	– [a]	65	6	12	17	– [a]
J	299	10	17	34	51	– [a]	35	13	27	40	– [a]	60	7	14	21	– [a]
K	325	15	13	26	39	– [a]	30	17	35	52	– [a]	55	12	23	35	– [a]
L	350	15	10	21	31	– [a]	30	16	33	49	– [a]	55	10	21	31	– [a]

[a]These values are specific to the product used to create the target and equal to the maximum C^* available at the hue angle and L^* specified. They are to be defined by the manufacturer of the product used to make the target.

each of the dye sets was computed. It is important to understand that although these companies make many different film products, within each company the image-forming dyes are common to many films. In fact the committee identified only five unique dye sets. These were the dyes used in Agfachrome, Fujichrome, Kodachrome, Ektachrome, and Konicachrome. (Today scanner targets meeting this standard are available on all but the Konica product.)

The minimum chroma of the five product gamuts was determined. Because these were theoretical plots, the committee picked a value of 80% in chroma of this gamut and defined it as a gamut limit that could be achieved by all products. Three lightness levels were selected at each hue angle to best represent the gamut at that hue angle. At each lightness level, three chroma values were identified – one corresponding to the 80% common gamut, one to two-thirds and one to one-third of that value. All products were expected to produce these 108 patches. A fourth value was specified at each hue angle and lightness level to be product unique. It is intended to represent the maximum practical chroma available from a particular dye set at the hue angle and lightness level specified.

Table 18.1 is the table of values resulting from this work that became the aims for the colour gamut area of the transmission target that were contained in ANSI IT8.7/1:1993.

Two additional elements were added to the basic Q60 design to complete the IT8 target definition. A 22-step grey scale, with specified values of L^*, was added at the bottom of the target. The values chosen were based on the earlier gamut mapping work and represented the range of neutral density that could be produced by all of the dye sets evaluated. To ensure that the characteristics of the specific product used to create a target were included, a product-specific neutral D_{min} and D_{max} were added as unnumbered patches at either end of the grey scale.

The second element added was the vendor optional use of columns 20–22. Here the vendor manufacturing a target was allowed to add any feature they deemed worthwhile. Each vendor has chosen to use this area differently. Kodak has the image of a model and several skin tone patches; Agfa and Fuji have both chosen to have patches of special colours in this area.

Provision was made for targets sized to fit 4×5 inch material, a single 35 mm frame, and a set of seven 35 mm frames. The set of seven 35 mm targets was intended to accommodate 35 mm scanners in the belief that the patch size required to make the target on a single 35 mm frame would be too small for colorimetric readings to be made accurately. The reading problem was solved by the manufacturers and most targets made have been either the 4×5 inch format or the single 35 mm format.

The committee also specified tolerances that were allowed in manufacturing the targets and data reporting requirements. Because the initial use of the targets was seen to be visual comparison between scanner input and reproduced images, the committee felt that the patches actually produced by the manufacturer of a target should fall within $10\Delta E$ of the specified value. In addition, the committee believed that systems would evolve that would make use of the actual values that were in a particular target. They therefore specified that 'For all targets the batch-specific mean value and standard deviation for each patch shall be available from the originator of targets manufactured in accordance with this standard.' Provision was also made for reporting data for 'calibrated' targets and a file format for the reporting of all data in digital form.

Reporting of this data has varied widely among the suppliers of targets. Kodak has been the exception, providing batch data for all targets manufactured on a publicly available website, ftp://ftp.kodak.com/gastds/q60data. These data have also included an estimate of the within-batch variability reported as the χ^2 parameter following the lead of Dr Dolezalek [1] as described in his 1994 TAGA paper.

The same procedure was followed in developing a reflection target for scanner characterization using photographic paper. The standard that described this target was ANSI/IT8.7/2 and its development and history are similar to that of the transmission target.

While neither standard makes provision for other than photographic products to be used in making targets, many involved believe that should the need arise, useful targets meeting the standard could be made using any colour marking technique. The standard specifically defines the actions that must be taken if the material being used to create the target cannot produce the colour specified for a particular patch. Further, the product unique areas of the target enable any material to be described regardless of its gamut.

18.2.2 CMYK output data set

The other initial area of target development was a data set to facilitate the colour characterization of CMYK output devices, again a relatively obvious need. Each scanner manufacturer used a different output target to develop the data required to compute colour separation data for their equipment. As a result the printer had to contend with a multiplicity of targets all aimed at the same data – how the plate preparation and printing converted CMYK data in the computer to printed colour.

Unfortunately, simply knowing the colour of the individual dots and the dot areas does not allow the printed colour of a combination of halftone dots to be easily predicted. Although there are several theoretical models, such as that described by Neugebauer, even these need extensive input data. In addition, starting in the late 1980s there was increased interest in interpolation procedures for use in creating look-up tables for use in graphic arts colour separation procedures.

The approach the committee chose was to develop a standard halftone data set (target) defined as CMYK dots on film or equivalent data in the computer. The goal was to include data appropriate for those using either computational techniques or interpolation techniques.

It was believed that having a standard set of input data would allow easier exchange of printing information among users and would also allow industry organizations to work together to provide representative data for specific process aims. Organizations like SWOP and SNAP in the United States had traditionally defined only process control aims of density and dot gain. Individual colour separation scanner vendors and colour proofing systems vendors usually had to approximate these conditions as best they could to develop their own characterization data. It was seldom that characterization data from any two tests had sufficient commonality of target data to allow them to be directly compared.

The development of the data set drew heavily on input of the major colour scanner vendors and their experience. It was decided that the target should be in two parts. The first was called the basic target and was aimed at those users who either used computational techniques, or simply needed a moderate set of data to refine more elaborate models based on their own printing tests. The basic target consisted of the 16 Neugebauer primaries (all combinations of the CMYK solids), 13 step halftone scales in C, M, Y, and K, and 114 overprints of various CMYK combinations including two-colour (R, G, B) scales and near neutrals.

The second part, or extended target as it was called, had a series of regular arrays in C, M and Y with varying amounts of black (K) added. The first two groups used six levels of each colour (0, 10, 20, 40, 70, 100% dot), first without black and then with 20% black. As the level of black increased the committee felt that the lighter tones of the chromatic colours were not as important. For the 40% and 60% levels of black, five levels were used by dropping the 10% dot in the chromatic colours. The highest level of black, 80%, used only the 0, 40, 70, and 100% levels of the chromatic colours. This provided 746 additional patches for a total of 928 patches. (Of the 928 patches only 836 are unique – the two-part nature of the target introduced 92 duplicate patches.) Although no layout for these patches is required, the standard does provide a default layout (shown in Plate 2.1) which is the most prevalent form in which the target has been printed.

Initially published as IT8.7/3, *Graphic technology – Prepress digital data exchange – Input data for characterization of 4-colour process printing*, and still generally known as IT8.7/3, it has also been approved and published ISO 12642 with the same contents and title.

A second CMYK data set, to be called IT8.7/4, is currently under development by CGATS. It is intended to supplement the IT8.7/3 target, particularly for packaging applications. It will have 950 patches with only 16 duplicate sets of patch values. More importantly it has many more patches in the highlight and shadow portions of the scales and many more light overprints. One of the goals of the design of this target is that it be able to support offset, gravure, flexographic, and metal decorating types of printing. The packaging industry wants to be able to use the same data set for all printing characterization work. In addition the committee is defining a set of fingerprinting data that will be contained within the characterization data set. Further, the goal is that all tone values in the process control and run targets will be a subset of the fingerprinting data set. This will allow consistency throughout package printing process.

18.2.3 Colour measurement and computation

The scanner and output target standards described above all use or generate colorimetric data. The measurement of colour and the computation of colorimetric parameters is clearly spelled out by the International Commission on Illumination, or the CIE as it is better known. Unfortunately, in some ways the CIE does too good a job. They describe how to measure colour under a wide variety of geometries and illuminants, and make no recommendations as to the specific conditions preferable for a specific application. And, in fact, the choice is often arbitrary. However, if one group chooses one arbitrary set of conditions and another group chooses a different set, they cannot meaningfully exchange data.

Therefore the standards described above all included the definition of specific conditions for the measurement and computation of colorimetric data. These were consistent with, and restrictions of, the basic procedures established by the CIE. It was natural therefore that these be collected and defined in a standard for graphic arts applications. The initial steps were taken by ANSI CGATS. CGATS had been formed in the US as an umbrella committee for graphic arts standards development and new projects were moved into CGATS from the IT8 committee. (CGATS may use either the CGATS or IT8 designation on its standards.)

The resulting standard CGATS.5, *Graphic technology – Spectral measurement and colorimetric computation for graphic arts images*, was quickly moved into ISO as 13655 with the same title. As with the target standards it drew upon the existing photographic viewing (ISO 3664) and densitometry standards (ISO 5 Series) for many of its requirements – in large part to ensure consistency with those standards.

For both reflection and transmission work the instrument geometry was the same as the densitometry standards – 0/45 (or 45/0) for reflection and 0/diffuse (or diffuse/0) for transmission, with the ISO 5 densitometry standards pointed to for all instrument tolerances. D50 was chosen to match the viewing standard and the 2° CIE standard observer was felt to be more suited to graphic arts images than the 10° observer. Black backing was chosen to allow consistency between density and colorimetric measurements and to allow a single set of spectral reflectance data to be used to generate both density and colorimetric data. (It should be noted that TC42 Photography is only now defining the procedure for computation of status density from spectral data.) The standard also specified that, in the measurement of fluorescent specimens, a D50 source should be used but at the same time acknowledged that it was probably unrealistic to expect this to be accomplished in portable instruments.

When it came to computations, rather than point to the various CIE tables of illuminant and observer (and the CIE computational procedure), the committee chose to include and require the use of weighting functions. These were taken, with permission, from ASTM E308 [2] which was the only published source of weighting functions that the committee was able to identify. Weighting functions for data spacings of 10 and 20 nm are provided from 340 to 780 nm.

ISO 13655, like ASTM E308, specifies that computation be made from 340 to 780 nm. It notes that if the measured spectral data begin at a wavelength greater than 340 nm, then all the weighting factors for wavelengths less than the first measured wavelength shall be summed and added to the weighting factor for the first wavelength measured. The data between the last measured value and 780 nm are handled similarly.

18.3 VIEWING STANDARDS

18.3.1 Hardcopy viewing

As has been mentioned several times, it is critical that viewing conditions and colorimetric measurement conditions are closely linked. Although ISO 3664:1975 was titled *Photography – Illumination conditions for viewing colour transparencies and their reproductions*, it was the only international standard for viewing conditions and as such was used by the graphic arts industry. TC42 had attempted to revise it on several occasions but was unsuccessful in reaching agreement. Another key document that was often referenced was ANSI PH2.30-1989,*For Graphic Arts and Photography – Color Prints, Transparencies, and Photomechanical Reproductions – Viewing Conditions*. It had been more recently updated and there was input from the printing and publishing community. In 1995, when PH2.30 was being slated for revision, a small group from TC130 suggested to the TC42 Secretariat and to the ANSI PH2 committee (both groups are supported by PIMA, the Photographic & Imaging Manufacturers' Association) that TC130 would be willing to help lead the revision of ISO 3664 and update it using the work already accomplished in PH2 as a starting point. This offer was accepted and a joint working group (JWG) was formed between TC42 and TC130. ISO 3664:2000, *Viewing conditions – Graphic technology and photography* was the result. (It also replaces ANSI PH2.30.)

There are several key changes from prior versions of the standard. The spectral power distribution of CIE D50 is specified in the wavelength region of 340 to 400 nm in consideration of fluorescence issues. To help ensure compliance of the spectral power distribution of a source to that of D50 in both the visible and UV, the use of CIE Publication 51 (*A method for assessing the quality of daylight simulators for colorimetry*) has been added to the colour rendering index (CRI) requirements and the tolerances on the chromaticity coordinates.

In addition, for reflection viewing, a second level of illumination has been added to the previously defined 2000 ± 500 lux used for critical comparison between two or more images. It is intended for practical appraisal of tone reproduction, inspection, etc. of a single image in isolation, and is specified to be 500 ± 125 lux. As in previous standards the surround and backing are specified for both reflection and transparency viewing.

New in this document is the addition of a section on 'Conditions for the appraisal of images displayed on colour monitors'. It specifies a monitor white point of D65 and specifies that the luminance level of the white displayed on the monitor shall be greater than $75 \, \text{cd/m}^2$ and should be greater than $100 \, \text{cd/m}^2$.

The specification on ambient conditions may be summarized as follows:

- Ambient illumination shall have a colour temperature equal to, or less than, that of the CRT display white point.
- The level of illumination when measured at the face of the monitor, or in any plane between the monitor and the observer, shall be less than 64 lux and, preferably, less than 32 lux.
- The user should avoid any sources of reflection or glare.
- A dark, neutral surround shall be provided for the image displayed.

Conditions for the display of prints in galleries, etc., addressed in earlier standards, are included in an informative annex rather than in the main body of the standard.

18.3.2 Softcopy viewing

ISO 12646, *Graphic technology – Displays for colour proofing – Characteristics and viewing conditions*, is currently in preparation. It provides recommendations for uniformity, size, resolution, convergence and refresh requirements, luminance levels and viewing conditions for a colour display used to simulate a hardcopy proofing system in Graphic Arts. It has largely been produced with regard to CRT displays, which are by far the dominant technology in use at the present time. However, displays using other technologies would be expected at least to meet the specifications provided.

The white of the display (i.e. when $R = G = B = 255$) should be set to the chromaticity of D50; namely $u' = 0.2092$, $v' = 0.4881$, within a circle of radius 0.01 from this point. The luminance level should be as high as practical with the display technology used but shall be at least $80 \, \text{cd/m}^2$ and should be at least $120 \, \text{cd/m}^2$.

The ambient conditions shall comply with those specified in ISO 3664, although with the lower limits for ambient illumination. However, for this application, in which comparison to hardcopy is assumed, a more restrictive illumination condition is desirable. For such an application the level of illumination shall be less than 32 lux and the surround shall be no more than 10% of the maximum luminance of the screen. It should also be noted that the ISO 3664 requirement means that the colour temperature of the ambient illumination should be less than, or equal to, that of D50. However, for this application the illumination should approximate D50 (particularly if the level of ambient illumination is towards the high end of the specification).

This specification also adds the additional constraint that the level of illumination when viewing a black screen (i.e. an image defined as $R = G = B = 0$) shall be less than 5% of that obtained when viewing a white screen (i.e. an image defined as $R = G = B = 255$) when measured at the plane of the observer. This additional constraint is to ensure that any reflected glare from the front surface of the display does not significantly reduce the perceived contrast.

18.4 DENSITOMETRY AND RELATED METROLOGY STANDARDS

While densitometry is not normally considered a 'colour' standard, its use is an integral part of the process control which allows both the photographic and printing industry to maintain the level of colour quality they demand. Several application standards have been prepared and the basic ISO 5 series of densitometry standards is currently undergoing a major revision.

18.4.1 Related metrology standards

Both TC42 and TC130 have prepared standards supplementing the basic ISO 5 series of densitometry standards and/or providing guidance in the use of both density and colorimetry.

ISO 13656:2000, *Graphic technology – Application of reflection densitometry and colorimetry to process control or evaluation of prints and proofs*, is a good example. It provides guidance in the use of both densitometry and colorimetry (and which to use when) for process control and evaluation of single and multi-colour proofing and printing in the graphic arts. It defines terms, specifies minimum requirements for control strips, specifies test methods, and specifies reporting procedures for the results.

ISO 14807, *Photography – Method for the determination of densitometer performance specifications*, is in the final stages of approval within TC42. It defines a common set of reporting parameters and describes the methods to be used in the determination and presentation of individual densitometer performance and manufacturer-reported performance specifications. It grew out of the observation that the densitometer customer is met with a plethora of claims and specifications, in a variety of formats, pertaining to densitometer performance. Furthermore, various manufacturers have often used different terminology for describing what is speculated to be the same characteristic.

With this in mind, this standard identifies three characteristics of performance: ISO repeatability, ISO stability and ISO bias estimate. Standardized methods for evaluating these characteristics are presented. Any or all three of these characteristics can be evaluated and used to describe the performance of an individual densitometer and will be useful in comparisons of the performance of densitometers. The first two of these characteristics, ISO repeatability and ISO stability, are evaluated in such a way that, by use of suitable periodic sampling of production, a densitometer manufacturer can report average or typical repeatability and stability as specifications for a particular class, type or model of densitometer.

However, ISO bias estimate cannot necessarily be meaningfully averaged over such a class, type or model, since by determining a mean bias estimate, any instruments that are biased positively will be offset by any that are biased negatively. Because of this, bias estimate for a class, type or model of densitometer (if determined as a simple arithmetic mean of the bias estimates determined for individuals of that class, type or model) is of limited (if any) value and should not be reported. If determined as such an arithmetic mean, it may only be meaningful if that entire class, type or model is fraught with a systematic design defect. There is currently no agreement as to the most meaningful way to provide an ISO bias estimate for a class, type or model of densitometer.

The standardized methods for determination of ISO repeatability and ISO stability provide manufacturers with a uniform basis for stating densitometer performance characteristics as specifications, thereby providing the customer with the most useful information. To clarify and provide mutual understanding, a list of definitions applicable to the performance characteristics has been provided.

ISO 15790, *Graphic technology – Reflection and transmission metrology – Documentation requirements for certified reference materials, procedures for use, and determination of combined standard uncertainty*, was prepared by a joint working group between TC42 and TC130 and has received ISO approval and is awaiting publication.

A certified reference material (CRM) is defined as well-characterized materials with values traceable to stated references. They may be used to calibrate or to determine the performance characteristics of measurement systems in order to facilitate the exchange of data and to assist in quality control. Their use will help to ensure the long-term adequacy and integrity of the measurement and quality control processes. Densitometers, colorimeters and spectrophotometers are widely used to make measurements for quality and process control in the graphic arts, photographic and imaging industries. The intent of this standard is to establish documentation requirements that describe characteristics of reflection and transmission certified reference materials which may be used for verifying performance of these instruments. In many areas (e.g., cyan, magenta, yellow colorants, etc.) no references are readily available that are traceable to national standards. This standard provides guidance in such circumstances by showing how to determine the reproducibility of results of measurement, even in the absence of CRMs.

Although the calibration reference materials provided with many reflection and transmission instruments used in graphic arts and photography are not identified as CRMs, they often meet the requirements of CRMs but simply lack the appropriate documentation.

One of the key applications of CRMs should be in the determination of uncertainty in measurement. However, there is very limited understanding of this approach within the industry. This standard therefore describes practical procedures to determine values that represent components of the uncertainty of measurements for the graphic arts, photography and other image technology industries. A computational procedure is also provided to combine these components to determine 'combined standard uncertainty'. For application of this simplified procedure, it is assumed that the input variables are independent of each other and have normal distribution, and that their standard deviations are each much smaller than the absolute magnitude of the corresponding input variable. While these assumptions are not always correct, they provide a reasonable basis for the practical use of a CRM.

This standard provides guidance and is a resource for both manufacturers and users of CRMs. General procedures are also identified for the use and maintenance of certified reference materials.

ISO 14981:2000, *Graphic technology – Process control – Optical, geometrical and metrological requirements for densitometers for graphic arts use*, was prepared by TC130 to supplement and extend the requirements specified in the ISO 5 series of densitometry standards specifically for graphic arts applications. The current activity to revise the ISO

5 series grew out of the realization that having two standards that could apply to the same instrument system was undesirable. The current understanding is that when the revision of ISO 5 is complete and the requirements of ISO 14981 have been incorporated, it will be withdrawn.

18.4.2 ISO 5

Although ISO 5 was the fifth ISO standard to be identified, and it has been revised many times, it was recently realized that within the body of the standard the definition of density and the requirements and tolerances for the instruments used to measure density are not separately identified. In addition, although widely used by the printing and other industries, ISO 5 has been largely a photographic standard.

After considerable discussion a joint working group was created in 1999 to address the revision of the complete ISO 5 series, which currently includes ISO 5-1:1991, *Photography – Density measurements – Part 1: Terms, symbols, and notations*; ISO 5-2:2001, *Photography – Density measurements – Part 2: Geometric conditions for transmission density*; ISO 5-3:1995, *Photography – Density measurements – Part 3: Spectral conditions*; and ISO 5-4:1995, *Photography – Density measurements – Part 4: Geometric conditions for reflection density*.

A key goal of this revision is to separate the definition of density from the requirements and tolerances on the instrumentation to measure density. This will allow separate requirements on instrumentation used in different application areas – specifically graphic arts and photography. Additional goals are to include the definition of polarization requirements (not used in photography) and to define the procedure for computing density from spectral data.

The computation of density from spectral data is a more thorny issue than is anticipated at first glance. The current standard (ISO 5-3) specifies status density in terms of spectral products defined at 10 nm intervals. If the spectral power of the influx spectrum is multiplied by the spectral response of the receiver (which includes the photodetector and all intervening components between it and the plane of the specimen – including filters) wavelength by wavelength, spectral products are obtained. Although the influx spectrum is defined to be CIE standard illuminant A, in those cases where there is no fluorescence in the specimen or in the optical elements, the standard states that it is not necessary to specify the spectral characteristics of the influx and receiver separately, as long as the correct spectral products are obtained. Thus the spectral products are analogous to points along a continuous function.

While it would be nice to treat these discrete values as weighting functions for computation from spectral reflection or transmission data, that would not be correct. The current approach is to interpolate the 10 nm data to 1 nm using a cubic spline function and define the 1 nm spectral products as the new definition of status density (none of the 10 nm data points change). Since intervals of 1 nm in spectral reflection or transmission represent about the finest practical interval used, these spectral products can also be treated as weighting functions at 1 nm intervals.

The revised standard will also include weighting functions at other data spacings (probably 5, 10, and possibly 20 nm) which will be computed from the 1 nm data. The approach being considered to create the weighting functions is that used for computation of tristimulus values defined in ASTM E308 and documented in ASTM E 2022, *Standard Practice for Calculation of Weighting Factors for Tristimulus Integration*. Unfortunately, many people have treated the current spectral products as weighting functions. If ISO 5-3 is revised as proposed, data computed using the spectral products as weighting functions will be wrong. Fortunately the errors in most cases will be small.

18.5 OUTPUT CHARACTERIZATION

In the current world of colour imaging, the relationship between output device drive values and the colour produced can be either relatively simple or very complex, and everything in between. For a colour monitor, the default relationship has been defined in IEC 61966-2-1: *Multimedia systems and equipment – Colour measurement and management – Part 2-1: Colour management – Default RGB colour space – sRGB*. This defines a relatively simple model of the relationship between RGB code values and monitor output in a given viewing condition.

For CMYK output devices the relationship is far more complex. Two approaches are currently being used to 'define' this relationship. One is the traditional specification of ink colour, substrate, and printing process control parameters. The other is referred to as the 'reference printing condition' concept.

18.5.1 Printing process definition

In the traditional world of printing the approach used to define the relationship between CMYK data and printed colour has been to define the type of printing (offset with negative or positive plates, gravure, flexography, letterpress, etc.), ink colour, screen ruling and substrate to be used, and the process control parameters of the printing such as density of the solids, tone value increase at several points, ink trap, etc. In the days before electronic data distribution and manipulation (distribution of press-ready film separations) for each class of printing, all presses had to print approximately the same.

TC130 has created a family of standards to define these parameters which includes the multi-part ISO 2846, *Graphic technology – Colour and transparency of ink sets for four-colour-printing*, and ISO 12647, *Graphic technology – Process control for the manufacture of half-tone colour separations, proof and production prints*. A complete listing of the various parts is contained in the standards reference list in Appendix 1 at the end of this chapter.

18.5.2 Reference printing conditions

The other approach being investigated is the definition of reference printing conditions for each class of printing – essentially for each paper class. In this approach a colour gamut is defined, based on practical printing experience, and a typical or reference within-gamut mapping is provided between CMYK data and printed colour. The printing process is calibrated to match the defined gamut and local process control used to maintain a predictable within-gamut CMYK to printed colour relationship (characterization). The conversion of the CMYK data prepared based on the 'reference' condition to that necessary for the actual press characterization is accomplished using colour management tools.

It is anticipated that five or six colour gamut definitions (reference printing conditions) will cover the full range of process colour printing from newsprint through heavyweight catalogue papers. It is also interesting to note that there is some speculation that gravure, for example, may be able to obtain a larger gamut on some papers than can offset. Thus the different gamuts may not be as directly tied to paper type as was initially thought.

If this approach proves feasible, as more and more data is available electronically it will greatly simplify the issues of prepress and proofing. It also provides a much greater separation between prepress and printing, allowing each more flexibility.

The first example of this approach was the documentation in CGATS TR 001:1995 of the characterization data (printed colour vs CMYK data) for publication printing as defined in the United States SWOP specification (Specification for Web Offset Publications). This characterization data was created from a series of press tests conducted by SWOP, with the support of CGATS, where the printing was accomplished close to the middle of the SWOP process control aims and the IT8.7/3 target was used for measurement of the printed colour. TR 001 is currently defined to be the electronic definition of SWOP printing.

In another series of tests, the CGATS committee, working with the United States SNAP (Specification for Newsprint Advertising Production) Committee, has determined that offset, letterpress, and flexographic printing on newsprint can all achieve the same gamut. A single reference printing and proofing condition is planned for newsprint and will be identified as CGATS TR 002.

In TC130 ISO 12647-7, *Graphic technology – Process control for the manufacture of half-tone colour separations, proof and production prints – Part 7: Reference printing conditions for electronic data exchange*, is being prepared to document a logical family of these reference conditions for process colour data definition. The successful completion and implementation of this standard will depend on continued understanding of electronic data manipulation, colour management, implementation of local process control, and a focus on printed colour.

One of the issues that this approach raises is that as process control parameters vary between presses and processes, image structure will also change even though the colour printed is the same. That is to say, dot sizes will not be constant between processes (if dots are even used) and different printing may use different combinations of process colour values to produce the same composite printed colour. The magnifying glass will no longer be the same process control tool it used to be.

Another option introduced by this approach is the concept of virtual CMYK data. Given a family of reference printing conditions and the associated characterization data, a colour management system could simply exchange three-component data (either source data such as scanner code values or a reference RGB) and the appropriate input (source to PCS) and output (PCS to device) profiles, and the CMYK data could be created as needed in the output processor of the proofer, platemaker, film setter, etc. Assuming the ICC solves the issues of CMM consistency, the CMYK data produced from a given set of input data and profiles will be essentially identical. This would allow the addition of any local transforms for press, proofer, etc. to be essentially invisible to the user and simply part of the final data computation without the need for additional intermediate files (and potential data quantization).

18.6 TEST IMAGES

Although test patches are important for the measurement of printing conditions, human observers find natural-appearing images far more useful for visual evaluation. Consistent use of the same images builds a base of understanding and allows comparison between systems that is not practical when image selection is random or personal. TC130 initially developed a set of CMYK images as ISO 12640:1997, *Graphic technology – Prepress digital data exchange – CMYK standard colour image data (CMYK/SCID)*. Included are a series of eight 'natural' images (pretty pictures) as well as 10 test and control elements or 'synthetic' images. The IT8.7/3 CMYK (ISO 12642) colour data set, described earlier, has been partitioned and is included as four of the synthetic images.

As these images only exist as electronic data, the data files are contained on a CD-ROM that is a normative part of the standard. Two versions of the data have been prepared. One has a data spacing of 16 samples per mm and a data range of 28 to 228 representing dot values of 0% to 100%. The other has a data spacing of 12 samples per mm with a data range of 0 to 255. All images are approximately 5×7 inches.

The CMYK/SCID images have proved to be so useful that TC130 is also developing two additional image sets that will become parts of this standard. (The CMYK/SCID images will become ISO 12640-1.) They are *Part 2: XYZ/sRGB encoded image data (XYZ/SCID)* and *Part 3: CIELAB encoded image data (CIELAB/SCID)*.

The XYZ/SCID images are restricted to the sRGB gamut and are intended for applications where monitor to print evaluations are desired. They are encoded as both 8-bit per channel sRGB data and 16-bit per channel CIEXYZ data. Included are eight natural images and seven synthetic images. The synthetic images include a series of test patches and a set of vignettes. The natural images have been deliberately chosen to be different from the natural images used in the CMYK/SCID images to avoid possible confusion of data.

The CIELAB/SCID images are still in the development stage. They will also include a series of natural images and a set of test or synthetic images. The natural images are being prepared by scanning a set of large gamut chromes and then further expanding their gamut to create a set of image data that has a larger gamut than any output device. These images can thus be used to evaluate different gamut and colour manipulation tools and will provide a common reference for such work.

The gamut limit for the image gamut expansion work was based on a maximum gamut study of real-world surface colours that combined known published data with a variety of speciality printing inks and paints. The resultant gamut limit in itself is useful and may be published as a separate report.

Closely allied to the test images, but aimed more at digital image capture devices, is a set of data called *Standard object colour spectral database for colour reproduction evaluation (SOCS)*. It is in preparation as an ISO Technical Report. This work is being largely accomplished by the Japanese delegation to TC130 and will contain both a 'standard' data set of several hundred sets of spectral reflectance data and the database of several thousand data sets from which the standard set was derived. The standard set will contain data in the following categories: photographic materials, offset prints, computer colour prints, paints (not for art), paints (for art), textiles, flowers and leaves, outdoor scenes, and human skin. All data will be made available on a CD-ROM.

18.7 OTHER RELATED STANDARDS WORK

There are many other standards efforts going on, both in the accredited standards committees and in industry groups, that will directly impact colour imaging. Many of these are reported on or alluded to in other chapters of this book. These include the work on large gamut RGB colour spaces being done in IEC TC100, on multimedia in ANSI PIMA, and in TC42. In addition, the work of the International Color Consortium (ICC) is critical to the ongoing standards effort supporting colour imaging. It is hoped that the specifications being developed by the ICC will be moved under the ISO umbrella so that they can be more easily referenced and specified.

We have already discussed many of the standards being developed by TC42 and TC130 in the area of densitometry, viewing conditions, etc. In addition both committees are actively involved in developing standards that support electronic data transfer and/or system characterization. While these may at first glance not look like colour standards, they all bear on the quality of the colour image delivered. It is difficult to define just what are colour imaging standards.

It is also important to note the work of CIE Division 8, Image Technology. CIE Division 8 was formed in 1998 in response to strong inputs from the imaging industry. Its terms of reference are 'To study procedures and prepare guides and standards for the optical, visual and metrological aspects of the communication, processing, and reproduction of images, using all types of analogue and digital imaging devices, storage media and imaging media.' It currently has 6 Technical Committees and will impact on future colour imaging standards. A full description of the work of Division 8 is not appropriate here, but the interested reader is directed to www.colour.org.

18.8 WHERE ARE WE GOING?

As the joint activities between TC42 and TC130 indicate, particular technology issues are no longer restricted to one application area. The model for future standards development will be cooperative activities between application groups to ensure that the standards developed will have broad applicability. Even more importantly, many of the key activities are outside the traditional ISO or IEC standards arenas. The real challenge for all of us involved in standards development is to find ways to co-ordinate the work of these diverse groups to minimize overlap or conflict and maximize the interoperability between the standards and specifications developed. We do not have the luxury of either time or budget to tolerate duplication of effort or competition between groups.

Another challenge to the accredited standards process (the ISO and IEC) is to find ways to build upon the work of industry vendors (*de facto* standards like Adobe's PostScript and PDF), trade groups, and consortia (like the ICC). We must find ways to bring the work of these groups into a common environment with the traditional standards where the various classes of standards can complement each other rather than be seen as competing. Fortunately, both ISO and IEC are working to develop procedures for fast-tracking industry or 'publicly available specifications' (PAS) to enable them to be recognized within the traditional standards environment.

As the pace of technology quickens, and as the available manpower and budgets tighten, we must continue to develop ways to move the standards process forward more quickly. Whether it is in a single company, a trade group, a consortium, or an accredited standards committee, the pacing item is always reaching agreement between competing interests to find a workable common solution. Once technical agreement is reached, creating a specification or standard moves quickly. The only time advantage that companies, trade groups and consortia have is that the smaller the group the more arbitrary the decisions and the fewer the inputs that must be considered. Unfortunately, this is often counter to broad applicability. This then is the conundrum – how to reach broad consensus on technical solutions quickly!

18.9 MORE INFORMATION

A listing of colour-related international standards is given in Appendix 1, and a selected bibliography of papers describing the evolution of the graphic arts colour-related standards in Appendix 2.

More information about these standards activities can be found at various websites. A short list of those directly involved include:

- Graphic arts: www.npes.org, www.din.de
- Photography: www.pima.net
- General: www.iso.ch, www.iec.ch

REFERENCES

1. Dolezalek, F. K. (1994) Appraisal of Production Run Fluctuations from Colour Measurements in the Image. *TAGA Proceedings*, 154–164.
2. ASTM E308-99, Standard Practice for Computing the Colors of Objects by Using the C. I. E. System, ASTM Standards, copyright American Society for Testing and Materials, (1916) Race St., Philadelphia, PA 19130, USA.

APPENDIX 1. PARTIAL LISTING OF COLOUR IMAGING STANDARDS

Document no.	Title
ISO 5-1:1991	Photography – Density measurements – Part 1: Terms, symbols, and notations
ISO 5-2:2001	Photography – Density measurements – Part 2: Geometric conditions for transmission density
ISO 5-3:1994	Photography – Density measurements – Part 3: Spectral conditions
ISO 5-4:1995	Photography – Density measurements – Part 4: Geometric conditions for reflection density
ISO 2846-1:1997	Graphic technology – Colour and transparency of ink sets for four-colour-printing – Part 1: Sheet-fed and heatset web offset lithographic printing
ISO 2846-2:2000	Graphic technology – Colour and transparency of ink sets for four-colour-printing – Part 2: Coldset offset lithographic printing
ISO 2846-3:DIS	Graphic technology – Colour and transparency of ink sets for four-colour-printing – Part 3: Gravure printing
ISO 2846-4:2000	Graphic technology – Colour and transparency of ink sets for four-colour-printing – Part 4: Screen printing
ISO 2846-5:CD	Graphic technology – Colour and transparency of ink sets for four-colour-printing – Part 5: Flexographic printing
ISO 3664:2000	Viewing conditions – Graphic technology and photography
ISO 12640-1:1997	Graphic technology – Prepress digital data exchange – CMYK standard colour image data (CMYK/SCID)
ISO 12640-2:CD	Graphic technology – Prepress digital data exchange – Standard colour image data – Part2: XYZ/sRGB encoded image data (XYZ/SCID)
ISO 12640-3:WD	Graphic technology – Prepress digital data exchange – Standard colour image data – Part3: CIELAB encoded image data (CIELAB/SCID)
ISO 12641:1997	Graphic technology – Prepress digital data exchange – Colour targets for input scanner calibration
ISO 12642:1996	Graphic technology – Prepress digital data exchange – Input data for characterization of 4-colour process printing
ISO 12645:1998	Graphic technology – Process control – Certified reference material for opaque area calibration of transmission densitometers
ISO 12646:CD	Graphic technology – Colour proofing using a colour display
ISO 12647-1:1996	Graphic technology – Process control for the manufacture of half-tone colour separations, proof and production prints – Part 1: Parameters and measurement methods

ISO 12647-2:1996	Graphic technology – Process control for the manufacture of half-tone colour separations, proof and production prints – Part 2: Offset lithographic processes
ISO 12647-3:1998	Graphic technology – Process control for the manufacture of half-tone colour separations, proof and production prints – Part 3: Coldset offset lithography and letterpress on newsprint
ISO 12647-4:WD	Graphic technology – Process control for the manufacture of half-tone colour separations, proof and production prints – Part 4: Gravure processes
ISO 12647-5:FDIS	Graphic technology – Process control for the manufacture of half-tone colour separations, proof and production prints – Part 5: Screen printing
ISO 12647-6:CD	Graphic technology – Process control for the manufacture of half-tone colour separations, proof and production prints – Part 6: Flexographic printing
ISO 12647-7:WD	Graphic technology – Process control for the manufacture of half-tone colour separations, proof and production prints – Part 7: Reference printing conditions for electronic data exchange
ISO 13655:1996	Graphic technology – Spectral measurement and colorimetric computation for graphic arts images
ISO 13656:2000	Graphic technology – Application of reflection densitometry and colorimetry to process control or evaluation of prints and proofs
ISO 14672:2000	Technical report – Statistics of the natural SCID images defined in 12640
ISO 14807:FDIS	Photography – Method for the determination of densitometer performance specifications
ISO 14981:2000	Graphic technology – Process control – Optical, geometrical and metrological requirements for densitometers for graphic arts use
ISO 15790:FDIS	Graphic technology – Reflection and transmission metrology – Documentation requirements for certified reference materials, procedures for use, and determination of combined standard uncertainty
ISO 17321-1:WD	Graphic technology and photography – Colour characterization of digital still cameras (DSCs) – Part 1: stimuli, metrology and test procedures
ISO 17321-4:WD	Graphic technology and photography – Colour characterization of digital still cameras (DSCs) using colour targets and spectral illumination
IEC 61966-2-1	Colour measurement and management in multimedia systems and equipment – Part 2.1: Colour management in multimedia systems – Default RGB colour space – sRGB

(*continued overleaf*)

Document no.	Title
IEC 61966-2-2	Colour management – Extended RGB colour space – sRGB64
IEC 61966-2-3	Colour management – Default YCC colour space – sYCC
IEC 61966-3	Colour measurement and management in multimedia systems and equipment – Part 3: Equipment using cathode ray tubes
IEC 61966-4	Colour measurement and management in multimedia systems and equipment – Part 4: Equipment using liquid crystal display panels
IEC 61966-5	Colour measurement and management in multimedia systems and equipment – Part 5: Equipment using plasma display panels
IEC 61966-7	Colour measurement and management in multimedia systems and equipment – Part 7: Colour printers
IEC 61966-8	Colour measurement and management in multimedia systems and equipment – Part 8: Colour scanners
IEC 61966-9	Colour measurement and management in multimedia systems and equipment – Part 9: Digital cameras
ANSI PIMA 7466:WD	Electronic still picture imaging – Reference input medium metric RGB color encoding (RIMM-RGB)
ANSI PIMA 7666:2001	Electronic still picture imaging – Reference output medium metric RGB color encoding (ROMM-RGB)
ANSI PIMA 7667:WD	Electronic still picture imaging – Extended sRGB color space – e-sRGB
ANSI CGATS TR001:1995	Graphic technology – Color characterization data for type 1 printing (1995)
ANSI CGATS TR002:WD	Graphic technology – Color characterization data for coldset printing on newsprint (draft)
ANSI IT8.7/1:1993	Graphic technology – Color transmission target for input scanner calibration (incorporated in ISO 12641)
ANSI IT8.7/2:1993	Graphic technology – Color reflection target for input scanner calibration (incorporated in ISO 12641)
ANSI IT8.7/3:1993	Graphic technology – Input data for characterization of 4-colour process printing (incorporated in ISO 12642)
ANSI IT8.7/4:WD	Graphic technology – Input data for characterization of 4-colour package printing

Notes: WD = Working draft stage of development
CD = Committee draft
DIS = Draft International Standard
FDIS = Final Draft International Standard
ISO = International Standard (ISO)
IEC = International Standard (IEC)

APPENDIX 2. BIBLIOGRAPHY

1. McDowell, D. Q. (1991) Critical Review of Colour Standards for Electronic Imaging, Standards for Electronic Imaging Systems: Proceedings of a conference held 28 Feb.–1 Mar. 1991, Critical Reviews of Optical Science and Technology. *SPIE CR37*, 40–53.
2. McDowell, D. Q. (1991) Summary of Colour Definition Activity in the Graphic Arts. *Proceedings – Image handling and Reproduction Systems Integration, SPIE*, **1460**, 29–37.
3. McDowell, D. Q. (1993) Summary of IT8/SC4 Colour Activities. *Proceedings – Device-Independent Colour Imaging and Imaging Systems Integration, SPIE*, **1909**, 229–235.
4. McDowell, D. Q. (1994) Colour Standards Activities in the Graphic Arts. *Proceedings – Colour Hardcopy and Graphic Arts III, SPIE*, **2171**, 174–182.
5. McDowell, D. Q. (1995) Graphic Arts Standards Update – 1995. *Proceedings – Colour Hardcopy and Graphic Arts IV, SPIE*, **2413**, 323–332.
6. McDowell, D. Q. (1996) Graphic Arts Standards Update – 1996. *Proceedings – Colour Imaging: Device-Independent Colour, Colour Hard Copy and Graphic Arts, SPIE*, **2658**, 138–146.
7. McDowell, D. Q. (1997) Graphic Arts Standards Update – 1997. *Proceedings – Colour Imaging: Device-Independent Colour, Colour Hard Copy and Graphic Arts II, SPIE*, **3018**, 148–155.
8. McDowell, D. Q. (1999) Graphic Arts Colour Standards Update – 1999. *Proceedings – Colour Imaging: Device-Independent Colour, Colour Hard Copy and Graphic Arts IV, SPIE*, **3648**, 294–303.
9. McDowell, D. Q. (1999) The Relationship Between Uncertainty in Reflectance Factor Data, and Computed CIELAB Values–Some Intuitive Tools. *Proceedings – Colour Imaging: Device-Independent Colour, Colour Hard Copy and Graphic Arts IV, SPIE*, **3648**, 291–293.
10. McDowell,D. Q. (1999) Colour Standards in ISO and IEC. In *presentation at International Colour Management Forum*, March 24–25, University of Derby, UK.
11. McDowell, D. Q. (1999) Reference Printing Conditions, What Are They & Why Are They Important? *The Prepress Bulletin*, **88**(6), 42–44. March/April.
12. McDowell, D. Q. and Johnson, A. J. (1999) Colour Management and Graphic Arts: Past Present, and Future, presentation at 68th Annual Meeting of Inter-Society Colour Council, May 6–7, Vancouver, BC, Canada.
13. McDowell, D. Q. (1999) Graphic Arts Colour Standards. *TAGA Proceedings*, 661–670.
14. McDowell, D. Q. (1999) Colour and the Graphic Arts. *American Ink Maker*, **77**(9), 28–32. September.
15. McDowell, D. Q. and Johnson, A. J. (1999) Colour Management and Graphic Arts: Past Present, and Future, presentation at Taipei International Conference on Graphic Communications, Oct 3–5, Taipei, Taiwan, R.O.C.
16. McDowell, D. Q. and Pate, L. (1999) Reference Printing Conditions and Colour Management for Proofing, PIRA International Conference Proceedings, Proofing – New Technologies & Best Practice for the Publishing & Printing Industries, Nov 15–16, London, England.
17. McDowell, D. Q. and Steele, L. C. (1999) Gravure Print Characterization on Packaging Substrates, Second Preliminary Report, Gravure, Winter, 44–51.
18. McDowell, D. Q. (2000) Colour Management: What's Needed for Printing and Publishing? *The Prepress Bulletin*, **89**(6), 9–17. March/April.
19. McDowell, D. Q. (2000) Colour Standards in Graphic Arts and Photography – Past, Present, and Future. *Proceedings – IS&T's NIP16: International Conference on Digital Printing Technologies*, October 15–20, pp. 546–551.
20. McDowell, D. Q. (2000) Do You Know Where Your Measurements Are? (with respect to their true value). *The Prepress Bulletin*, **90**(3), 20–22. November/December.
21. McDowell, D. Q. (2001) Digital Image management, An Overview of International Standards. *PIRA International Conference Proceedings, Digital Image Management*, pp. 29–30. London, England. March.

Author biographies

Roy Berns is Hunter Professor of Color Science, Appearance and Technology at the Munsell Color Science Laboratory, Rochester Institute of Technology.

He received a PhD in colour science from the Rensselaer Polytechnic Institute in 1984, and has since presented invited papers and lectures at many events, including those organized by the CIE, AIC, ISCC, RPS, IS&T and SID.

Currently, Professor Berns is active in the following research areas: colour tolerance equations; colorimetric and spectral device characterization; colour appearance psychophysics; and colorimetric and spectral colour reproduction of paintings. In the CIE, he is active on committees dealing with spectrophotometric performance (TC 2-28), the effects of rod intrusion on metameric colour matches (TC 1-43), and colour tolerance equations (TC 1-47).

Ed Giorgianni is an imaging scientist in the Imaging Research and Advanced Development Division at Eastman Kodak Company. He has many years of practical experience in designing photographic, electronic and hybrid colour-imaging systems. He holds numerous patents in the fields of colour management and imaging technology. Among his inventions are the digital colour-encoding methods used on many commercial imaging systems, including the PhotoCD system. In addition, Ed is an award-winning instructor and frequent lecturer at technical symposia and universities, and is a co-author of *Digital Color Management* (Addison-Wesley).

Phil Green is a member of the Colour Imaging Group at the London College of Printing, and Course Director of the college's postgraduate programme in Digital Colour Imaging.

He worked in the printing industry from 1975, joining the London College of Printing in 1986. He received an MSc in Interactive Systems Analysis from the University of Surrey in 1995, and is currently completing a PhD at the University of Derby. He has authored a number of graphic arts textbooks, including *Understanding Digital Color* (GATF), and *Digital Photography* (Pira International). He is active in CIE TC8-03 Gamut Mapping and his current research interests are gamut mapping and colour difference.

Larry Hanlon is an R&D project manager at Hewlett-Packard Laboratories in Palo Alto, California. Following a post-doctorate at the University of Pennsylvania studying synthetic metals, he joined HP Laboratories in 1977 to develop photoreceptors for the first HP laser printers. He has since managed technology developments involving displays, e-beam lithography resists, and electronic packaging and interconnect solutions for the advanced microprocessors in HP's PA-RISC computers. Since returning to HP Labs in 1992 to work on colour imaging, Larry has managed projects in colour reproduction and digital photography. His teams have delivered key technologies that have enabled HP's inkjet colour copier and AIO business, initiated the sRGB colour data exchange standard, and enabled true photographic quality thermal inkjet printing.

Jack Holm is a senior scientist working on strategic technology in the CTO office of Hewlett-Packard's Imaging and Printing Systems. He is HP's primary standards representative for digital imaging, participates in ISO TC42 (Photography), ISO TC130 (Graphic Technology), IEC TC 100 (Multimedia) and ISO/IEC JTC1 SC29/WG1 (JPEG), and serves as liaison from ISO TC42 to the International Color Consortium (ICC) and CIE.

He holds a BS in Physics from Texas A&M University and an MS in Imaging Science from RIT. He has been active in digital photography research for over a decade, and was involved in the development of the HP/Pentax digital cameras. Previously he served as a digital photography consultant, and on the faculty of the Rochester Institute of Technology School of Photographic Arts and Sciences.

Paul M. Hubel is a Principal Project Scientist at Hewlett-Packard Laboratories, working on colour algorithms for digital photography and digital photofinishing. He was the principal architect for the imaging pipeline of current Hewlett-Packard and Pentax digital cameras. His recent work has been on illumination estimation and colour correction for digital cameras and image processing methods to optimize digital photofinishing.

Dr. Hubel received his DPhil from Oxford University in Engineering Science in 1990. His doctoral thesis was on Colour Reflection Holography. He has published over 25 technical papers, and currently holds 4 patents.

Tony Johnson is a Professor of Colour Imaging and member of the Colour Imaging Group at the London College of Printing. He is also Technical Secretary of the ICC.

Professor Johnson first joined the printing industry as an apprentice hot metal compositor. He studied printing technology at LCP, colour physics at Imperial College, London and mathematics with the Open University. After 10 years' research into colour reproduction with Pira he was appointed as research manager of Crosfield Electronics Ltd in 1983 and became a Principal Lecturer at LCP in 1997.

He is active in the international standards community, including ISO TC130 and CIE TC8-03 and TC8-05. His current research interests are colour management, colour difference and the colorimetry of graphic arts media.

Naoya Katoh is a Senior Staff Scientist at Personal-Imaging Company, Mobile Network-Company, Sony Corporation.

He received his B. Eng. degree in precision mechanics in 1987 from Kyoto University, Japan. In the same year he joined Sony Corporation where his main research focused on the development of novel printing methods. He received an MS degree in color science in 1997 from the Munsell Color Science Laboratory, Rochester Institute of Technology, NY, USA, and Ph.D, Engineering in 2002 from Chiba University, Japan. His current research at Sony Corporation focuses on the color image processing and the digital photography.

He is a member of the Society for Imaging Science and Technology (IS&T) and The Institute of Image Information and Television Engineers (ITE). He received the IS&T's Charles E. Ives Award in 1994. He is an active member of IEC/TC100/TA2, ISO/TC42/WG18, and CIE/Division 8 on color management related standards.

James King is a Principal Scientist at Adobe Systems Incorporated, and is one of the people responsible for the development of new products and new features for existing Adobe products. Jim King joined Adobe in 1988 and until 1996 was the founder and Senior Director of Adobe's Advanced Technology Group (ATG). He is now a member of that group.

Earlier, Dr King was manager of I/O Systems Laboratory (IOSL) at the IBM Almaden Research Center and at the IBM T. J. Watson Research Center. He received a PhD in Computer Science from Carnegie-Mellon University. He is a member of the ACM, IEEE Computer Society, the Seybold Conference Advisory Board, and the IS&T (Information Sciences and Technology). He is the inventor of several patents.

Wolfgang Lempp is Director of Technology Development at the Computer Film Company, concentrating on development projects for digital film mastering and digital cinema technology. He graduated in theoretical physics at Munich University and worked in the film industry for 20 years, initially as special effects technician and then visual effects supervisor. One of the Computer Film Company's original team members and key contributor to development of the pioneering digital film facility, he received a technical academy award in 1996. At CFC he has developed image processing tools, workflow and colour management applications, film scanners and film recorders, and image processing hardware. For two years he was System Architect and Director of Technology at the Babelsberg F/X Centre in Germany, a fully integrated digital audio-visual production and post production facility for film, TV and new media.

M. Ronnier Luo is a Professor of Colour Science at the Colour & Imaging Institute at the University of Derby. He received his PhD in Colour Physics from the University of Bradford in 1986. He has numerous publications in the areas of colour measurement, colour difference, colour appearance modelling, and colour reproduction. He is the Chairman of the Colour Measurement Committee (CMC) of the Society of Dyers and Colourists (SDC), CIE TC 1-52 Chromatic Adaptation Transforms and CIE TC 8-2 Colour Difference Evaluation in Images. He was the recipient of the 1994 Bartleson Award for his work in colour science.

Lindsay MacDonald is Professor of Multimedia Imaging at the Colour & Imaging Institute, University of Derby. For 18 years he was with Crosfield Electronics Ltd (now Fujifilm Electronic Imaging), where he designed and wrote the software for the world's first computer-based page composition system in 1977.

Professor MacDonald is a Fellow of the British Computer Society (FBCS), the Institution of Electrical Engineers (FIEE), the Royal Photographic Society (FRPS), and the Royal Society for the Arts (FRSA). He is co-author or co-editor of a number of books, including *Computer Generated Colour* (Wiley), *Display Systems: Design and Applications* (Wiley),

Colour Imaging: Vision and Technology (Wiley) and *Colour Image Science: Exploiting Digital Media* (Wiley).

Marc Mahy is a Principal Colour Scientist at Agfa-Gevaert N.V. He received an MS degree in physics at the Katholieke Universiteit Leuven (Belgium) in 1986 and a PhD in sciences at the same university in 1994 for his study on colour spaces for visual and physical image processing. Recently, he became a member of Agfa's Colour Technology Centre (CTC) of the graphic arts division to manage a team for the development of software technology for colour imaging. Dr Mahy is the author of several publications on colour management and is a member of several international standardization organizations such as ISO, CIE and ICC.

David Q. McDowell graduated from the University of Rhode Island in 1957 with a BSc degree in Engineering Physics. He recently retired from Eastman Kodak Company, after 42 years, where he was a Senior Technical Associate in the Professional Imaging Division. In retirement, he is continuing his involvement with national and international standards activities.

McDowell is the chairman of the ISO Steering Committee for Image Technology (SCIT), chairman of the US Technical Advisory Group (USTAG) to ISO TC130 (Graphic technology), acting chair of TC130/WG2 (Prepress Data Exchange), chair of ISO TC42/JWG21 (Revision of ISO 5), and secretary of CIE Division 8 (Image Technology). He is also IS&T Standards Chair and Standards Editor of the IS&T Newsletter. He is a long-time member of both IS&T and TAGA and is a past president of TAGA.

Ján Morovic is a lecturer in Digital Colour Reproduction at the Colour & Imaging Institute at the University of Derby, where he is module leader for two modules on the MSc in Imaging Science. He was born in Bratislava in 1974. After first completing a BA at the London College of Printing, in 1998 he received a PhD in Colour Science at CII for his research project 'To Develop a Universal Gamut Mapping Algorithm'. Dr Morovic also serves as chairman of the CIE's technical committee 8-03 on Gamut Mapping. His research interests include image characteristics, gamut mapping and colour reproduction.

Leonardo Noriega is a Research Assistant at the Colour & Imaging Institute at the University of Derby, working on the digitization of colour film stock. He received an MSc in Computer Science at the University of Kent at Canterbury in 1991, an MSc in Machine Perception and Neurocomputing from the University of Keele in 1993, and a PhD from Nottingham Trent University in 1998. His work career has encompassed various aspects of engineering, including two years working in the Computer Science department of the Polytechnic University of Valencia, and a further two years working at the School of Industrial Engineers at the University of Murcia (now the Polytechnic University of Cartagena) in Spain. He has been a Research Fellow in the Department of Communications and Neuroscience at the University of Keele and in the Department of Textiles at UMIST before joining the University of Derby in January 2000.

Peter Rhodes is a Senior Research Fellow in the Colour & Imaging Institute at the University of Derby. In addition to teaching MSc students Electronic Image Communication and Colour Notations Systems, his research interests include computer-mediated colour fidelity and communication for which he received his PhD in 1995 from Loughborough University.

Danny Rich is director of the Sun Chemical (GPI) Color Research Laboratory, where his current responsibilities include visual and instrumental tolerancing, corporation-wide instrument reproducibility, standards and calibration and lighting engineering for Sun Chemical worldwide. He completed a PhD at Rensselaer Polytechnic Institute on 'The Perception of Moderate Color Differences in Surface-Color Space' in 1980. He was Manager of Research for Applied Color Systems from 1984 where he did research on colour simulation, instrument design and optical metrology and calibration, and was later Manager of Advanced Colorimetry and Metrology for Datacolor International. Dr Rich has published on all aspects of colour science and engineering, including visual perception, instrumentation, and mathematical modelling.

Kevin Spaulding is a Research Associate in the Imaging Science Division at Eastman Kodak Company. He received a BS in Imaging Science from the Rochester Institute of Technology in 1983, and MS and PhD degrees in Optical Engineering from the University of Rochester in 1988 and 1992, respectively. He has been with Eastman Kodak Company since 1983.

He is also Technical Secretary for the CIE TC8-05 committee, which is tasked with defining standards for the unambiguous communication of colour information in images. He is the author of more than 30 technical papers, and an inventor of more than 60 patents/patent applications. His research interests include digital colour encoding, colour reproduction, digital halftoning, image quality metrics, and image processing.

Sabine Süsstrunk is Assistant Professor for Imaging and Visual Representation in the Audiovisual Communications Laboratory, Communication Systems Department, at the Swiss Federal Institute of Technology (EPFL) in Lausanne, Switzerland. Her main research areas are colour imaging, image quality metrics, and digital archiving. From 1995 to 1999, she was the Principal Imaging Researcher at Corbis Corporation in Bellevue, Washington. From 1991 to 1995, she was an Assistant Professor in the School of Photographic Arts and Sciences at the Rochester Institute of Technology (RIT). She received an MS in Graphic Arts Publishing from the Rochester Institute of Technology (RIT) and is an active member of ISO/TC42/WG18 and JWG20 – Electronic Imaging standards groups.

Arthur Tarrant studied physics at the University of London and first worked on colorimetry at the National Physical Laboratory. He then moved to Battersea College, which later became the University of Surrey. His work there has included spectroscopy, optics, instrumentation, lighting and consumer science. His main research has been on the spectral composition of daylight, scattered light in optical instruments, colour names and visual colour matching instruments. He is an honorary fellow of both Surrey and Derby universities. He has received the CIBSE Lighting Award and the Bronze Medal of the Society of Light and Lighting, both for distinguished services to lighting.

Ingeborg Tastl is a digital colour imaging scientist at Hewlett-Packard Laboratories, working in the area of digital imaging and printing since April 2001. Before that her focus was in the area of digital photography while working at Sony's US Research Laboratories and at the École Nationale Supérieure des Télécommunications in Paris. Colour science, imaging science and computer graphics and their applications have been her research interests since she received her MS degree and her PhD degree in computer science from Vienna University of Technology, Austria. She is a member of IS&T, and the SID's Technical Program Co-Chair for the 9th Color Imaging Conference.

Dawn Wallner has recently retired from Sun Microsystems, for whom she was a representative to the ICC. She has 25 years of software programming and design experience, and has worked on satellite image manipulation, graphics databases for flight simulation, libraries for 3D graphics and image processing, and integration of colour management into operating systems.

Index